Occupational Health and Safety in the Food and Beverage Industry

A safe and healthy working environment is a vital aspect of the food and beverage processing industry. *Occupational Health and Safety in the Food and Beverage Industry* provides key information on food and beverage manufacturing accident prevention, disease, injury management, and safer steps for employees to get back to work and discusses food security, safety, biosecurity, defense, ethics, food safety and quality including food adulteration.

Features

- Discusses fundamentals of occupational health and safety in the food and beverage industry
- Highlights standards and legislations as related to occupational health and safety for food and beverage processing sectors
- Covers hazards, elements, accident prevention, various hazards present in food and beverage sectors, and their disease and injury management
- Explores ethical issues in agri-food processing sectors and their effects on sustainability
- Introduces importance, organization, and management for food and beverage processing sectors to prevent losses

The book is intended for professionals in the fields of occupational health and safety, food engineering, chemical engineering, and process engineering.

Occupational Health and Safety in the Food and Beverage Industry

Ebrahim Noroozi and
Ali R. Taherian

CRC Press
Taylor & Francis Group
Boca Raton London New York

CRC Press is an imprint of the
Taylor & Francis Group, an **informa** business

Front cover image: Siberian Art/Shutterstock

First edition published 2023
by CRC Press
6000 Broken Sound Parkway NW, Suite 300, Boca Raton, FL 33487-2742

and by CRC Press
4 Park Square, Milton Park, Abingdon, Oxon, OX14 4RN

CRC Press is an imprint of Taylor & Francis Group, LLC

© 2023 Ebrahim Noroozi and Ali R. Taherian

Reasonable efforts have been made to publish reliable data and information, but the author and publisher cannot assume responsibility for the validity of all materials or the consequences of their use. The authors and publishers have attempted to trace the copyright holders of all material reproduced in this publication and apologize to copyright holders if permission to publish in this form has not been obtained. If any copyright material has not been acknowledged please write and let us know so we may rectify in any future reprint.

Except as permitted under U.S. Copyright Law, no part of this book may be reprinted, reproduced, transmitted, or utilized in any form by any electronic, mechanical, or other means, now known or hereafter invented, including photocopying, microfilming, and recording, or in any information storage or retrieval system, without written permission from the publishers.

For permission to photocopy or use material electronically from this work, access www.copyright.com or contact the Copyright Clearance Center, Inc. (CCC), 222 Rosewood Drive, Danvers, MA 01923, 978-750-8400. For works that are not available on CCC please contact mpkbookspermissions@tandf.co.uk

Trademark notice: Product or corporate names may be trademarks or registered trademarks and are used only for identification and explanation without intent to infringe.

Library of Congress Cataloging-in-Publication Data

Names: Noroozi, Ebrahim, author. | Taherian, Ali R., author.
Title: Occupational health and safety in the food and beverage industry /
Ebrahim Noroozi, Ali R. Taherian.
Description: First edition. | Boca Raton : CRC Press, 2023. | Includes
bibliographical references and index.
Identifiers: LCCN 2022043217 (print) | LCCN 2022043218 (ebook) | ISBN
9781032300368 (hardback) | ISBN 9781032300375 (paperback) | ISBN
9781003303152 (ebook)
Subjects: LCSH: Food industry and trade--Safety measures. | Food industry
and trade--Health aspects. | Beverage industry--Safety measures. |
Beverage industry--Health aspects.
Classification: LCC TP373.5 .N67 2023 (print) | LCC TP373.5 (ebook) | DDC
664--dc23/eng/20221006
LC record available at https://lccn.loc.gov/2022043217
LC ebook record available at https://lccn.loc.gov/2022043218

ISBN: 978-1-032-30036-8 (hbk)
ISBN: 978-1-032-30037-5 (pbk)
ISBN: 978-1-003-30315-2 (ebk)

DOI: 10.1201/9781003303152

Typeset in Times
by Deanta Global Publishing Services, Chennai, India

This book has been dedicated to:

Our families

and

All victims of OHS in food and beverage manufacturing

Contents

Preface...xix
Foreword xxi
Authors' disclaimer..xxv
About the Authors...xxvii
Acknowledgments...xxix

Chapter 1   Introduction: Health and Safety Organization 1

    1.1   Introduction 1
    1.2   Safety: a comprehensive responsibility 1
    1.3   Organization 2
    1.4   The safety department...2
    1.5   Production departmental responsibility 2
    1.6   The production line/laboratory responsibilities 3

Chapter 2   Fundamentals of Health and Safety Management 5

    2.1   Introduction 5
    2.2   Investment in health and safety management pays off............6
         2.2.1   Losses and due diligence..7
    2.3   Elements of a safety management system 8
         2.3.1   Core elements of health, safety, and
                 environmental management 8
    2.4   Modern safety management evolution......................8

Chapter 3   Health and Safety Legislation, Audit, and Standards 9

    3.1   Introduction 9
         3.1.1   What are guidelines?......................................9
         3.1.2   Why have food standards become so important? 10
         3.1.3   The need for audit 10
    3.2   Food hygiene legislation...10
         3.2.1   REACH: the new European chemical legislation 11
         3.2.2   International standards on health and safety.............11
    3.3   Safe food handling guidelines.................................11
         3.3.1   Food standards and audit..12
                3.3.1.1   What are food standards? 12
         3.3.2   The Canadian food safety approach........................13
                3.3.2.1   The Integrated Inspection System (IIS)......14
                3.3.2.2   Legislative levels in Canada 14
         3.3.3   Standards and their importance for audit.................15

vii

viii   Contents

	3.3.4	Standard-setting process and principles	15
	3.3.5	What do auditors look for?	16
	3.3.6	Developing food safety and quality management systems	16

3.3.6.1 Hazard Analysis Critical Control Points (HACCP) 16
3.3.6.1 Inspection 18
3.3.6.2 Total Quality Management System (TQMS)/Good Manufacturing Practices (GMP) 18
3.3.6.3 Quality control (QC) 18
3.3.6.4 Quality assurance (QA) 18
3.3.6.5 Good Manufacturing Practice (GMP) 18
3.3.6.6 ISO 22000 19

3.3.7 An international risk-based food inspection system .. 19

3.4 Recall and traceability 20
    3.4.1 Why is traceability needed? 21

Chapter 4 Accident Prevention Elements 23

4.1 Introduction 23
    4.1.1 Accident/incident definition 23
    4.1.2 Frank Bird accident ratio pyramid 24

4.2 Accident prevention and causation models 24
    4.2.1 Hazard identification and risk minimization 25

4.3 Accident/incident investigation, analysis, and costs 25
    4.3.1 Legislations 26
    4.3.2 Responsibilities 26
    4.3.3 Procedures 27
    4.3.4 Communication/training/evaluation 27

4.4 Transitional employee policy 27

4.5 Contractor safety program 28
    4.5.1 Responsibilities 28
    4.5.2 Procedure 28
    4.5.3 Communication/training/evaluation 29
    4.5.4 Relevant remarks 29

4.6 Emergency response and preparedness 30
    4.6.1 How to develop and implement a plan 30
    4.6.2 Facilities: fixed and movable equipment 30
    4.6.3 Laboratory facilities: design and equipment 31

4.7 First aid and cardiopulmonary resuscitation (CPR) 31

4.8 Health and safety policy 31
    4.8.1 Health and safety forms 31
    4.8.2 Accident and inspection report form 32

4.9 Industrial hygiene 33

4.9.1	Definition	33
4.9.2	Purpose	33
4.9.3	Procedure	33
4.9.4	Health and safety committee/joint health and safety committee (JHSC)	33
4.9.5	Conducting inspection	34
	4.9.5.1 Inspection tools and preparation	34
	4.9.5.2 Inspection ethics	34
4.9.6	Personal protective equipment (PPE)	34
	4.9.6.1 PPE requirements	35
	4.9.6.2 The hazard assessment	35
	4.9.6.3 Body protection wear (face, hand, eyes, jaws, etc.)	36
	4.9.6.4 PPE where chemical hazards are involved	36
	4.9.6.5 PPE devices for chemical hazards	36
	4.9.6.6 Impact hazard	37
	4.9.6.7 PPE devices for impact hazards	37
	4.9.6.8 PPE devices for dust hazards	37
	4.9.6.9 Dust hazard	37
	4.9.6.10 Optical hazard	38
	4.9.6.11 PPE for optical hazard: lens requirements	38
	4.9.6.12 Foot protection	38
	4.9.6.13 Prevention	39
	4.9.6.14 Head protection	39
	4.9.6.15 Respiratory protection	39
	4.9.6.16 Employee training	40
4.9.7	Legislation	40
	4.9.7.1 Purpose	41
	4.9.7.2 Responsibilities	41
	4.9.7.3 Procedures	41
	4.9.7.4 Communication and training	41
	4.9.7.5 Evaluation	42
4.9.8	Preventive maintenance	42
	4.9.8.1 Purpose	42
	4.9.8.2 Contractors	42
	4.9.8.3 Preventive maintenance program components	43
	4.9.8.4 Communication and training	43
	4.9.8.5 Evaluation	44
4.9.9	Purchasing policy	44
	4.9.9.1 Responsibilities	44
	4.9.9.2 Procedures	44
	4.9.9.3 Communication and training	45
	4.9.9.4 Evaluation	45
4.9.10	Work refusal	47

x

Contents

| | | 4.9.10.1 | Communication/training/evaluation | 49 |

4.9.11 Work stoppage 49
4.9.12 Working alone 49

    4.9.12.1 Workplace inspection 50
    4.9.12.2 When to inspect 50
    4.9.12.3 Who should be inspected? 51
    4.9.12.4 How to inspect? 51
    4.9.12.5 What to inspect? 51
    4.9.12.6 Inspection topics 52
    4.9.12.7 Responsibilities 54
    4.9.12.8 Procedures 54
    4.9.12.9 Communication and training 55
    4.9.12.10 Evaluation 55
    4.9.12.11 Due diligence 55
    4.9.12.12 Field work and external safety
               considerations 55
    4.9.12.13 Safety promotion and recognition 56
    4.9.12.14 Commendation is an important element 57

Chapter 5   Occupational Hazard Origin and Preventive Measures 59

Introduction 59
    Historical perspective 59
5.1   Hazard prevention strategy/hygiene 60

    5.1.1 Water sources of contamination 61
    5.1.2 What you can do? 61

5.2   Biological hazards 62

    5.2.1 Foodborne illnesses by pathogenic microorganisms 63
        5.2.1.1 Magnitude of foodborne illness 66
        5.2.1.2 Prevention 66
        5.2.1.3 Integrated Food Safety System (IFSS) 67
    5.2.2 Hazards associated with live or dead animals 67
    5.2.3 Food safety priorities 67
    5.2.4 Food allergy, sensitivity, and food intolerance 69
        5.2.4.1 Food allergy prevention 70
    5.2.5 Food safety of biotechnology-derived products 71
    5.2.6 Skin disease/dermatitis 72
        5.2.6.1 High-risk occupations 72
        5.2.6.2 Prevention 73
    5.2.7 Respiratory illnesses 73
    5.2.8 Hearing problems 73
        5.2.8.1 Preventive management 74
    5.2.9 Food hygiene safety and sanitation concern 74
        5.2.9.1 Musculoskeletal injury 75
        5.2.9.2 Working with animals 76

Contents xi

5.2.9.3 Transient emissions 77
5.2.9.4 Burn 77
5.3 Chemical hazard 78
5.3.1 Natural toxicants/prevention 80
5.3.2 Chemical hazard prevention 80
5.3.3 Workplace Hazardous Material Information
System (WHMIS 2015) 81
5.3.3.1 Controlled products 82
5.3.4 Future Transition to Globally Harmonized
System (GHS)/WHMIS after GHS 82
5.3.5 Fire safety 83
5.3.5.1 Fire classes 84
5.3.5.2 Fire extinguisher classes 85
5.3.5.3 Type and size of fire extinguishers 85
5.3.5.4 Inspection and maintenance of fire
extinguishers 86
5.3.5.5 Fire prevention management 86
5.3.5.6 Emergency fire procedures 87
5.3.5.7 Fire prevention special hazard 88
5.3.5.8 Combustible dust and explosion hazards 88
5.3.5.9 Chemical storage safety 90
5.3.5.10 Laboratory safety 92
5.3.5.11 Chemical inventory management,
segregation, and storage 92
5.3.5.12 Special chemical hazards 95
5.4 Physical hazards 99
5.4.1 Physical hazards in food and drink manufacturing 100
5.4.2 Common physical hazards in food and drink 100
5.4.3 Classification of physical hazards 101
5.4.4 Developing an effective physical hazards plan 101
5.4.5 Current regulations around the world 102
5.4.6 Prevention of physical hazards in food and drinks 102
5.4.7 Preventive management 103
5.4.8 Slips, trips, and falls 104
5.4.9 Identifying and assessing fall hazards 107
5.4.10 Falls from height and principles of working at
height 107
5.4.11 Fall hazards assessments 108
5.4.12 Preventive measures; falls from height 108
5.4.13 Ladder hazard prevention management 109
5.4.14 Scaffold safety 110
5.4.15 Workplace transport 110
5.4.16 Forklift truck 111
5.4.17 Struck by objects 112
5.4.17.1 Preventive management; struck by objects ... 113

5.4.18	Food processing machinery, tools, and equipment safety		114
	5.4.18.1	Main causes of injury	114
	5.4.18.2	General machine hazard safety	114
	5.4.18.3	Food processing machine safety prevention management	115
	5.4.18.4	Siting of fixed equipment	115
	5.4.18.5	Electrical safety	115
	5.4.18.6	Maintenance (including cleaning)	115
	5.4.18.7	Protective clothing	116
	5.4.18.8	Specific equipment/machine/tools hazard	116
5.4.19	Hazard prevention		117
5.4.20	Commercial mixers		117
	5.4.20.1	Identification and assessing mixer hazards	117
	5.4.20.2	Mixers hazard prevention management	117
	5.4.20.3	Prevention management; transport hazards	120
5.4.21	Lockout (energy control)		124
	5.4.21.1	Multiple workers/contractors and lockout/tagout	125
	5.4.21.2	Lockout/tagout exercise	125
	5.4.21.3	Hoists	126
	5.4.21.4	Hazard's priorities	126
	5.4.21.5	Glassware, sharp, and bottling hazards	126
	5.4.21.6	Hazard prevention	126
	5.4.21.7	Handling sharp objects	126
	5.4.21.8	Glass/bottling hazards prevention management in processing areas	127
	5.4.21.9	Fragile devices	127
	5.4.21.10	Radiation/and food irradiation hazards	127
	5.4.21.11	The food irradiation process and hazard	128
5.5	Noise hazard/hearing problem		129
	5.5.1	Hearing problem and noise level	133
	5.5.2	Food and beverage industry noise hazards	134
	5.5.3	Identifying and assessing noise hazards	134
	5.5.4	Noise preventive management and control	134
	5.5.5	Hearing conservation program (HCP)	135
	5.5.5.1	Engineering	135
	5.5.5.2	Administrative	135
	5.5.6	Stunning electrical equipment	135
	5.5.6.1	Hazard prevention	135
	5.5.6.2	Cleaning and storage	136
	5.5.6.3	How to use work equipment safely	136
	5.5.6.4	Vibration	136

Contents   xiii

| | | 5.5.6.5 | Pressure vessels and pipelines | 137 |

5.6 Electrical safety 139
    5.6.1 Arc flash hazard 139
    5.6.2 Protection against electrical hazards....................... 140
    5.6.3 Electrical hazard prevention.................................... 141
    5.6.4 Preventing power tool hazard................................. 142
    5.6.5 Power cord safety 143
        5.6.5.1 Inerting and purging 143
        5.6.5.2 Hazard preventive management................ 144
        5.6.5.3 Compressed gases................................... 144
5.7 Thermal hazards.. 148
    5.7.1 Cryogenic hazard prevention................................. 149
    5.7.2 Confined space hazard 149
    5.7.3 The risks 150
    5.7.4 Hazardous atmospheres... 150
        5.7.4.1 High-hazard atmosphere........................... 151
        5.7.4.2 Moderate-hazard atmosphere 151
        5.7.4.3 Low-hazard atmosphere........................... 151
    5.7.5 Hazard alert.. 151
        5.7.5.1 Oxygen: too little or too much.................. 151
        5.7.5.2 Toxic atmospheres 151
        5.7.5.3 Explosive atmospheres............................. 152
        5.7.5.4 Oxygen... 152
        5.7.5.5 Flammable material (Fuel) 152
        5.7.5.6 Ignition sources....................................... 153
5.8 Physical hazards 153
    5.8.1 Loose and unstable materials................................. 153
    5.8.2 Slip, trip, and fall hazards 154
    5.8.3 Falling objects 154
    5.8.4 Moving parts of equipment and machinery 154
    5.8.5 Electrical shock... 154
    5.8.6 Substances entering through piping 155
    5.8.7 Poor visibility 155
    5.8.8 Temperature extremes 155
5.9 Noise... 155
    5.9.1 Risk of drowning... 155
    5.9.2 Confined space entry program 156
    5.9.3 Identifying and assessing confined space hazards 156
    5.9.4 Construction 156
    5.9.5 Location.. 157
    5.9.6 Contents... 157
    5.9.7 Work activity 157
    5.9.8 Hazardous atmosphere 157
        5.9.8.1 Preventive management and hazard
                  control 157

xiv Contents

| | | 5.9.8.2 | Pre-entry precautions | 157 |

5.9.8.2 Pre-entry precautions 157
5.9.8.3 Testing the atmosphere 157
5.10 Purging 157
5.11 Blanking-off procedures 158
5.12 Clean out 158
5.13 Lockout procedures 158
5.14 Fire safety 158
5.15 Electrical safety 158
5.16 Emergency procedures 158
5.17 Entry precautions 158
Personal Protective Equipment 158
5.18 Access and egress 159
5.19 Administration 159
5.20 Training 159
5.21 Work permit 159
5.22 Specific hazards and elements 160
5.22.1 Welding 160
5.22.2 Hazard associated with nanotechnology 160
5.22.3 Temporary and seasonal workers 160
5.22.4 Sanitation and sanitary occupational hazards 161
5.22.5 Preventive management and control 161
5.22.6 Psychosocial hazards/stress 161
5.22.7 Environmental and stress hazard 162
5.22.8 Ergonomics hazards 163

Chapter 6 Occupational Injury and Its Management in Food and Beverage
Industry 165

6.1 Introduction 165
6.1.1 Introduction: common occupational injuries 166
6.1.2 Main causes of injury and accidents in food and
beverage manufacturing 166
6.1.3 Meat, poultry, fish, and petfood industry 168
6.1.4 Grain and flour milling industries 171
6.1.5 Bread, cake, and biscuit manufacturing 172
6.1.6 Chocolate and sugar confectionary manufacturing 173
6.1.7 Fruit and vegetable processing industries 174
6.1.8 Dairy, milk, and cheese manufacturing 175
6.1.9 Oils and fat processing industries 177
6.1.10 Canning 178
6.1.11 Egg production 180
6.1.12 Flavor manufacturing 180
6.1.13 Honey production (apiculture/beekeeping) 181
6.1.14 Tobacco manufacturing 182
6.1.15 Catering and food service industry 182

Contents

xv

	6.1.16	Spice manufacturing 182
	6.1.17	Food additive 183
	6.1.18	Chilled and frozen products 183
	6.1.19	Health and safety temperature requirements 183
	6.1.20	Working in chill units and freezers 183

6.2 Bottled water alcoholic and non-alcoholic drink and beverage industry.. 183

    6.2.1   Introduction 183
    6.2.2   Evolution of the industry 184
    6.2.3   Main causes of injury 186
    6.2.4   Soft drinks industry.. 187
    6.2.5   Brewing industry 187
            6.2.5.1   Main cause of injury 188
    6.2.6   Potable spirits industry... 189
    6.2.7   Fire and explosion 189
    6.2.8   Fruit juice production and frozen concentrate 189
    6.2.9   Soft drink concentrates 190
    6.2.10 Coffee manufacturing 190
    6.2.11 Tea manufacturing... 191
    6.2.12 Spirit manufacturing hazards and prevention 192
    6.2.13 Wine industry 192
    6.2.14 Hazard prevention in beverage and drink industries.... 192
    6.2.15 Injury management and prevention in food and beverage sectors 193
            6.2.15.1 Some general OHS practices 194
            6.2.15.2 Some specific OHS practices................... 195
    6.2.16 Sugar-beets industry... 197
    6.2.17 Working conditions 198
    6.2.18 Hazards and their prevention 198

6.3 Injury management and occupational rehabilitation 199
    6.3.1   Rehabilitation 200

Chapter 7 Food Security, Safety, Biosecurity, and Defense 203

7.1 Food security 203
    7.1.1   Introduction 203
    7.1.2   Effects of food insecurity and risk factors 204

7.2 Bioterrorism and agri-food..204
    7.2.1   Prevention measures..204

7.3 Food safety and sanitation...205

7.4 Food defense, biosecurity, and terrorism 207

7.5 Food production security..208
    7.5.1   Food tampering 208
            7.5.1.1   Food tampering signs and preventive measure 208

xvi   Contents

7.6   Food fraud and adulteration 209
      7.6.1   Food fraud definition..209
           7.6.1.1   Historical perspective: the fight against food fraud and adulteration......................209
           7.6.1.2   Food crime..211
           7.6.1.3   Food fraud or error 211
           7.6.1.4   Food fraud vulnerability............................211
           7.6.1.5   Food adulteration in the 21st century 212
      7.6.2   Development of food laws and regulations 212
           7.6.2.1   Food legislation...212
      7.6.3   Preparing a food defense plan 213
      7.6.4   Food biosecurity and terrorism 213
           7.6.4.1   Comparative risks of food and water as vehicles for terrorist threats 214
      7.6.5   Cybercrime..214
      7.6.6   Prevention and response systems in the food industry..215
7.7   Future directions...215
7.8   Food preparedness during emergencies and disasters...........215
7.9   Emergency food services..215

Chapter 8   OHS from Ethics to Sustainability an Agri-food Concern 219

8.1   Introduction 219
8.2   Sustainability an agri-food concern 220
      8.2.1   Sustainability pillars 221
      8.2.2   Hazardous waste management.................................222
      8.2.3   Sustainability in the laboratory 223
8.3   Ethical issues related to OHS and sustainability...................224
      8.3.1   Purposes of the code of ethics.................................224
      8.3.2   Ethics and waste management 224
      8.3.3   Is safety an ethical issue? 225
           8.3.3.1   Food sovereignty.......................................225
           8.3.3.2   Food ethic 225

Chapter 9   The Future of OHS in Food and Beverage Industries....................229

9.1   Introduction 229
9.2   Emerging trend, priority areas, and an OHS model..............230
      9.2.1   Emerging trends 230
      9.2.2   Priority areas 231
      9.2.3   An OHS Model..231

Contents

xvii

| | | 9.2.3.1 | Continuous Improvement Management System (CIMS) | 231 |

9.2.3.1 Continuous Improvement Management System (CIMS) 231
9.2.3.2 Program core elements to include 232
9.2.3.3 Hazards to be considered 232
9.2.3.4 Injury Management Policy (IMP) 232
9.3 Shaping the future 232

Appendix 1: Case Studies 235

Appendix 2 241

References 243

Index 283

Preface

This book has 9 chapters in 282 double-spaced pages (including Appendixes and References), 41 tables, and 7 figures as well as more than 20 case studies in Appendix 1.

The book content covers occupational health and safety (OHS) organization, fundamental management, accident and occupational injury prevention, adulteration, food safety, food defense, biosecurity, ethics, and sustainably and wraps up with a chapter on the future of occupational health and safety professionals in easy language to understand scientific information of OHS model, correlating with 838 references.

Occupational health and safety has been one of the leading sciences that undoubtedly touches every discipline being practiced. Exploring the unknown in science and technology means exploring the unknown in hazards. Things can happen at rates much faster than a human can react to prevent an accident, and yet we are expected to experiment, research, develop, discover, and produce in such an environment and do well and safely without damage to people, property, process, and environment.

From its roots more than 500 years ago to the present day, capitalism expanded from Western Europe to the USA and then to much of the rest of the world.

This expansion has not gone uncontested; resistance has been both direct and indirect, including political, religious, and social protest and even revolution but its impact has also been felt in development in many disciplines including occupational health and safety standards over time.

The economic cost associated with occupational health and safety has raised more attention to many other issues besides fatalities and injuries such as market loss, lawsuits, higher insurance premium, cost of accident, loss of reputation, loss of professionals, reduce worker morale and business efficiency, improve public moral and PR, increase financing and regulatory compliance, and boost corporate and social responsibility.

Since 1900 we have noticed major political, social, economic, scientific, and industrialization milestones, information technology, artificial intelligence, and other developments in food and beverage processing sectors such as nonthermal processing and health food (organic products, nutraceuticals, functional or holistic foods, etc.), food defense and biosecurity, information technology, artificial intelligence and cybercrimes, mental health issue, and climate change. Climate change is expected to lead to increased bacterial, viral, and pathogenic contamination of water and food by altering the features of survival and transmission patterns through changing weather characteristics, such as temperature and humidity, and therefore continue to have a major impact on OHS.

OHS hazards being chemical, biological, physical, or psychological have become an important topic to deal with since the last century and continue to grow and diversify, facing challenging issues such as COVID-19 at a global scale to date.

The global food and beverages market has been nearly $5,943.6 billion in 2019, having increased at a compound annual growth rate (CAGR) of 5.7% since 2015.

The market is expected to grow at a CAGR of 6.1% from 2019 and reach $7,525.7 billion in 2023.

So far, there has not been an adequate source of OHS information in the food and beverage processing industry sectors with many of its manufacturing sectors still being considered one of the high hazards in manufacturing sectors. Therefore, updating sources of knowledge and information for prevention, hazard reduction, and management not only in the workplace but also for the environment is a necessity.

This has also been indicated in a recent US study as the first use of the farm-to-table model to assess occupational morbidity and mortality, and these findings highlight specific workplace hazards across food system industries impacting both consumers and producers. This model unifies diverse industries into a common chain.

In this way, it reshapes our understanding of not only food safety concerns through contaminated food but also the costs that society bears for occupational injury, illness, and death that occur in the process of producing and delivering food to consumers.[718]

The next decades will witness the emergence of a world market and greater effort toward internationalization and harmonization of OHS standards for further protection of people, product, process, property, as well as the environment, in particular.

The rate of change in occupational health and safety has been rising substantially in recent decades in both processing and particular laboratories. Every indication is pointing out the fact that changes will continue to accelerate in the 21st century.

Foreword

Occupational health and safety (OHS) has been one of the leading sciences and undoubtedly touches every discipline being practiced. Exploring the unknown in science and technology is similar to exploring the unknown in hazards.

In the USA, the term OHS is referred to activities that can happen within or outside of the working environment.[1]

The food and drink sector provides safe, quality, healthy, and affordable food to millions of people worldwide.

Despite structural changes in the past decades, the sector remains a large source of manufacturing output and employment, particularly in developing countries where the industry grew rapidly. Officially available statistics suggest that more than 22 million workers were employed worldwide in food and drink manufacturing in 2008. These figures may increase significantly if jobs throughout the entire food production system are counted.

The new trends will have a considerable impact on accidents and safety in various sectors of the economy, particularly business and industry.[60] In recent years health and safety organizations responsible in major industrialized countries have identified to better handle industry's problems; in this regard it is better to address the issues separately for each sector such as aerospace, construction, petrochemical, marine, mining, pharmaceutical, forestry, cosmetic, agriculture, or food industry as the case may be. Accidents of various types and forms have been reported.

Between 1962 and 2002 there were 13,332 work-related highway death in the USA, almost 1.3 million are killed, and between 20 and 50 million are injured on the world roads according to WHO which is the leading death of people between 25 and 44 years of age.[3]

A 2018 US Bureau of Labor and Statistics has also reported various injuries and fatalities related to various industries including agri-food, fruits and vegetable, cattle and dairy production, potato and oil seed farming, orchards, even hunting and fishing, etc. including food manufacturing separately which provided OHS-related activities.[696]

A recent US 2020 census also indicates that food manufacturing workers have a 60% higher rate of occupational injury and illness than workers in other industries. The risk of occupational death is 9.5 times higher for food industry workers than in non-food jobs.

Besides 24% of food workers have been injured at their current job, and 17% of workers were injured in their first year. Researchers found that injuries from slips, trips, and falls were highest in food processing, storage, and retail.[668] Such statistical data helps food and beverage companies to plan ahead for their OHS program strategy.[669]

The food industry with many of its divisions such as slaughterhouse, poultry, meat processing, fish and seafood, fruits and vegetable, oil, dairy, flour milling and

xxi

grain, confectionary and bakery, condiments and seasonings, breweries, sugar, flavor, and others carries in general a high accident rate compared to other sectors.

Major political, social, and economic changes and scientific development such as that in low temperature, high pressure, and nanotechnology[324] has brought in major changes with hazards associated since 1900 in each decade.[325]

Each of these sectors like in some other industries carries their own specific ways of research, development, production, and marketing before a product reaches the consumer.

The National Safety Council (NSC) in the USA has been commissioning professionals at various times (1985 to present) to evaluate, identify, and recommend measures to reduce or eliminate accidents.

The first group committee members reported emerging trends to be based on their importance, persistence, significance confidence building measures and concluded that the following five issues could be ranked highest on the basis of these criteria[60]:

1. Ergonomics
2. Considering health, safety, and ergonomic issues at the concept and design stage
3. Incorporating safety into the management system
4. Repetitive motion trauma, including carpal tunnel syndrome
5. Back injuries with five other major areas that have been identified to be closely related to
   a) occupational health and wellness,
   b) public health and safety issues,
   c) competence health and safety professionals,
   d) changing nature of the workforce, and
   e) responses to changing technology.

Based on the results, NSC concluded the following factors were referred as emerging trend for OHS improvement in workplace: health and safety management; adequate and qualified OHS professionals, measured cut back during financial crises, ergonomics issues; better loss control management, cybercrime and transportation safety for future evaluation as priority areas to be improved.[60]

The World Health Report 2007 "A Safer Future" emphasized that the new watchwords are diplomacy, cooperation, transparency, and preparedness in order to ensure global public health security in the 21st century.[309]

The global food and beverages market is expected to grow from $5,838.8 billion in 2020 to $6,196.15 billion in 2021 at a compound annual growth rate (CAGR) of 6.1%.[646]

The World Health Organization (WHO) has been asked by its member states to provide advice and assistance on strengthening of national systems to respond to food safety.

As indicated by International Labor Organization (ILO), despite structural changes in the past decades, the food and beverage sector remains a large source of manufacturing output and employment, particularly in developing countries where the industry grew

Foreword

rapidly and although, overall, working conditions have gradually improved in the food and drink sector, there are still a number of challenges to overcome.[707]

The various chemical, physical, biological, and psychological hazards plus the addition of other hazards due to globalization, new technologies (IT, AI), new work culture and climate change affect, etc. continue to point to the existence and increase of OHS hazards in food and beverage industries as associated with each level of food production (farming, fishing, aquaculture); processing (refining, slaughtering, and packing); and distribution or storage (transporting and warehousing) to retail (grocers, wholesaler, restaurant, and ready-to-eat food server); and services which have had a lower rate of severe injuries than that of all non-food private industries that need to be addressed.[718]

A recent study in USA indicating in a first farm-to-table model in assessing OHS morbidity and occupational mortality rate for 'food system industries' have been significantly higher than 'non-food system industries' where the annual economic cost in the USA has been estimated to be approximately $250 billion.[718]

Applying the farm-to-table model is a novel construct within OHS and has the potential to reshape the understanding of how market forces in the food industry may impact producers and consumers in food chain hazard and can assist to reshape our understanding of the burden of "foodborne" illness to include not just pathogens and toxins that are transmitted to consumers through contaminated food but also the costs that society bears for occupational injury, illness, and death that occur in the process of producing and delivering food to consumers.

The study also indicated a lack of adequate data and not including the public sector establishments such as government and military.

Underreporting also occurs because of the employer, employee, physician, and survey design factors. In addition, the Bureau of Labor and Statistics (BLS) is not collected for farms with fewer than 11 employees and may exclude migrant or seasonal workers, which may lead to a substantial undercounting.[721] Besides, the study only includes traumatic causes of death and only those that occurred during a single shift or workday. Thus, it fails to capture occupational fatalities caused by long-term exposures or work-related illnesses. Work-related illnesses are associated with an estimated 49,000 deaths annually.

Therefore, it is likely that the true burden of injury and illnesses in food system industries may be higher than that we have reported here.[718] Other topic of discussion includes climate change and rising temperatures, resulting in higher food- and water-borne diseases, lower harvest, food price and adulteration increase,[825,826] as well as OHS stress and cyber-crime related factors from farm to table.[728–737]

The goals of OHS programs include fostering a safe and healthy work environment. OHS may also protect co-workers, family members, employers, customers, and many others who might be affected by the workplace environment. However, additional safety measures will be required by users, and they should consult pertinent local, state, and federal laws, legal counsel, and other professional consultation prior to initiating any accident prevention program circumstances.

The next decade will witness the emergence of a competing world market and greater effort toward internationalization and harmonization on OHS standards for the protection of workers, process, food product safety, food defense and security, as well as the environment as a whole will be required.[309]

The information provided in this book is intended to serve as a starting point for good occupational health and safety practices and does not provide specific minimal legal standards. This information could provide basic guidelines for accident prevention. Therefore, it cannot be assumed that all necessary warning and precautionary measures have been provided.

Authors' disclaimer

Although the information and recommendations in this publication have been compiled from sources believed to be reliable to the best of our ability, the authors make no guarantee as to, and assume no responsibility for, the correctness, recommendations, and completeness of such information provided. Additional occupational health and safety measures may be required. Users should consult pertinent local, state, provincial, federal, national, and international laws, regulations, and other legal standards that may be required prior to initiating any accident or other similar program prevention measures. The information provided is intended to serve as a starting point and basic information for good OHS practices and may not provide specific or minimum legal standards.

Therefore, it cannot be assumed that all necessary warning and precautionary measures have been provided and need to be clarified in advance by workers and employers, and as it has also been indicated in the book, the employers are responsible for:

- Performing a "hazard assessment" of the workplace to identify and control physical and health hazards.
- Identifying and providing appropriate personal protective equipment (PPE) for employees.
- Training employees in the use and care of the PPE.
- Maintaining PPE, including replacing worn or damaged PPE.
- Periodically reviewing and evaluating the effectiveness of the PPE program.
- In general, employees should properly wear PPE, attend training sessions on PPE, care for, clean, and maintain PPE, and inform a supervisor of any defective needs, repair, replacement, and/or maintenance that may be required.

About the Authors

Ebrahim Noroozi obtained his BSc degree with first class honors from the Institute of Food Science Engineering in Tehran, Iran, in 1972 and his MSc degree with honors in Food and Dairy Chemistry at McGill University, Montreal, Canada, in 1978.

He has published and/or collaborated in more than 100 research, books, industrial reports, and publications as well as a number of newly developed products. He has been providing extensive presentations, workshops, and lectures in OHS in the food and beverage processing industry nationally and internationally since 1978. He joined the Food Science Department at McGill University in Canada in 1987, where he is presently Manager of the Department's Teaching Laboratories and has been acting professionally as Food Scientist (CFS), Occupational Health & Safety Professionals (CRSP), as well as Professional Chemist (P. Chem.) in the field.

Professional accreditation, certifications, and awards:

- Board of Canadian Registered Safety Professionals as Certified Registered Safety Professionals (CRSP), Quebec RSC Chair & Ambassador, Quebec, Canada, 1995 to present
- The International Food Science Certification Committee (IFSCC) at the Institute of Food Technologists (IFT) as Certified Food Scientist (CFS), 2013 to present
- Ordre des Chimistes du Quebec as Professional Chemist (P. Chem.), 1989–1999
- Winner of more than a dozen national and international awards including Canada's Governor General Sovereign Medal for Volunteers (SMV) 2021
- Commissioner of Oath, Ministry of Justice, Quebec, Canada, 1987 to present

Ali R. Taherian joined McGill University in 1991 after serving food process industries for 15 years to pursue his master's and PhD educations.

He then joined Agriculture and Agri-Food Canada as a research professional while serving McGill University as an adjunct professor. He published more than 50 peer-reviewed articles, chapters, and books and provided over 500 presentations and workshops on food safety and security worldwide.

He also co-supervised and graduated over 12 master's and PhD students from different countries.

Acknowledgments

The first Canadian edition of *Occupational Health and Safety in the Food and Beverage Industry* has been in the making based on our professional and educational experiences in academic, industry, and government organizations in the past several decades. We look forward to readers for their continued input for any improvement of the book in the future.

The authors would like to thank the following reviewers for their constructive suggestions, ideas, and comments for the manuscript proposal on behalf of CRC Press:

Dr Varoujan Yaylayan, Professor (Flavor Chemistry) & Chair, Department of Food Science, McGill University, Montreal, Canada

Ms Lee-Anne Lyon-Bartley, BASc, CRSP, NEBOSH, CP-FS, Executive Vice President, Health, Safety, Environment & Quality, Dexterra Group Inc., Ontario, Canada

Dr Reza Zareifard, PhD, Scientist, Food Research & Development Center, Agriculture & Agri-Food Canada, Montreal, Canada

Dr Yixiang Wang, PhD, Professor (Food Packaging), Department of Food Science, McGill University, Montreal, Canada

Ms Judith McNulty-Green, The National Institute of Occupational Safety and Health (NIOSH), Washington, DC, USA

As well as special thanks for intellectual support and encouraging assistants from the following professionals, colleagues, and many friends:

Mr Howard McGraw, CRSP, Former Director of Occupational Health & Safety, Maple Leaf Canada, Industrial Accident Prevention Association (IAPA), Ontario, Canada

Mr Wayne Wood, MSc, CIH, Former Director, Environmental Health & Safety, McGill University Montreal, Canada

Claude Lahaie, Former Manager, Waste Management Program & Associate Director, Emergency Measures & Fire Prevention and presently Associate Director, Building Operations, McGill University, Montreal, Canada

Mr Joseph Vincelli, MSc, CRSP, Operational Manager, Environmental Health & Safety, McGill University, Montreal, Canada

Dr. Soraya Azarnia, Ph.D., Food Scientist, Professional Consultant, and University Laecturer

Mr Peter Fletcher, CRSP, Executive Secretary, Board of Canadian Registered Safety Professionals (BCRSP), Ontario, Canada

Late Dr Habibollah Hedayat, MD, PhD, Former Professor (Food Science & Nutrition), Father of Iranian Food Science & Nutrition, Former Dean and Founder of Faculty of Food Science Engineering, Tehran, Iran

Late Dr Shahab Vaeza Zadeh, DVM, Former Professor (Food Science Engineering), Faculty of Food Science Engineering, Tehran, Iran

Late Mr Das Nabin, MSc, Former Technical Chemist, John Abbott Collège, Ste Anne de Bellevue, Quebec, Canada

1 Introduction: Health and Safety Organization

1.1 INTRODUCTION

In the continually expanding context of occupational health and safety (OHS), the word safety refers to physical, chemical, biological, and psychological hazards that could impact hazardous conditions in the workplace.[6]

Over the last century, organizations and institutions have developed a growing capability to create catastrophic accidents. Someone's error often causes a slip, trip, or fall. The slips, trips, falls and other kinds of errors may result in the destruction of people, organizations, or systems of organizations. As a result, the general issue of risk integration or reliability enhanced in organizations is becoming increasingly more important over time.[546]

Ultimately, knowledge of hazards and their elements is the earliest step for health and safety improvement. The International Board for Certification of Safety Managers (IBFCSM) defines a threat as "Any solid, gas, or liquid with the potential to cause harm when interacting with an array of initiating stimuli, including human related factors."[12]

In order to reduce the risks associated with all threats, the Canadian health and safety legislation issued that employers are required to have a written health and safety program in their workplace. Documentation of occupational health and safety policy could help the effectiveness of the program and mirrors the unique needs of a workplace. The document should be regularly reviewed and updated in order to assist in the application of policy in the workplace. This policy should be signed by a senior manager or president for further establishment of the commitment to the health and safety of the workplace.[814]

In short, assignment of responsibility and a well-written OHS policy is the first step in hazards prevention which requires the following essential elements: (1) identification; (2) safe work procedures; (3) training; (4) inspection; (5) accident/incident investigation; (6) health and safety committee meetings; (7) records and statistics; and (8) medical back-to-work program.

1.2 SAFETY: A COMPREHENSIVE RESPONSIBILITY

Today, safety is the responsibility of various divisions within an organization instead of being limited to the safety department, management, or production line. Due to the advent of regulations, many other units also have specific responsibilities which can directly affect the health and safety program. Among these units include but are not limited to environment, health, and safety (EHS) office; human resources; legal/

DOI: 10.1201/9781003303152-1

2 Occupational Health and Safety in the Food and Beverage Industry

regulatory office; risk management; planning, engineering, and production; purchasing; fire and emergency measures and facilities.[12]

When it comes to safety, therefore, everyone has a responsibility to play. The following provides a general view of such responsibilities for employers, supervisors, and workers for their various responsibilities in the organization's mission.[519]

1.3 ORGANIZATION

Nowadays, science-based human resources perk the attention of different organizations.[44] They include both motivation/incentives and formation of proper organizational structure. They also focus on the training of new employees in order to take a more active role in monitoring OHS needs specific to a company or organization.[12]

1.4 THE SAFETY DEPARTMENT

A typical hierarchy of controlling accidents in descending order of effectiveness has been referred to an effective safety department for hazard elimination and reduction, engineering application, training/certification, accident prevention/incident and investigation, personal protective equipment (PPE), food permit license, hazardous waste management, industrial hygiene, medical hygiene/first aid and cardiopulmonary resuscitation (CPR), spill response, fire and emergency measures, biological, chemical, and physical safety programs, record keeping, and evaluation as well as liaison with OHS/EHS regulatory officials.[12]

1.5 PRODUCTION DEPARTMENTAL RESPONSIBILITY

The local department is an entity within an organization partaking in reduced available resources but should be encouraged and provided with means to progress in safety practices. The leadership or chair of the departmental safety committee should provide a natural channel for communication with and within the personnel and provide a means of enforcing safety policies of the departmental concerns. No safety department is sufficiently large enough to be present everywhere. Safety personnel need assistance from local production, sectors, or department to implement safety practices and policies.

They may also develop their tailor-made occupational health and safety (OHS) designs to implement general OHS policies in their sections.

In the manufacturing environment, the production department should assign someone as the safety coordinator who may also act as a liaison to the organization's safety committee as it may be a food manufacturing or other organization. It is required and preferred upon the situation that each production department has its safety committee with representation from lower to higher management at this committee.

Their primary function is to advise individuals and make recommendations on improvements for safe operation. They should be supported for their physical and staffing needs. They may decide to have audits, training, setting standards, and/or

Introduction: Health and Safety Organization

similar OHS procedural promotion for accident prevention as well as improving life quality.

1.6 THE PRODUCTION LINE/LABORATORY RESPONSIBILITIES

The critical factors when considering laboratory safety have been referred to infectious agents, economics, awareness of risks, and management consideration.[331]

The personnel in the production line, the laboratory, or the smallest unit in an organization will ultimately carry the final health and safety responsibility. Although, the supervisor may not carry similar attentivity as workers but is responsible for setting the highest standards of performance for all the workers. Therefore, the supervisor must make sure that all the personnel are fully aware of the associated risks in their sections.

Laboratory safety is one of the most important part of health and safety in academic, research institutions, and food industries. The lab safety program should be constantly improved to match the needs of day-by-day changes considering experiment modifications, chemical storages, waste segregation, inventory, and so on. Timely practices are also needed to encourage and attain environmental and public health protection and improvements in lab safety program.[332]

2 Fundamentals of Health and Safety Management

2.1 INTRODUCTION

Today, people often long for the good old days. But their yearning is only for the real or imagined positive parts of the past. In the past health and safety were vastly different than now where the industrial revolution was in its infancy, machines and equipment were unguarded, and unskilled and untrained working conditions and lack of OHS regulation existed.

Today, there have been significant changes in accepting any type of fatal and/or nonfatal accidents as well as other incidents and overall losses to find out the root causes of the problem. Such changes have of course been due to many new regulations as well as the importance of skilled and educated workers and their familiarity with OHS in the workplace.

In fact, today accidents have been identified as being categorized as a personal risk rather than an employer's responsibility or the state compared to those a century ago. Many proactive safety measures in collecting accident reports have led to better identification of the root causes.

Many workers and employers failed to notice and report accidents due to carelessness, clumsiness, fatigue, or individual differences which could be the root causes of accidents. However, today all these factors become the focusing points of educational and training proposes in order to address the prevention measures.[551]

In order to minimize accident probability, the first step became the need of encouraging the habit of accident prevention.[551] Therefore, this required constant and correct use of engineering and process control measures such as providing personal protective equipment (PPE), if possible, less and/or minimizing hazardous substances, supplies, machineries, or methodology. In order to achieve such objectives, it was realized that we need to have management responsibility and support in areas such as product safety and quality, production quantity, customer service satisfaction, and target date.[547]

Therefore, accident prevention needs an effective OHS program supported by higher management level who is ultimately responsible for health and safety in an organization.

Of course, the presence of a trained and qualified safety professional with the authority to act is a must. A good accident prevention program should consider constant training and reminders of common and possible hazards involved to which people may be exposed with particular attention to the materials, equipment, machine, process, and facilities with which they work.

DOI: 10.1201/9781003303152-2

The rapid pace of new information, technology, and changing regulations in the 21st century demand the ongoing training even for experienced workers.[14]

To assess your needs and priorities in food and drink production, you need to consider the facilities involved. One recent survey reviewed by the Food Products Division of the Industrial Accident Prevention Association (IAPA) in Ontario, Canada, concluded that 87% of those surveyed thought organizations would benefit by developing an effective occupational health and safety (OHS) program guide in complying with regulations as well as established standards such as Good Manufacturing Practices (GMP) or Hazard Analysis Critical Control Points (HACCP) as part of their OHS programs.[16]

2.2 INVESTMENT IN HEALTH AND SAFETY MANAGEMENT PAYS OFF

Investment in OHS is not only to do research and find out the current needs but also to address the financial merits as well as provide insight into which sectors and types of interventions are needed for future economic evaluations.[16]

Such losses are considerably high. In one report from the USA, the total number of fatal and nonfatal injuries was more than 5,600 and 8,559,000 with a total cost of $6 and $186 billion, respectively.[7,19]

First, protecting workers from injury and illness is not only a law and a legal issue but also required morally on behalf of everyone. Second, even during hard financial times, the organization needs to prioritize its interventions and target areas in its OHS program.

Once you know where the priorities are, then even if you have to cut back at least you do so where there is the least possibility of losses involved to be considered in your OHS program.

Overall, it is essential for any organization to evaluate the cost and losses due to the lack of an adequate OHS program.

A closer look at the issue that affects any organization reveals such direct or indirect benefits by considering some of the following factors[17]:

- More business, better business: In fact, many companies check their supplier's exemplary safety records, and financial institutions will consider duplicate safety records before providing loans for investment.
- Better quality: Many businesses, large and small, have found that the quality of their products improved after they corrected safety problems. Examples of factors contributing to quality improvement in an organization are training, effective communication, a system for ensuring standards, and employee involvement as part of the organization's OHS program.
- Motivating employees: Organization's active commitment to safety lets workers know that they matter most to keep them in their jobs by preventing them from both injury and illness.

Fundamentals of Health and Safety Management

- Reduced costs: It has repeatedly been demonstrated that for every dollar lost in direct costs due to an injury/illness, workplace experiences 3–10 times losses of indirect cost.
- Other benefits: Investing in health and safety would be avoiding criminal prosecution by regulations, settlement payment for injury or death, legal fees for defense against the claim, reduced equipment and operating cost, emergency equipment expenditure, salvage of damaged equipment, vehicles recovery cost, operation loss, loss of experience and trained workers, training cost for replacement, lost time of personnel other than those involved in the accident, the investigation, rescue, or salvage, such as public relation and management people, insurance and medical premium cost increase, public confidence loss, prestige and moral degradation, payment of emergency services such as police, fire department, security, and regulatory agencies and much more.[326]
- Also, negligence in health and safety can cause direct and indirect cost/losses such as fatal injury and suffering, motivation loss among employees, reduction of profit or bankruptcy, legal fees and court battle costs due to negligence and criminal offenses, process, product, machinery or property damage, insurance cost rise such as medical, fire, theft, etc., environmental losses due to air, land, and water contamination or pollution, market share losses due to bad reputation from consumers, contactors, granting agencies or financial institutions for providing loan, etc., ethical value loss, psychological issues on present employees, attracting new workers and professionals, training cost for replacement workers, and cost associated with emergency services such as police, fire department, ambulatory services, or regulatory agencies.[326]

It should be noted there are other losses, such as co-workers or supervisor's time and production losses, worker's rehabilitation, and adaptability after returning to the job. In fact, investment in health and safety programs is like insurance for your organization.[466]

In Canada, factors to consider for the accuracy and reliability of injury rates could be jurisdictional size, injury under-reporting, differences in legislation, and interpreting in comparing fatality and injury rates as reported by the University of Regina.[612]

In Canada, there have been some statistical reports on various hazards and injuries in the food, beverage, and tobacco processing industry.[613,614,327]

Nationally in Canada, there has been a report of 67 fatalities and various other injuries mainly due to motor vehicles, floor and ground surfaces, machinery, environmental conditions, and non-pressurized container.[216]

2.2.1 Losses and due diligence

As recently witnessed in Canada, company owners and managers are increasingly at risk of being taken to criminal court and facing huge fines and imprisonment, even if they did not mean to do anything wrong, did not know they were doing anything wrong, or did not know that an employee was committing a prohibited act.

In one significant new law "Safe Food for Canada Act," the accused could be liable on both charges of fine indictment and imprisonment of an offense.[462]

2.3 ELEMENTS OF A SAFETY MANAGEMENT SYSTEM

Any successful implementation of an OHS program requires an effective safety management system that overall covers the protection of people, process, property, and environment as a whole and must include the following seven steps[463]:

a. Safety plan
b. Process and procedures
c. Training
d. Verification/monitoring
e. Supervision
f. Reporting
g. Evaluation

2.3.1 CORE ELEMENTS OF HEALTH, SAFETY, AND ENVIRONMENTAL MANAGEMENT

Employers have legal obligations for an effective and successful OHS not only to be designed based on their needs but also to provide leadership for its successful implementation.[464]

2.4 MODERN SAFETY MANAGEMENT EVOLUTION

Historically, many factors have changed since the industrial revolution of 1900s where equipment and machines were unguarded, workers were unskilled and untrained, work hours were much longer, educational levels were lower, employers and employees were careless, and safety laws were limited. In fact since the last century evolution of safety has been impacted by many factors such as union formation, consumer demands, technology, workforce changes, safety laws, environmental issues, ethics, and so on.[353] OHS has gradually stepped away from finding whose fault it has been and rather tries to find the root cause for correction by getting the employees and employers involved in workplace health and safety which has been beneficial as a loss control measure and prevention issue to both.[465]

We have also witnessed, as organizations and workforce may reshape and/or disappear, sometimes so do some OHS knowledge and skills along the way, and that is why such developmental changes in the process may require continuous improvement to keep up with changes.[690]

3 Health and Safety Legislation, Audit, and Standards

3.1 INTRODUCTION

Protection of consumers from the consumption of unsafe and contaminated food is an essential part of food production. In this regard, engagement of the preventive measures integrates the greatest amount of expenditure for the food producers.

Gradual increase in investment for consumer protection came along with the evolvement of agricultural development and urbanization in the 9th century first in the Fertile Crescent (present Iran, Turkey, and Syria). Throughout history, dietary laws have sustained spiritual identity in the face of invasion and forced conversions by foreign tribes and religions where societies have permanently attached symbolic value to different foods in their culture.[351]

In ancient Babylon, for example, the "Code of Hammurabi" prescribed punishment of overseers for injuries suffered by workers. For instance, if a worker lost an arm due to an overseer's negligence or oversight, the overseer's arm was taken – to match the worker's loss.[352]

In recent years the engagement of safety professionals has continued to rise worldwide and created vast opportunities for OHS employment due to the upsurge of epidemical diseases such as COVID-19.[366]

Because food legislation will apply to (the processing of) a substance if this substance can be qualified as food, it is essential to know whether the substance is a food or not.[817] In addition, the Food and Drug Administration (FDA) has, recently, announced the Food Safety Modernization Act (FSMA) Produce Safety Rule (PSR) which is now available in Chinese and Portuguese languages. PSR was previously translated into Spanish for the broader global audience and access.[686]

3.1.1 WHAT ARE GUIDELINES?

Guidelines are departmental documents for the interpretation of the regulation or legislation. They could be driven by legislation and are often used to advise how one might comply with a regulation. Guidelines do not partake the law enforcement.

DOI: 10.1201/9781003303152-3

3.1.2 WHY HAVE FOOD STANDARDS BECOME SO IMPORTANT?

The importance of standardization relies on an equal share in food control for all the trade partners. In other words, each partner has to put in a similar effort to assure the food meets market expectations. The increased need to meet a wide range of standards by trade partners to ensure "equivalence" in food control or achieve market expectations has elevated the commercial importance of food standards. The rising actual and perceived economic impact of meeting these standards has necessitated investigations into the relevant costs and benefits. Transparency within the standard-setting and compliance process is therefore crucial to international trade.

Audit systems are now used by government agencies that monitor these agreements, and therefore, audit system has been recognized as the mechanism to ensure effective food control systems have been implemented and maintained.[59]

3.1.3 THE NEED FOR AUDIT

Audits are conducted by trained professionals. Those facilities that participate in a program of in-plant audits are thoroughly examined and receive technical assistance in all areas affecting pesticide use, product integrity, regulatory compliance, and proper handling of ingredients. For more than 50 years, the American Institute of Baking has offered training and certification to people providing food audit services.[18]

The role of audit is therefore of increasing importance, and the relevant skills needed by food "inspectors" need to be defined and agreed upon at national and international levels. Auditing food control systems using standard methods is now recognized as a challenge in the complex world of food protection in the 21st century.[59]

3.2 FOOD HYGIENE LEGISLATION

It has been reported that in the USA, more than 3,000 state, local, and tribal agencies have a primary responsibility to regulate the retail food and food service industries.[31]

They are responsible for the inspection of over one million food establishments, restaurants, and grocery stores, as well as vending machines, cafeterias, and other outlets in healthcare, schools, and correctional facilities.

FDA strives to promote the application of science-based principles in retail and food service settings to minimize foodborne illnesses. It assists regulatory agencies and industries by providing a model Food Code, scientifically based guidance, training, program evaluation, and technical assistance.[31] In the UK, food hygiene legislation is closely related to the legislation on the general requirements and principles of food law. The legislation lays down the food hygiene rules for all businesses, applying practical and proportionate controls throughout the food chain, from primary production to sale or supply to the food consumer.[31]

The Canadian Food Inspection Agency (CFIA) has held broad consultations with frontline inspectors, consumer associations, industries, and government stakeholders

Health and Safety Legislation, Audit, and Standards   **11**

to develop a single and consistent inspection approach. The original intent was to find an applicable model for inspection of all food-related activities, whether related to human, animal, or plant health or the environment.[33]

3.2.1 REACH: THE NEW EUROPEAN CHEMICAL LEGISLATION

Registration, Evaluation, Authorisation and Restriction of Chemicals (REACH) came into effect on June 1, 2007. As a competent agency for the health assessment of chemicals and products, BfR played a major role in shaping the new REACH regulation, which applies equally to all member states of the European Union.[488]

3.2.2 INTERNATIONAL STANDARDS ON HEALTH AND SAFETY

The food, drink, and tobacco committee section of the International Labor Organization (ILO) in Europe and Central Asia has also been active in promoting OHS standards for ISO 45001 in 2018. ISO 45001 specifies requirements for an occupational health and safety (OH&S) management system. It also offers guidance for preventive measures and proactive improvement of OH&S performance.[679,680]

OHSAS 18001 was first introduced by British Standard Institution (BSI) in 1999 with the main focus on controlling the risk of hazards. The main difference between ISO 18001 and ISO 45001 is the reactive and proactive approaches, respectively. In the reactive approach, the problem is emphasized after happening, while the proactive approach focuses on the elimination of the event before its occurrence.

In China, food safety issues fall under National Medical Products Administration which is a restructured FDA. Under this act, food companies and local people's governments carry the main and integrated responsivities.[687]

3.3 SAFE FOOD HANDLING GUIDELINES

The followings are some guidelines recommended for proper safe food handling[34]:

- Avoid working with food and drink if you are sick with illnesses such as cold sore, infected cuts, etc.
- Purchase food from certified suppliers with valid permits from health authorities.
- Have complete procedural standards from receiving to preparation and storage.
- Apply a standard food safety method such as Hazard Analysis Critical Control Points (HACCP) or Good Manufacturing Practices (GMP), etc. in your production.
- Follow proper personal hygiene and pest control systems in place.
- Wash and sanitize raw fruits and vegetable products before serving.
- Sanitize utensils, equipment, and machine in the processing line.
- Handle food products with proper utensils when feasible and wear gloves.

- Keep cold food cold (below 4 °C) and hot food hot (above 60 °C), the temperature from 4 to 60 °C is called a "Danger Zone."
- Follow proper OHS practices to avoid biological, chemical, and physical hazards.

3.3.1 FOOD STANDARDS AND AUDIT

3.3.1.1 What are food standards?

Food standards are a set of criteria that a food must meet to be suitable for human consumption. For example, food source, composition, appearance, freshness, permissible additives, and maximum bacterial content fall under these criteria. In addition, ensuring consistency in manufacturing food products from safety, nutritional, and management system perspectives is a vital measure which may be required by law or by the market.

The existence of standards goes back to ancient Athens where beer and wine were inspected for their purities. Today, most modern food standards and food control systems date back to the late 1800s. In the USA, the primary federal law governing food additives is Food, Drug and Cosmetic Act, dating back to 1938. Over time, various issues related to labeling, health claims, color additives, direct or indirect additives, irradiation, and, in recent years, biotechnology-derived products become topics of interest to industry, government, and consumers.[35]

The legal food standards cover a wide range of food criteria from specific products like meat, fish, and egg to crosscutting such as the horizontal general hygiene directive. Briefly, the existence of clear standards is crucial for complex food networks to safeguard the efficiency of food production within numerous legal requirements.

In Canada, the Standard Council of Canada (SCC) carries the responsibility of leading and enabling the use of national and international standards. The main emphasis of this council is on the improvement of the workplace's psychological health and safety by considering risk mitigation, cost-effectiveness, recruitment, retention, as well as organizational excellence and sustainability.[688] For example, the development and use of Hazard Analysis Critical Control Points (HACCP) in support of the World Trade Organization and General Agreement on Tariffs and Trade (GATT) is maintained by Standard Council of Canada.

The international standard-setting body – Codex Alimentarius was also set up by the FAO/World Health Organization (WHO) Food Standards program established in 1963 to protect consumer health and safety.[59]

One of the primary tools to control the implementation of quality standards and ensure the appropriateness of intended use is Good Manufacturing Practices (GMP).[40] In fact, GMP is the foundation of any effective food safety program, including HACCP. The GMP Advantage program addresses the hazards associated with the personnel and the processing environment. The food production facilities can apply for GMP certification once the 60 GMP standards and the accompanying monitoring procedures are implemented. The application should be forwarded to the Canadian General Standards Board (CGSB) for an audit and GMP certification.[41]

Health and Safety Legislation, Audit, and Standards

However, food standards, once set up, are not usually implemented due to deficiency of resources, resistance to change, lack of commitment, and increased training requirements.[59] In addition, the commercial value of such an implementation has not been promoted enough in the industry due to fears of the implication cost, or it has been addressed poorly by the government. Moreover, food supply shortage and price increase could render the safety of food to become secondary to its availability.[36]

Nowadays, the food industry faces diverse challenges due to an increase in both level of regulations related to the pandemic and consumer concerns for food safety. Accordingly, manufacturers have been actively working to develop suitable systems for controlling the risks, methods of assessments, and food quality improvement in the market.

Traditionally, the regulators did set the standards with regulation setting and enforcement of standards. Most of the Latin American Countries of the Western hemisphere base their food standards on Roman law, while the English-speaking countries have their basis in Anglo-Saxon Law.[117]

Along with the growth of the food industry, the government's approach to food inspection has evolved. The trade commercialization of food commodities worldwide has also placed more pressure on the manufacture, storage, and distribution of quality food products around the world, where the economic cost may put food safety and quality in the danger zone for consumption. Thus, the institutional roles in risk-based management must be clarified. It should also be noted that food safety faces similar challenges for institutional, scientific, business, and regulatory compartments. For instance, a food product such as a loaf of bread is made in the USA using seven different ingredients supplied by diverse countries of origin all around the world.[53] Consequently, all national and international bodies are looking for a new and pioneering method of assessment to attain safe and healthy food products.

Regulators, on the other hand, are more focused on auditing the food safety systems engaged by food businesses and moving toward proactive and preventive measures, rather than inspection. The food industries, now, carry the greater responsibility for assuring the health and safety of food products. This has led auditing to become a vital issue for the food industries. The food businesses must, therefore, focus on self-auditing or third-party accreditation to improve their performance, rather than the effectiveness of regulators' auditing.

Another crucial issue which affects international trading is the transparency of the standard-setting. The regulator must provide clear information regarding the standard and its use, as well as make sure it has been understood by all trading partners. It has been known that, in the long term, the cost of the regulators' new approach to assessment of the industry control system is less than that of the traditional approach. Moreover, the new approach provides a greater pledge for assuring safety in the food system.[58]

3.3.2 The Canadian food safety approach

In Canada, the Canadian Food Inspection Agency (CFIA) carries the responsibility of consumer protection from food fraud and consumption of food involving danger or

14   Occupational Health and Safety in the Food and Beverage Industry

risk. This organization was founded in 1997 and is the combination of three departments including Agriculture and Agri-Food Canada, Fisheries and Oceans Canada, and Health Canada. The main action of the CFIA is assuring the safety of food using tools such as inspection, sampling, and data collection and verifying no compliance. Activities of CFIA cover the vast area of all federal food inspection from farm to fork and include animal feeds, food production chain, distribution, and retail stages. The agency assures the safety of food, animal health, and plant protection.[59]

The use of traceability systems by Canadian federal, provincial, and territorial governments, in partnership with industry, provides a unique approach to assure safety and wholesomeness of food and feed from farm to fork in Canada. The traceability systems are based on three essential elements: animal identification, animal movement, and premises identification. The ability to rapidly trace an animal throughout its life cycle is essential to isolating animal health emergencies, and it can help limit the economic, trade, environmental, and social impacts of such emergencies.[57]

The CFIA has several successful programs including: 1) the quality management program (QMP), which is a mandatory program for fish and seafood products. This program covers both exports and inter-provincial trades in order to verify whether the food safety regulatory conditions are encountered. 2) The QMP for the fish importers (QMPI), who have voluntarily chosen to assume additional responsibilities under this program. 3) The Food Safety Enhancement Program (FSEP), also a voluntary program, designed to encourage the development and maintenance of HACCP-based systems. This program which is a federally registered agri-food processing center covers processed dairy, meat and poultry, fruits and vegetables, and shell and egg products.

3.3.2.1 The Integrated Inspection System (IIS)

Presently Canadian system is considered among the best in the world. The IIS, naturally, has evolved to meet the challenges of new hazards. It is considered the first corporate business plan of the CFIA which was developed as a mechanism to guide the evolution of all agency inspection programs under a consistent approach for all food commodities. This program considers that the industry is responsible for controlling its products and processes in agreement with recognized standards. The government, on the other hand, is responsible for verifying the effectiveness of the industry's control systems and intervening when necessary. The primary objectives of IIS are to provide efficient, uniform, and transparent inspection strategies for a food safety system in order to protect consumers and deliver the national and international requirements for accessing Canadian food, animal, plant, and forestry products. The strength of the IIS food control system lies in the integration of internal and external parties and the cooperation of all stakeholders for attaining the compulsory results.

3.3.2.2 Legislative levels in Canada

Health and safety regulations that may indirectly affect the food processing facilities within one nation could be originated from various levels[61]:

1. Federal legislations include Workplace Hazardous Material Information System (WHMIS); Transport of Dangerous Goods, or Canada Labor Code Regulation.
2. Provincial such as also WHMIS, Occupational Health and Safety Act, Environmental Regulation, and Sanitation and Waste Disposal Regulation.
3. Municipal such as Fire Codes, Building Codes, and Sanitation and Waste Disposal Regulations.

3.3.3 Standards and their Importance for Audit

The establishment of standards for private companies, generally, involves both standard-setting coalitions and private firms. The main aim of standard-setting is to enable supply chain management by progressively considering globalization and competitive international food market. The increase in consumer awareness of food safety along with the rise in global and complex supply chains is the main drive for setting effective standard systems to guarantee the wholesomeness and safety of food products.

The standard-setting is also crucial for trade partners in order to practice equivalence in food control to achieve market expectations. It is of great importance to be lucid and transparent in setting the standard, in such a way, that it does not impose any barrier in international trade for developing countries.

With this regard, Health Canada has provided the guidelines to clarify interpretation of legislation and regulations. Health Canada has also developed and enforced different regulations to assure the health and safety of Canadians.[59]

However, food safety, based on the general agreement, should not be used as a competitive tool in the marketplace, although this may not be the fact constantly.[25]

3.3.4 Standard-Setting Process and Principles

1. **Simplification**: There must be a continual process in simplifying complex processes and making them easily understandable. Institute of Food Technologists (IFT) Good Manufacturing Practice Guidelines is a document for food manufacturers on producing safe and quality food commodities.[59]
2. **Cooperation**: The standards should be reached by agreement among all parties involved, as the economic and social needs may be different.
3. **Implementation**: The standard must be achievable and accessible for the implementation of original World Health Organization (WHO) documents on food safety systems, such as HACCP, which did have principles of the method but lacked sufficient information until recent publication, where it has been simplified.
4. **Selection**: Subjects that standards may cover in the food sector can be varied and wide and must be chosen carefully. It is, therefore, important to make sure that all the aspects of standards are considered to ensure the relevancy and durability of the standard.

5. **Revision**: Usually, standards are reviewed every three years. Currently, significant food safety hygiene rules are being studied by the EU, which makes the food operators throughout the chain responsible for food safety.[59]

6. **Determination of compliance**: After the determination of crucial criteria for a process or product, the specifications must comprise an explanation of the suggested or required examinations and approaches. The complexity of the compliances for essential control in a food processing plant should also be considered, especially when dealing with thousands of ingredients which may end up in hundreds of different products.[59] In such cases, therefore, devoted management along with continuous monitoring is required to ensure compliance determination with labeling food safety and environmental standards.

    For instance, new legislation on genetically modified organisms (GMOs) for products such as xanthan gum should be understood for both the raw materials and the medium of growth.[59]

7. **Legal requirement:** Legal requirement is conditional based on the national belief in the necessity for legally enforcing standards.

3.3.5 What do auditors look for?

Auditing is an essential act for the verification of appropriate food safety practices in different food and beverage processing amenities. The auditing report enables assisting the selection of suppliers that have taken all reasonable steps to protect customers from harm or validate internal policies and procedures.[26]

Auditing covers both regular and non-regular types. In regular type, new or existing suppliers are checked to confirm continued compliance with the formerly accepted quality management system or "due diligence." The non-regular auditing is done during serious interruption of supply, variability in product quality, high consumer complaint, negative microbial trend, media reports, or product recall.

The audit carried out by retailers or their agents is considered a legal audit. The auditing may start with a tick sheet which has been designed in advance, not necessarily, fitting the retailer's requirements. In the case where both the retailer and supplier are involved in the preparation of auditing queries, the non-fit or pointless procedures may be eliminated toward mutual advantages of both business and consumer.[59]

3.3.6 Developing food safety and quality management systems

Additional food safety and quality systems useful to food and beverage industries are provided below:

3.3.6.1 Hazard Analysis Critical Control Points (HACCP)

In regard to food safety management procedures, a variety of systems have been developed by the food standards agency. These systems help the food and beverage sectors to ensure their products are safe and wholesome for consumption.[27]

Health and Safety Legislation, Audit, and Standards

The most recognized system for assuring the safety and quality of food products is the Hazard Analysis Critical Control Points (HACCP). The use of the HACCP system was originated by the Pillsbury Company when a systematic control for the safety of food products was a demand by National Aeronautical and Space Administration (NASA) for space flights.

The systematic control of food safety is a completed package in comparison with the simple traditional methods in which only sampling and analyzing of finished products were used to identify the presence and level of hazards. The approach of Pillsbury was based on understanding the hazards which could run food unsafe followed by the development of control measures to prevent the hazards from occurring prior to consumption.

It is worth noting that successful systematic control of food safety requires the knowledge of the HACCP system and engagement of the ownership and all operational staff as a unified team.

The effectiveness of systematic control of hazards is due to the proper implementation of prerequisite systems in order to identify and monitor all critical control points (CCP). It has been reported that lack of knowledge among CCP monitors for understanding the importance of their role has resulted in poor monitoring and record keeping and finally failing to take proper action. The poor control of food spoilage is not only a waste and unsustainable but also costly with adverse consumer confidence.[467]

Briefly, HACCP focuses on controlling food safety hazards associated, specifically, with the actual food product or ingredients, or the specific food manufacturing process, which requires comprehensive knowledge of the facility, components, and products.[121]

The province of Manitoba in Canada offers 60 GMP program standards and the accompanying monitoring procedures. Once these programs are implemented, the manufacturer can apply to Canadian General Standards Board (CGSB) for an audit and GMP certification, and the facility will be recognized by Manitoba Agriculture, Food and Rural Initiatives.[122]

It is essential to remember that regardless of the type of food safety program you plan to implement, you must always ensure all required regulations that may apply to the facility, commodities, and processing are encountered. It should be noted that GMP and HACCP do not replace food safety regulations but instead complement them.

Programs such as HACCP or GMP include the following eight areas: (1) premises (building exterior, building interior, sanitary facilities, water quality and supply, ventilation); (2) transportation and storage program (receiving, transportation, storage, non-food chemicals/ingredients, returns, and reworking on products within the plant); (3) equipment (design and installation, preventive maintenance, and calibration); (4) personnel training (general food hygiene training, technical training, cleanliness and conduct, communicable diseases/injuries, controlled access for personnel and visits); (5) sanitation and pest control; (6) sanitation of equipment, premises, processing, and storage areas to minimize the risk of product contamination and prevent threads from unclean equipment and surroundings; (7) recall and product traceability; and (8) record keeping.

3.3.6.1 Inspection

Food inspection is an act of controlling the quality, and assuring all aspects of standardization, uniformity, and wholesomeness of food products has been met. The purpose of inspection is measurement, investigation, and monitoring of different features of a product to approve the elements which have attained the established standards.

In the case of the National Aeronautical and Space Administration (NASA), which was referred to earlier, the organization appointed the Pillsbury Company to develop a zero-tolerance food safety system. Later on, the developed controlling measures turned into the HACCP system. The engineering principle called Failure Mode and Effects Analysis (FMEA) was used as the basis of the system of hazard analysis and control.[558] In simple terms, HACCP is a logic system covering all the stages of food production from growing stage to consumption and includes all the intermediate processing and distribution activities.[559]

3.3.6.2 Total Quality Management System (TQMS)/ Good Manufacturing Practices (GMP)

The organization is growing increasingly everywhere with a conscious of the competitive potential for quality, which is an issue due to the contractual definition of standards which, in contrast, previously were vague and unmonitored.[364]

Total quality management (TQM) is, fundamentally, defined as a management approach to the long-term achievement of customer satisfaction. In order to accomplish the TQM system, the engagement of all the members of an organization in enhancing the processes, products, services, and culture is essential.[28]

3.3.6.3 Quality control (QC)

Quality control is the part of GMP, which covers sampling, specification, testing, documentation, and release procedures. It ensures that the essential and relevant examinations for the raw, intermediate, and finished products are maintained and that the food product is ready for release to the market. Therefore, the food quality should be confirmed by essential laboratory tests and documentation, considering the incorporation of all necessary activities affecting product quality.[452]

3.3.6.4 Quality assurance (QA)

Quality assurance is a prevalent perception that covers all matters that independently or cooperatively impact the food quality. QA is the total of the organization's arrangements with the objective of ensuring that a product meets the required quality for its intended use. Therefore, it integrates the GMP and all other factors that are outside the scope of these guidelines.[452]

3.3.6.5 Good Manufacturing Practice (GMP)

Food plants may operate under federal regulations or various state and local codes. The regulations are designed to prevent the production of food ingredients or food products that may lead to contamination with filth, hazardous substances, and adulteration. Whether operating under the Food and Drug Administration (FDA) or

Health and Safety Legislation, Audit, and Standards

the United States Department of Agriculture (USDA), food plants are governed by federal regulations. Furthermore, other concerned agencies such as Environmental Protection Agency (EPA), Clean Air Act, Federal Water Pollution Control, and Occupational Safety and Health Administrator (OSHA) for worker safety and local public health agencies may be involved.[556]

The GMP covers a sequence of general principles that are essential for observation during manufacturing and should not be considered as a guideline on how to manufacture a product. The quality program and manufacturing process may consider different issues to justify the GMP requirements. Different activities such as sanitation, inspection, microorganism control, and chemical and physical hazards all may play an important role to be addressed.[556] Therefore, it is the company's responsibility to determine the most effective and efficient quality process.[123]

In the USA, the legal standard for the responsibility of all engaged in drug manufacturing is high, and the actual harm from contamination of a drug product has to be proven.[557]

According to FDA, the preventive actions for controlling the problems encountered in food safety include: (1) training: covers ongoing and targeted training for different areas such as allergen control, cleaning and sanitation procedures, incoming ingredient reception protocol, and monitoring for employees, management, and suppliers; (2) auditing: time-by-time auditing and inspections of the facility and raw material suppliers either internally or by external companies; (3) documentation: documenting all training activities, raw material handling policies and activities, cleaning and sanitation, receiving records, and use of sign-off logs; and (4) validation/evaluation: evaluating the effectiveness of training and creation of liability; validation of cleaning via conducting the necessary testing.

3.3.6.6 ISO 22000

The new ISO 22000 food safety standard formally integrates HACCP within the structure of a quality management system.[558]

ISO 22000 is an international food safety standard which has been established by the International Organization for Standardization. It is an outline for a Food Safety Management System (FSMS) which includes the elements of Good Manufacturing Practices (GMPs), Hazard Analysis Critical Control Points (HACCP) principles, and ISO 9001:2000. Implementation of this standard system indicates local, national, or international markets are loyal to food quality and safety standards. Briefly, ISO covers the following five steps: step 1: learn about ISO 22000; step 2: compare the current system to the requirements; step 3: plan the project; step 4: design and document the system; and step 5: train employees and an internal audit team.[46] It is also essential to train an internal audit team and adequate auditors to maintain the auditing schedule.

3.3.7 AN INTERNATIONAL RISK-BASED FOOD INSPECTION SYSTEM

Food safety and quality and consumer protection against fraud to fundamental human rights are promoted and supported by the Food and Agriculture Organization of the United Nations (FAO) and the World Health Organization (WHO). It also has major social, food security, and economic implications for all member countries.[4] The national food control system needs to be established with modern and effective legal and regulatory requirements to address food safety throughout the food chain. Obviously, a mechanism must be in place for enforcement to be effective in this regard. Food safety regulations must be based on risk and harmonized with Codex Alimentarius and other relevant international standards.[11]

As mentioned, the HACCP system throughout the food chain has clear benefits to enhance food safety and prevent foodborne diseases; however, it will take some time to establish prerequisite conditions and consider effective factors for food inspection to reduce or eliminate food safety hazards. The guidelines for the implementation of the HACCP system can be found in FAO/WHO guidance to governments and the application of HACCP in small and or less developed food businesses.[41]

3.4 RECALL AND TRACEABILITY

Preferably, each food producer must have a recall plan as a part of the FSMS. At a minimum, the recall plan should cover procedures for submitting information to the FDA/FSMS. The plan should also cover the notification for the removal of the contaminated product from the supply chain and the evaluation of the recall's impact.[689]

Food traceability is now an essential aspect of a due diligence defense, for example, in providing courts with copies of product recall procedures and records. Simply, blaming the supply chain is widely seen as an insufficient defense in assessing whether a business has been negligent in its food safety responsibilities.

Recall and traceability can be used as a communicative tool for the reconstruction of the food product history, allowing active control of the ways foods are produced for all stakeholders, governments, industries, and consumers as a whole.[560]

Traceability of raw materials and product recall has long been known as an important aspect of food safety. Based on Food Safety Act, the food industry regulations in most countries required to[548]:

a. Hold accurate records of all ingredients within a product,
b. Demonstrate procedures that provide traceability of components and products,
c. Provide product recall and emergency incident procedures, and
d. Ensure that the "due diligence" and precautionary principles apply.

Consumer information describes that recalls and advisories arise when a product is discovered to carry potential health risks or pose safety hazards to consumers using those products. The advisories, warnings, and recalls can apply to almost any product, including household products, toys, cosmetics, medication, food, electronics, and vehicles.[47]

3.4.1 WHY IS TRACEABILITY NEEDED?

Traceability in food is defined as the capability to trace back all processes involved in producing a food product from farm to consumption and disposal. The knowledge of the traceability system allows to identify the source, time, and producer of a food product from raw material to consumption and discarding. The process of traceability also enables the identification of risks, withdrawal of contaminated food, and delivery of precise information to the civic, thus diminishing interruption to trade.[49] The effectiveness of traceability or the ability to trail any food throughout all the stages of production lies in the availability of common requirements across all who are involved in the production of a food product.[29,56]

The number of recalled food products in Canada increased from 144 to 172 from 2007 to 2009, respectively.[45] Also, an average of 220 recalls per year for various reasons, such as allergens, microbials, extraneous materials, and the like, have been reported in Canada.[670]

Thus, while food companies might benefit from traceability and government may eventually demand it, food chain traceability is largely about building relationships with consumers and giving them what they want – the ability to trust that they know what they are eating.[30]

Recall of food products is the endpoint of defense in a food safety emergency situation and is considered a major economic loss each year.[45]

The initiation of food safety inquiry and recall process depends upon the existence of a potential for contaminated food after distribution in the market.[19]

Although the Canadian food supply has been known as the safest in the world, the zero-risk food safety system does not exist. Thus, food may develop contamination at any point in the production system due to the physical, chemical, or biological hazards.[20]

The CFIA classifies recalls, based on the level of health risk associated with the recalled food product. The record of all recalls (Class I, II, and III), including those that did not trigger a public warning, is also available.[50]

A provision of the federal food safety law passed last year requires that all players in the country's food supply chain be able to quickly trace from whom they received a food product and to whom they sent it. They all have to maintain that information in digital form, creating deep wells of information that, in some cases, consumers could tap into through their computers or cell phones. The recall may be done on voluntary bases or by request of the state laws. The FDA has three levels of recall classifications that can be found in Table 3.1.[55]

Every food processor or provider should develop a recall plan of action to retract the food product from the marketplace quickly and efficiently. The traceback plan is different from the recall plan in that it helps to trace back the food product to the farm or field it was harvested initially.[328]

The FDA was granted sweeping new food recall authority. Food Safety Act passed final approval by the USA House of Representatives on December 21, 2010. Food Safety Act is still the most important single aspect of the food safety bill that

TABLE 3.1

FDA recall – classification of food products

Recall – classification	Hazard level involved
Class I	An emergency situation involving the removal of products from the market which could lead to an immediate or long-term, life-threatening situation and involve a direct cause–effect relationship, such as *Clostridium botulinum* in the product
Class II	A priority situation in which the consequences may be immediate or long term and may potentially be life-threatening or hazardous to health, such as *Salmonella* in food
Class III	A routine situation in which even life-threatening consequences (if any) are remote or nonexistent. The products are recalled because of adulteration or labeling violation, and the product does not have any health hazards

covers the requirement for a comprehensive and detailed food recall traceability record-keeping program.

Food Safety Act compliance will be hugely important for every food manufacturer, transport, warehouse, and sales facility. One may contact the trained staff to learn more.[51]

The need for a robust recall information management and traceability solution has expanded dramatically from a simple knowledge of what items you sent to whom, to understanding the items, logistics path, lot details, manufacturing genealogy, etc. for all products and ingredients along the supply chain, past or present.[51]

An effective traceability system also helps to ensure that the food supply chain is not compromised from a food security standpoint and provides an excellent way to document regulatory compliance with bioterrorism rules.[21]

4 Accident Prevention Elements

4.1 INTRODUCTION

The existence of violation is evidence of a problem in one of the three areas in which mandatory occupational health and safety (OHS) requirements stipulate action by the operator(s). Any violation, accident, or incident indicates an examination, installation, or correction problem.[561]

4.1.1 ACCIDENT/INCIDENT DEFINITION

"Accident" like "safety" is difficult to be defined by the rank and file. Webster's Third New International Dictionary defines it as an "Event or condition occurring by chance or arising from unknown or remote causes" and also as "sudden event or chance occurring without intent or volition through carelessness, unawareness, ignorance, or a combination of causes producing an unfortunate result."[119]

An **accident** may be defined as an undesired event that results in harm to people, damage property, or loss of process.[468]

An **incident** can be defined as an undesired event which under slightly different circumstances could have resulted in harm to people, damage property, or loss of process.[468] Injury can be defined as physical or psychological injury, disease, or illness where a personal injury could lead to death.[48]

In case of injury, one must obtain first aid or medical treatment even if the injury is minor. While many minor injuries heal without treatment, a few result in serious prolonged disability that could have been prevented had the employee received treatment when the injury occurred.[54] About 8,900 workers in the hotel and restaurant industries are injured on the job every year in British Columbia. More than half of these workers have to take time off work because of their injuries. The types of injuries range from severe cuts, burns, and scalds to strains, sprains, and broken bones. Some hotel and restaurant workers have even been killed on the job.[9]

More than 5,000 injuries within the UK food and drink manufacturing sector are reported to the Health and Safety Executive each year, representing about a quarter of all the reported injuries in manufacturing. Manual handling and slips and trips are the most common causes of these injuries.[22]

Since 1990, the number of injuries reported in food and drink industries in the UK has dropped by 38%, and the number of injuries reported per 100,000 employees has also dropped by 26% – but still stands to be twice the rate of other manufacturing industry sectors. In addition, hundreds of workers in the UK food industry are likely to suffer an over-three-day accident during their working lifetime. These are

DOI: 10.1201/9781003303152-4

the results of an annual survey on occupational injuries and ill health in the food and drink industry.[62]

Another survey consisting of 63 companies and 85,000 staff showed 2 deaths, 133 major injuries, and 767 over-three-day injuries in the sector in 2011/2012. This represents a sample size of approximately 17% of the food and drink industry workforce.[63]

The US Census Bureau reported that the following general categories of causes resulted in fatal work injuries due to transportation – 43%; assaults/violent acts – 16%; contacts with objects – 17%; falls – 12%; exposure to harmful substances or environment – 8%; fires – 3%; and others – 1%.[15]

There is no question that accident hidden costs to industry and society are so many and, in fact, more than what most organizations realize. Although the changes in social responsibility is slower than that of technology, but major disastrous industrial accident in 20th century led to the formation of many laws and legislations to protect people, process, product, property, and environment.[118]

The programs designed to curb such hazards and losses are the result of careful planning and mixing the right elements to get designed objectives such as through experience, experiment, analysis, and previously observed human error or natural disaster.[119]

For example, "mercury," a toxic metal, undergoes a global distribution where it leaves the earth's surface, both terrestrial and aquatic, in the form of mercury vapor. Certain volcanic actions are important sources of its release into the atmosphere, but anthropogenic sources are also important as they contribute to almost half of these toxic vapors. Human activities like smelter industries, cement manufacturers, crematoriums, municipals, and other factory wastes are to mention a few.[120]

4.1.2 FRANK BIRD ACCIDENT RATIO PYRAMID

In a major study referred to the 1–10–30–600 or "Accident Ratio Study" in the USA, an analysis involving data collection from over 1,753,498 accidents reported by 297 cooperating companies, representing some 21 different industrial groups, employing 1,750,000 employees who worked for three billion man-hours during the exposure period analyzed in 1969, it was concluded that for every 600 incidents with no visible injury or damage (near accident or close calls), there were 30 property damage accidents of all types and ten minor injuries less than serious, and finally resulted in one major or serious injury including disabling and serious injuries.[352]

4.2 ACCIDENT PREVENTION AND CAUSATION MODELS

Numerous accident and loss causation models have been introduced in recent years. Large parts are complex and difficult for many to understand. However, one of the best reported models which is relatively simple to understand and comprise the key points that enable the user to consider few critical facts for accident prevention and

Accident Prevention Elements

TABLE 4.1
The ILCI Loss Causation Model

Cause	Effect (Contributing Factors)
1. **Lack of control**	Inadequate program standards compliance with standards such as management training, planned inspection, emergency preparedness, engineering control, purchasing control, and safety meeting
2. **Basic cause**	a. Personal factors: inadequate physical/physiological or psychological capacity, physical, mental, or physiological stress, lack of knowledge, skill, and improper motivation
	b. Job factors: inadequate leadership or supervision, engineering, purchasing, maintenance, tools and equipment, and standards. Wear and tear or misuse
3. **Immediate causes**	a. Substandard acts: improper lifting, improper use of equipment, failing to use PPE, horseplay, removing safety devices, failure to secure
	b. Substandard conditions: defective tools, noise exposures, fire and explosion hazards, poor housekeeping, radiation or temperature extreme, inadequate warning signs, ventilation, or illumination
4. Incident	Contact with energy or substance such as struck against or by something, fall, caught in, on, or between, contact energy sources such as electricity, heat, cold, chemical, become overstress or overload
5. **Loss**	a. Personal harm: major, serious, or minor injury or illness.
	b. Property damage: catastrophic, major, serious, or minor
	c. Process loss: catastrophic, major, serious, or minor

loss management and control the problems could be **The ILCI Loss Causation Model**, with the contributing factors tabulated in Table 4.1.[468]

4.2.1 HAZARD IDENTIFICATION AND RISK MINIMIZATION

Some of the most common hazards faced by food processing workers include pushing and lifting heavy bins, tubs, barrels, pumps, hoses, and mixers, using knives, working with hazardous chemicals, and entering confined spaces.[519]

4.3 ACCIDENT/INCIDENT INVESTIGATION, ANALYSIS, AND COSTS

This chapter lists notable **industrial disasters**, which are disasters caused by industrial companies, by accident, negligence, or incompetence. They are a form of an industrial accident where great damage, injury, or loss of life is caused.

Other disasters can also be considered industrial disasters if their causes are rooted in the products or processes of industry. There are many causes of accidents such as poor engineering design, lack of safety training, inadequate or defective machine, etc. to reduce costs of construction and fabrication which lead to unexpected design failures. Table 4.2 lists some food industry disasters.[38]

TABLE 4.2
Historical Accident Examples in Food Manufacturing

Data	Location	Cause of Accident
1878	Fluor mill, USA	Fluor dust explosion
1919	Molasses Co., USA	Tank explosion
1979	Mill, Germany	Flour dust explosion
1991	Chicken processing, USA	Fire
1998	Grain elevator, USA	Dust explosion
2008	Sugar refinery, USA	Dust explosion
2013	Poultry factory, China	Fire, ammonia leak

In Michigan, USA, the average lost workdays per injury amounted to 17.9 days. However, within all the causes of accidents, the greatest number of average days lost was due to cutting and slicing machines with an average of 167.7 days lost per injury.[32]

There are no doubt that accidents have major effects on human injuries and death. For instance, the following types with a total cost of $96.9 billion involving various accidents have been reported earlier in the USA: motor vehicle accidents $47.6, work accidents $33.0, home accidents $10.8, and public accidents $7.2.[354]

4.3.1 LEGISLATIONS

The existence of harsher Occupational Health and Safety Acts in Canada and around the world has provided an opportunity to improve upon and/or prevent an unplanned, undesired event that could cause injury or damage to people, processes, products, properties, or the environment. Incidents are near-accidents, and so-called accidents are not random events but rather preventable events. With proper hazard identification and evaluation, management support, preventive and corrective pleasures, monitoring and training, unwanted events, therefore resulting losses, damages, or injuries can be reduced or prevented.

4.3.2 RESPONSIBILITIES

All accidents and incidents should be investigated, regardless of the severity of injury or damage. For purpose of incident prevention, investigations must be fact-finding, not fault-finding; otherwise, they can do more harm than good. This is not to say responsibility should not be fixed where personal failure has caused injury, nor should that such persons be excused from the consequences of their actions. It does mean the investigation itself should be concerned only with facts. The investigating individual, board, or committee must not be involved with any disciplinary measures resulting from the investigation.

Accident Prevention Elements

4.3.3 Procedures

The investigation approach should consist of the following procedures: (1) identify responsibility and who should conduct the investigation and when; (2) specify investigation requirements specific to identification, assessment, and control issues; (3) ensure there are investigation methods, analysis, and reporting; (4) specify record-keeping procedures; (5) ensure follow-up and evaluation; and (6) review for continual improvement.

4.3.4 Communication/Training/Evaluation

Accident investigation should be communicated to all employees and identify the employees' role in this regard. The responsible person for investigation should be trained for this purpose and improve the knowledge or skills required accordingly. Training may include but not limited to skills in communication, investigation, interviewing, cause analysis, computer, and reporting.

Training components of any activity should include training objectives, outcomes expected from the training, competency standards for trainers, resources necessary to complete the training, training schedules for training and retraining, and record-keeping of completed training.

Evaluation should include interviews, loss data analysis, accident/incident report review, program/ performance audit, and management performance reviews.

4.4 TRANSITIONAL EMPLOYEE POLICY

Some organizations may have a transitional policy for injured employees after they have recovered and return to work after accidents.

All employers have an ethical, fiscal, and legal obligation to assist injured employees with their return to meaningful employment. Participation in the transitional program must be mandatory for all affected managers and employees.

The transitional program may only be refused if the employee provides a medical certificate to this effect that shows how the proposed work being proposed would pose risk to his/her health, adversely affect his/her recovery, or pose a danger to co-workers.

The key objective of any transitional work is to allow the employee to reduce/ combat the effects of physical deconditioning, aid alienation from the workforce, reduce the potential for permanent impairment, and employ the benefits of on-the-job work hardening.

To facilitate timely reintegration back to his/her pre-injury role in a constructive and progressive fashion may not exceed 90 days; however, prevention is easier and faster treatment/recovery. Working safely, using proper ergonomics, and following procedures on a regular basis do not add time to your day but add quality to your life.[8]

4.5 CONTRACTOR SAFETY PROGRAM

A contractor is defined as "the provider of services or work," and a receiver of these services is defined as "employer." The purpose of the contractor safety program is to ensure that potential risks in this process are minimized or totally – eliminated.

In this regard both the hazards associated with the type and nature of work being performed by the contractor as well as those being present at the worksite must be clearly identified and prevention measures be adapted as briefly discussed below.[469]

4.5.1 RESPONSIBILITIES

This should apply to all persons involved in the process of providing and supplying contracted services which include maintenance and repair organizations, new construction organizations, facilities, equipment and materials, cartage services, and temporary manpower services. These also include everyone involved in the bidding, selection, monitoring on-site, and closeout documentation of contracted services.

Responsibilities for ensuring the development, implementation, modification, and delivery of the contractor safety program should be established by including senior management, department managers, purchasing, maintenance and human resources staff, and contracted services management and staff. Suppliers, contractors, and sub-contractors should be expected to comply with the contractor controls requirements and all legislative requirements.

4.5.2 PROCEDURE

The employer should inform the contractor in advance of establishing a contract of the hazard involved, designated substances, and their controls activity, contractor obligations listed previously have been communicated and met, all contractors/employees are covered by the contractor's insurer, including the contractor's sub-contract work, contractor's clearance certificate every 3 months, copy of tender information, health and safety policy, and appropriate elements.

Contractors should provide a clearance certificate from their insurer, verifying coverage and bid documents, a copy of the rules to all employees used for the contract in fulfilling the contract and to enforce compliance to those rules, be informed of hazardous work permit required, be required to hold the sub-contractors accountable for all health and safety requirements to which they and their employees are held, ensure and provide evidence, that all employees have received Workplace Hazardous Material Information System (WHMIS) or similar hazardous training that may be required, make their employees available for this training prior to any employee beginning work on the project, ensure all their employees comply with company rules and other legislative, provide a list of any substances they will be introducing into the workplace, and Material Safety Data Sheet (MSDS) on those substances at least a week prior to start of the contract, and meet the contract controls activity obligations while engaged by the employer.

Accident Prevention Elements

In addition, the procedure should specify responsibilities for who is to do what and when, general and specific health and safety issues, delivery method or mode, record-keeping procedures, include documentation of dates and signature required, and ensure follow-up and evaluation of the contractor effectiveness controls and expected outcomes and to be reviewed for continuous improvement.

4.5.3 COMMUNICATION/TRAINING/EVALUATION

All employees and prospective contractors should be aware of the contractor controls activity standards. This can be accomplished through the contractor bidding or hiring process, rule books, notices, orientation training, or other effective means.

A week prior to contract work start, a meeting should take place to review emergency planning arrangements, hazardous work permit procedures, contract employee supervision agreements, accident notices and investigation needs, safety hazard information exchange arrangements, as well as health and safety resolution procedures.

Those responsible for meeting the standards should be trained in the requirements of the contractor controls activity.

The training should include (1) legal requirements (Occupational Health and Safety Act (OHSA), WHMIS, building code, fire code, etc.), (2) contract writing skills, and (3) purchasing standards knowledge and codes of ethics.[16]

Upon completion of the project, the health and safety coordinator should meet with the contractor to review the activities and experiences in fulfilling the contract.

An evaluation of compliance begins with comparing performance against the standards and procedures. Data collection can result from incident/accident investigation reports, interviews, loss data analysis, activity/performance audits, management performance reviews, etc. Employers should review the contractor controls activity performance annually and make recommendations for continuous improvement of the activity.

4.5.4 RELEVANT REMARKS

Successful accident prevention requires:

1. Study of all working areas to detect and control or eliminate hazards.
2. Study of all operating procedures and administrative controls.
3. Education, training, and discipline to minimize human factors.
4. Complete accident/incident investigation and analysis.
5. The main objective of accident investigation is to prevent the accident by revealing the facts and planning to prevent its recurrence.
6. All accidents including near-miss, minor, and especially those which are fatal or have caused serious injuries. They must be investigated, and employees should involve employees and contractor training and communication.
7. Analysis of accident/incident investigations can help management identify sources of problems, reveal the need for engineering or technical changes,

30 Occupational Health and Safety in the Food and Beverage Industry

and correct the inefficiencies, disclose unsafe practices which becomes the target for evaluation and planning of safety program's effectiveness.

4.6 EMERGENCY RESPONSE AND PREPAREDNESS

The actions taken in the initial minutes of an emergency are critical to warn employees.

The emergency response could be of natural origin such as fire, flood, earthquake, volcanic eruption, etc., technological such as hazardous material incidents, electricity outages, water problem, etc., and human origin such as crime, medical crisis, violence, etc.

It is reported more than 120 billion dollars is spent on emergency accidents by National Safety Council in the USA, and consultants estimate that over 60% of the companies with major disasters will go bankrupt.[476] Food processors should be prepared to respond to emergencies such as fires, explosions, chemical spills, or natural disasters.

4.6.1 HOW TO DEVELOP AND IMPLEMENT A PLAN

Follow steps can assist small- and medium-sized food processors in developing an emergency response plan[519]:

1. List all possible events (e.g., serious injuries, fires, explosions, or disasters)
2. Identify the major consequences associated with each event (such as casualties, equipment damage, or facility damage)
3. Determine the necessary measures to deal with those consequences (e.g., first aid, notification of medical authorities, emergency authorities, and so on)
4. Determine what resources will be required (medical supplies or rescue equipment)
5. Store emergency equipment where it will be accessible in the event of an emergency
6. Ensure that workers are trained in emergency procedures with the location of equipment
7. Hold periodic drill practices to ensure that employees will be ready
8. Communicate the plan to everyone involved

4.6.2 FACILITIES: FIXED AND MOVABLE EQUIPMENT

The safety features of any building may be movable such as fire extinguishers or fixed such as emergency lights, hoods, wash stations, or fire hoses. In order to facilitate the design and construction of a safe building, laboratory, or plant, fire and other building codes have been established in most localities which govern new construction and renovations to the existing buildings. Generally, under these codes, research laboratories come under the classification of a business use occupancy or

Accident Prevention Elements

possibly as a hazardous use occupancy, each of which incorporates different safety features.[361]

4.6.3 LABORATORY FACILITIES: DESIGN AND EQUIPMENT

The design of a laboratory facility depends upon both function and program needs. There is a growing need for "generic" laboratory spaces which are readily adaptable to different needs. Factors normally considered are as follows[362]: engineering and architectural principles, building classification, type of construction, area and height limitations, access (exit way features, corridors, stairs, doors, emergency lights, signs, etc.), construction and interior finish, and laboratory classification (low, moderate, high, and substantial risk). The risk level and needs will be defined in one of the following factors: (1) standard practices required (inventory, storage, MSDS, Safety Data Sheet (SDS), and hazardous waste), (2) special practices such as using hoods, personal protective equipment (PPE), and toxic gases, (3) special safety equipment (hoods, fire extinguisher, etc.), (4) laboratory facility factors such as easily cleanable, bench top resistant property, electrical codes, plumbing, ventilation, lighting, and workspace.

4.7 FIRST AID AND CARDIOPULMONARY RESUSCITATION (CPR)

In Quebec its human rights chapter legally requires that in modern society assistance as such must be provided and a person who refuses to come to the assistance of an injured person may be held liable. All employers are liable to provide training as specified to their employees including provision on first aid supplies and equipment.[271]

4.8 HEALTH AND SAFETY POLICY

A policy is a statement of intent and commitment to plan for coordinated management action. The health and safety policy should reflect the employer's positive attitude to a safe and healthy workplace. The policy should reflect the values of the organization and mandate the successful implementation of a health and safety program. This program must be written, reviewed at least annually, and posted in a conspicuous spot in the workplace.

The employer is responsible for ensuring the development, implementation, modification, and maintenance of the policy which is to be reviewed at least annually and/ or modified as needs or legal requirements change mandated by health and safety policy.

4.8.1 HEALTH AND SAFETY FORMS

There are various types of forms and different questionnaires that could be considered for different purposes based on the need and requirement of the company's activity in order to proceed with any health and safety follow-up improvement.

32 Occupational Health and Safety in the Food and Beverage Industry

An academic institution with research and scientific laboratories, hospitals, or industrial food and beverage processing sectors may require more detailed information on the nature and cause of the accident in order to plan accordingly.

4.8.2 Accident and Inspection Report Form

An accident report investigation should be based on the following information for further evaluation: form being labeled according to the type of hazard investigated (accident, audit, etc.), proper identification where the accident, incident, or occupational disease has happened such as victim's name, time and date, location, name of the investigator and witness, who is being copied on, etc., nature and extent of injury, damage or hazard occurred, first aid or outside medical attention received, name and address of the physician, etc., description of activity during the accident, cause of the accident, and/or what contributed to it (equipment, human error, lack of training, engineering design, etc.) and what employee was doing during the event, signature of the victim if does not agree to medical attention and is being released, body part affected, corrective action taken, etc. follow-up recommendation on preventive measures and person or department responsible for getting it done. The form has to be custom-made based on the need of an organization and the nature of the hazard involved (Table 4.3).

These forms come in many templates and can be customized based on the organization's need to cover critical health and safety issues. An audit/inspection form should be updated with laboratory/location ID information card, providing the name of the lab supervisor, emergency contacts and type of hazards in the lab, Material Safety Data Sheet, chemical inventory/WHMIS, lab environment, list of first aid providers, chemical safety storage, equipment/material safety storage, emergency measures, waste management, PPE, housekeeping, fume hoods, sustainability issues, etc. The form has to be custom-made based on the need of an organization and the nature of the hazard involved (Table 4.4).

TABLE 4.3
Example of an Accident/Incident Report Form

Section	Inclusion
A. Accident/incident	Injured/incident, identification, date
B. Event description	What, how, where, and when happened and action taken?
C. General information	Supervisor's name, material damage, contact information
D. Preventive measures	Root cause of the event, corrective action taken, did the worker had training, type of injury, damage to property, supervisor signature, medical and follow-up action taken, future corrective measure, signature

Accident Prevention Elements

TABLE 4.4
Example of an Inspection Report Form

Information	Description
General	Location address, inspector name, contact information, etc.
Hazard involved/preventive measures	Section, nature of accident/incident, injury type or damage, action taken, follow-up, corrective measure recommended for future
Acceptable/unacceptable	Signature and date

4.9 INDUSTRIAL HYGIENE

4.9.1 DEFINITION

Prior to the 20th century, physicians had made observations of health in working persons; particularly in Europe, but it was not until the 1930s or 1940s that engineers began to take an active role in finding ways to control the hazards found in industry.[333]

Industrial hygiene is a recognized science and art devoted to the recognition, evaluation, and control of those environmental factors or stresses – chemical, physical, biological, and ergonomic that may cause sickness and impaired healthy or significant discomfort to employees or the citizens of the community.[329]

4.9.2 PURPOSE

Monitoring in the workplace is often called exposure monitoring or air monitoring.

These terms refer to the detection and assessment of agents to which workers may be exposed in the workplace environment such as dust, fumes, smoke, gases, and vapors.[329] The purpose of monitoring is to determine the actual exposure levels of the agents and, where necessary, to determine where controls may be required.

4.9.3 PROCEDURE

A proper and developed monitoring strategy should include what should be sampled, who should do it, how to sample, where to sample, when to sample, and how long and how many to sample, ensuring proper involvement of Health & Safety committee members.

The industrial hygienist is trained to anticipate, recognize, evaluate, and control in particular chemical, physical, biological, and ergonomic hazards in the workplace.[470]

4.9.4 HEALTH AND SAFETY COMMITTEE/JOINT HEALTH AND SAFETY COMMITTEE (JHSC)

An Occupational Health and Safety Act is built on the principle that workers and employers must work together to identify hazards and resolve problems. Employee

34   Occupational Health and Safety in the Food and Beverage Industry

participation is most commonly found in the form of JHSC or health and safety representatives.

Section 23 of the Occupational Health and Safety Act of the USA requires any employer with more than 20 regular employees to establish a health and safety program. This means that workplaces requiring a committee will also require a program.

In Canada, a joint health and safety committee or the appointment of health and safety representatives is either mandatory or subject to the ministerial decision in all Canadian jurisdictions. Certain types of workplaces may be exempt from this requirement, depending on the size of the workforce, industry, incident record, or some combination of these factors.[671] The Canada Labor Code requires employers under federal jurisdiction to appoint a health and safety representative for each workplace with fewer than 20 employees. Employers under federal jurisdiction with 300 or more employees across Canada are required to establish a policy health and safety committee.[672] Consult the most up-to-date applicable legislation to find out the requirements for your workplace.

JHSC conduct inspections as part of their function. They give equal consideration to accident, fire, and health exposures. By periodically visiting areas, members may notice changes in the process and condition of the workplace environment.[471]

4.9.5   Conducting inspection

4.9.5.1   Inspection tools and preparation

Regular inspections are vital to planning and prevention of a number of hazards. Inspection tools can include clipboards, forms, pens/pencils, lockout/tag-out supplies, measuring tape, ruler, flashlight, laptop, ladder, or other supplies based on the inspection such as cameras, tape recorder, electrical equipment testing, sampling devices, sampling containers, calipers, micrometers, stopwatch, and personal protective equipment.

Furthermore, inspection should be planned for times when it could be the best opportunity to see the operation and/or condition which needs to be inspected.

4.9.5.2   Inspection ethics

Each regulatory agency may have specific requirements and rules for inspecting an organization's facilities. In general,[471]:

- Call at regular business hours
- Present credentials
- Identify the purpose of inspection and its length
- Request records and other documents as may be required
- Other related issues

4.9.6   Personal protective equipment (PPE)

When other control measures such as engineering controls and standards and administrative practices can't eliminate or reduce the hazard, personal protective

Accident Prevention Elements

equipment will be required during workplace practices.[473] The personal protective equipment (PPE) is designed to inform employers, supervisors, and workers about PPE in general and the PPE required by regulation for specific industries and hazards. They also describe when and how different types of PPE must be worn.[42]

PPE devices are designed to minimize a worker's accidental exposure to a hazard and should not be used as primary means of protection.[472]

PPE such as gloves and special clothing can be effectively used to avoid or minimize exposure to hazardous elements. When selecting gloves, aprons, boots, and sleeves, special attention should be given to their resistance to the source of hazards such as chemicals.[472]

PPE is designed to protect workers from serious workplace injuries or illnesses resulting from contact with chemical, radiological, physical, electrical, mechanical, or other workplace hazards. Besides face shields, safety glasses, hard hats, and safety shoes, protective equipment includes a variety of devices and garments such as goggles, coveralls, gloves, vests, earplugs, and respirators. Protection parts could include from eyes and face, head, hands and arms, legs and feet, lungs and breathing systems, and other body parts.[474]

4.9.6.1 PPE requirements

To ensure the greatest possible protection for employees in the workplace, the cooperative efforts of both employers and employees will help to establish and maintain a safe and healthy working environment.[475]

In general, employers are responsible for:

- Performing a "hazard assessment" of the workplace to identify and control physical and health hazards.
- Identifying and providing appropriate PPE for employees.
- Training employees in the use and care of PPE.
- Maintaining PPE, including replacing worn or damaged PPE.
- Periodically reviewing and evaluating the effectiveness of the PPE program.
- In general, employees should properly wear PPE, attend training sessions on PPE, care for, clean, and maintain PPE, and inform a supervisor of any defective needs, repair, replacement, and/or maintenance that may be required.

4.9.6.2 The hazard assessment

The hazard assessment should begin with a walk-through survey of the facility to develop a list of potential hazards in the following basic hazard categories: impact, penetration, compression (roll-over), chemical, thermal, harmful dust, light (optical) radiation, and biological.

In addition to noting the basic layout of the facility and reviewing any history of occupational illnesses or injuries, things to look for during the walk-through survey include sources of electricity, sources of motion such as machines or processes where movement may exist that could result in an impact between personnel and equipment, sources of high temperatures that could result in burns, eye injuries or fire, types of chemicals used in the workplace sources of harmful dust sources of light radiation, such as welding, brazing, cutting, furnaces, heat treating, high-intensity lights, etc., the potential for falling or dropping objects, sharp objects that could poke, cut, stab, or puncture, and biologic hazards involved.[475]

Documentation of the hazard assessment is required through a written certification that includes workplace identification, name of the person doing the assessment, date, and identification of the document certifying completion of the hazard assessment.

Everyday use of prescription corrective lenses will not provide adequate protection against most occupational eye and face hazards, so employers must make sure that employees with corrective lenses either wear eye protection that incorporates the prescription into the design or wear additional eye protection over their prescription lenses. It is important to ensure that the protective eyewear does not disturb the proper positioning of the prescription lenses so that the employee's vision will not be inhibited or limited.[475]

4.9.6.3 Body protection wear (face, hand, eyes, jaws, etc.)

Gloves are used in the lab for various purposes and should be selected even within each group mentioned below for particular characteristics of the process or product involved. If one involves in cutting food such as meat with knives or machine, resistant gloves for this purpose must be worn, and so for chemicals and fluids, radiation, thermal effect, sharp objects, etc. It is best to consult with MSDS or Safety Data Sheet (SDS) when it involves handling chemicals for best protection in choosing the right type of gloves as well as protective clothing. Employers should also ensure employees are abiding by PPE use as required as well as application of engineering control and guide for the prevention and reduction of accidents and incidents where and when applicable.

4.9.6.4 PPE where chemical hazards are involved

- A large percentage of eye injuries are caused by direct contact with chemicals. These injuries often result from an inappropriate choice of PPE that allows a chemical substance to enter from around or under protective eye equipment.
- Serious and irreversible damage can occur when chemical substances contact the eyes in the form of splashes, mists, vapors, or fumes.
- When working with or around chemicals, it is important to know the location of the emergency eye wash station and how to access it with restricted vision.

4.9.6.5 PPE devices for chemical hazards

- Goggles are primary protectors intended to shield the eyes against liquid or chemical splashes, irritating mists, vapors, and fumes.
- Face shields are secondary protectors intended to protect the entire face against exposure to chemical hazards.

Accident Prevention Elements

4.9.6.6 Impact hazard
- The majority of impact injuries result from flying or falling objects or sparks striking the eye. Most of these objects are smaller than a pin head and can cause serious injuries such as punctures, abrasions, and contusions.
- While working in a hazardous area where the worker is exposed to flying objects, fragments, and particles, primary protective devices such as safety spectacles with side shields or goggles must be worn. Secondary protective devices such as face shields are required in conjunction with primary protective devices during severe exposure to impact hazards.

4.9.6.7 PPE devices for impact hazards
- Spectacles are primary protectors intended to shield the eyes from a variety of impact hazards.
- Goggles are primary protectors intended to shield the eyes against flying fragments, objects, large chips, and particles.
- Face shields are secondary protectors intended to protect the entire face against exposure to impact hazards.

4.9.6.8 PPE devices for dust hazards
- Dust is present in the workplace during operations such as woodworking and buffing. Working in a dusty environment can cause eye injuries and presents additional hazards to contact lens wearers. Either eyecup or cover-type safety goggles should be worn when dust is present. Safety goggles are the only effective type of eye protection from nuisance dust because they create a protective seal around the eyes.
- Primary protectors are intended to protect the eyes against a variety of airborne particles and harmful dust.

4.9.6.9 Dust hazard
- Dust is present in the workplace during operations such as woodworking and buffing and in food industry working with materials such as powdered dairy, spices, flour, etc. in a dusty environment which can cause eye injuries and presents additional hazards to contact lens wearers.
- Either eyecup or cover-type safety goggles should be worn when dust is present. Safety goggles are the only effective type of eye protection from nuisance dust because they create a protective seal around the eyes.
- PPE for dust hazards will be regular or special goggles are primary protectors intended to protect the eyes against a variety of airborne particles or harmful dust.
- An explosive dust cloud will be formed, when the dust is suspended in the air to the extent that the concentration of dust per unit volume of clouds drops into the explosive range which will be discussed later on.[562]

4.9.6.10 Optical hazard

PPE selection: Optical radiation laser work and similar operations create intense concentrations of heat, ultraviolet, infrared, and reflected light radiation. A laser beam, of sufficient power, can produce intensities greater than those experienced when looking directly at the sun.

- Unprotected laser exposure may result in eye injuries including retinal burns, cataracts, and permanent blindness. When lasers produce invisible ultraviolet or other radiation, both employees and visitors should use appropriate eye protection at all times.
- Determine the maximum power density, or intensity, lasers produce when workers are exposed to laser beams. Based on this knowledge, select lenses that protect against the maximum intensity. The selection of laser protection should depend upon the lasers in use and the operating conditions. Workers with exposure to laser beams must be furnished with suitable laser protection.

4.9.6.11 PPE for optical hazard: lens requirements

When selecting filter lenses, begin with a shade too dark to see the welding zone. Then try lighter shades until one allows a sufficient view of the welding zone without going below the minimum protective shade.

4.9.6.12 Foot protection

Protective footwear worn in the workplace is designed to protect the foot from physical hazards such as falling objects, stepping on sharp objects, heat and cold, wet and slippery surfaces, or exposure to corrosive chemicals. Employers are encouraged to initiate a footwear protection safety program and train employees on foot safety and the importance of protective footwear as personal protective equipment.[127]

Some common problems in lack of proper footwear include severely aching feet, blisters, calluses, corns, rheumatism, arthritis, malformations of toes, fallen arches (flat feet), bunions, sprains, sweaty feet, and fungal infections (athlete's foot). Such problem can occur due to long periods of standing, hard flooring, poorly fitted footwear, high heels, pointed shoes, lack of arch support, too loose or too tight footwear, hot and humid environment, strenuous work, footwear with synthetic (non-porous) uppers.

Types of injuries include crushed or broken feet, amputation of toes or feet, trapped feet between objects or caught in a crack, falls of heavy objects, moving vehicles (lift trucks, bulldozers, etc.), conveyor belts (feet drawn between belt and roller), punctures of the sole of the foot, loose nails, sharp metal or glass objects which can result in cuts or severed feet or toes, laceration due to chain saws, rotary mowers, and unguarded machinery, burns due to contact with molten metal splashes, chemical splashes, fire, flammable or explosive atmospheres, electric shocks due to static electricity or contact with sources of electricity, sprained or twisted ankles, fractured or broken bones because of slips, trips or falls, slippery floors, littered walkways, incorrect footwear, and poor lighting.

Accident Prevention Elements

4.9.6.13 Prevention

- Identify workplace hazards and their correction
- Apply job rotation, job duty versatility, and team works will reduce foot pressure
- Use proper workstation design (footrail, footrest, etc.) so workers use alternate feet
- Where possible work standing or sitting at will
- Separate mobile equipment from pedestrian traffic
- Install safety mirrors and warning signs
- Use color contrast and lighting in complicated areas like stairs, ramps, etc.
- Use anti-slip flooring or anti-fatigue matting
- Provide training for using and maintaining, buy proper footwear considering the type of hazards in the workplace, type of work, body condition, type of footwear, consult physicians, podiatrists, and orthopedic practitioners, bathing feet daily and drying them thoroughly, check feet frequently for corns, calluses, cracks, keep your feet warm, trim your toenails straight across, slightly longer than the end of the toe, and prevent foot problems by visiting your podiatrist as part of your annual health check-up

4.9.6.14 Head protection

Head injuries could be serious and may even lead to death. Therefore, protecting the head is critical while working in certain locations and manufacturing sites in general hard hats protect you by providing the following features[126]:

- A rigid shell that resists and deflects blows to the head.
- A suspension system inside the hat that acts as a shock absorber.
- Some hats serve as an insulator against electrical shocks.
- Shield your scalp, face, neck, and shoulders against splashes, spills, and drips.
- Some hard hats can be modified so you can add face shields, goggles, hoods, or hearing protection to them.

4.9.6.15 Respiratory protection

A respirator is a protective device that covers the nose and mouth or the entire face or head to guard the wearer against hazardous atmospheres. Respirators may be[477]:

- Tight-fitting – that is, half masks, which cover the mouth and nose, and full-face pieces that cover the face from the hairline to below the chin; or
- Loose-fitting, such as hoods or helmets that cover the head completely.

In addition, there are two major classes of respirators:

- Air-purifying, which removes contaminants from the air

40 Occupational Health and Safety in the Food and Beverage Industry

- Atmosphere-supplying, which provides clean, breathable air from an uncontaminated source. As a general rule, atmosphere-supplying respirators are used for more hazardous exposures.

In the control of those occupational diseases caused by breathing air contaminated with harmful dusts, fogs, fumes, mists, gases, smokes, sprays, or vapors, the primary objective shall be to prevent atmospheric contamination.[128]

This shall be accomplished as far as feasible by accepted engineering control measures (e.g., enclosure or confinement of the operation, general and local ventilation, and substitution of less toxic materials). When effective engineering controls are not feasible, or while they are being instituted, appropriate respirators shall be used.

A respirator is a protective device that covers the nose and mouth or the entire face or head to guard the wearer against hazardous atmospheres. Respirators may be. Tight-fitting – that is, half masks, which cover the mouth and nose and full-face pieces that cover the face from the hairline to below the chin; or loose-fitting, such as hoods or helmets that cover the head completely.

There are also two types of self-contained apparatus: (1) air-supplying self-contained breathing apparatus (SCBA) for oxygen deficiency situations like where odor and toxic gases are present and (2) air-purifying applicable to regular lab applications.

The main elements for consideration are:

- Select respirator to situation requirements and consult MSDS
- Written operating procedures and training needed
- Assignment of respirators to individuals for their personal use if needed
- Evaluation of the facial fit for all respirator wearers
- Regular cleaning and disinfection of reusable ones
- Regular inspection of reusable respirators
- Storage of respirators in convenient and sanitary locations (sealed zip lock bag)
- Medical surveillance for those required to wear respirators to verify cardiopulmonary health and ability to function under increased breathing resistance

4.9.6.16 Employee training

Employers are required to train each employee who must use PPE. Employees must be trained to know at least the following: when PPE is necessary; what type of PPE is necessary? How to properly put on, take off, adjust, and wear the PPE, and the limitations of the PPE. And the proper care, maintenance, useful life and disposal of PPE.

4.9.7 LEGISLATION

Please verify any corresponding legislation within your state in this regard.

Accident Prevention Elements

4.9.7.1 Purpose

- PPE should be used when other controls are impractical, can't provide adequate protection from hazards, and must be available in emergencies.
- As PPE is important and the last defensive measure against hazards, the activity must emphasize proper training and compliance with standards.
- Permanent, contractors, and seasonal employees being part-time or full-time must be covered in the program and detail the nature of all PPE equipment selected.
- PPE must be identified and available based on specific needs.

4.9.7.2 Responsibilities

- The task of PPE responsibilities, development, implementation, modification, and its maintenance activity should be established by the employer and administered by immediate supervision.
- Employees must ensure the availability of the equipment and supervise its usage, and employees must adhere to regulations in this regard and wear the required ones.
- All levels of management should be accountable for the implementation and compliance with the requirements of the PPE activity.
- Employees need to comply with the PPE and legislative requirements for the standards and procedures.

4.9.7.3 Procedures

The procedures in this regard should address where PPE is required to provide fitting and wearing, maintenance and cleaning, storage, disposal, and enforcing the usage.

Procedures should also specify:

- Who has to use, when, and where?
- General and specific job requirements including health and safety-related issues
- Types of PPE to be used
- Record-keeping, documentation, dates, and signature needed
- Follow-up and evaluation for the PPE effectiveness of activity and outcome.

4.9.7.4 Communication and training

PPE standards should be communicated to all employees through the department supervisor, hiring process, plant audit and inspection, rule books, notices, orientation training, and other effective means.

Training should be provided to all employees required to wear and use PPE and should include the procedures for purchasing, issuance, reasons for PPE use, fitting and wearing, storage, maintenance, repair and sanitary cleaning, replacement, and disposal as well as record-keeping on all above issues.

4.9.7.5 Evaluation

Annually review the PPE activity performance and recommendation for continuous improvement of PPE.

4.9.8 PREVENTIVE MAINTENANCE

4.9.8.1 Purpose

- Preventive maintenance must be on regular bases regardless of if any defective problem has been identified or not. This will effectively eliminate and/or minimize wear and tear and monitor any corrective measures that may be required.
- Machines or processes that do not operate properly are considered safety hazards. Unscheduled downtime due to equipment failure is the most dangerous time for people in plants.
- Identification and analysis of tools, equipment, and machines used as part of a work process and making rational decisions to eliminate the problem are the key factors in safety practices as a preventive maintenance program.
- The heart of every preventive maintenance program is regularly scheduled attention to the machines, equipment, and processes covered by the program. This includes regular lubrication, inspection, and maintenance work unique to each machine. It is important to initially concentrate on production machinery; however, you should eventually include buildings and building systems such as roofs and utilities.[272]

Improving reliability will result in lowering maintenance costs which will lead to fewer spare parts and fewer time losses.[478]

4.9.8.2 Contractors

There should be a complete understanding and procedure to be followed by outside contractors. It is important for all parties to know who is responsible for what, timelines, deliverables, and standards to be met. The steps to be followed are:

1. Take equipment inventory to verify type, function, serial number, location, and availability of service manual. The master record for each machine, equipment, component, or process should indicate hazards around the machine, materials to be used, special tools or procedures required, sketches or drawings that show lubrication and inspection points, location and nature of utility feeds, machine shut-offs and lockouts
2. Assess each equipment by identifying the critical components
   Refer to the manufacturer's recommended service guide, legal issues, and previous history of the equipment.
3. Assess how critical is the equipment to production as well as health and safety

Accident Prevention Elements

This will help to assess the frequency and severity of criticality establishment and determine the priority and frequency of work that has to be done.

4. Develop the preventive maintenance program

Gathering information on each piece of machine and equipment and its components will assist the preventive program for maintenance scheduling.

4.9.8.3 Preventive maintenance program components

At minimum, it involves formal scheduling of work, clear statement of work, as well as records of work completed. It should include:

- A "systematic process" to determine which equipment/systems require preventive maintenance.
- "Equipment Records" in need of preventive maintenance, with serial numbers, descriptions of work done by whom and when.
- "Established maintenance methods" which may include specific types of lubrication or adjustments, condition or monitoring such as visual inspection for wear, damage, noise level, reliability analysis, mechanical integrity test, and so on.
- "Clear performance responsibilities." Every individual active in the preventive maintenance program should have clearly defined roles and responsibilities as well as clear lines of communication and reporting structure.
- "Training of maintenance staff" must be an ongoing activity to ensure an acceptable level of competence despite employee turnover and equipment change. Pay special attention to lock out/tag out procedures to prevent the startup of machinery during maintenance.
- "Information reporting system" which is important to have a procedure to report the date of maintenance, what was done, personnel involved, etc. after the work was completed.
- "Record-keeping system" which is valuable to record and file the information generated for future reference.
- "A priority system." The preventive maintenance program should prioritize the work to be done so that resources can be directed to the most critical work.
- "A follow-up procedures" which is important to include a follow-up procedure for all concerns raised by maintenance work.
- "Regular audits." The entire preventive maintenance program should be reviewed on a regularly scheduled basis to ensure conformance to company standards and the overall adequacy of the program.

4.9.8.4 Communication and training

All maintenance personnel have to be trained in their specific needs and responsibilities. Every supervisor must be aware of the requirements of the plan and what work is being done in their area of responsibility. Every operator must have been informed of the work being performed on their equipment.

4.9.8.5 Evaluation

Assign a coordinator to schedule, oversee, and keep records of the work done, measuring and evaluating the actual preventive maintenance work being done against the program requirements. The results should indicate whether the required work and schedules are being met successfully.

4.9.9 PURCHASING POLICY

The purpose of purchasing is to minimize health and safety hazards by controlling materials, equipment, and services being provided for the organization.[151]

All purchased materials and equipment should be evaluated for health and safety concerns prior to introduction into the workplace.

Individual purchasers are already reporting positive benefits from adopting such an "intelligent customer" approach to suppliers. One industry survey has found equipment safety to be the commonest concern of prospective buyers of food equipment.[151]

4.9.9.1 Responsibilities

The purchasing department should work closely with those responsible for health and safety to minimize health and safety hazards. It is also important to review the purchase of all potentially hazardous substances for health, safety, and environmental concerns.[273]

- For ensuring the development, implementation, modification, and delivery of the purchasing activity should be established by "Senior Management."
- All "persons involved in the purchasing and receiving" of goods and services should be responsible for ensuring the purchasing standards and procedures are being complied.
- The "employer" should ensure the purchasing standards and procedures are created and utilize the techniques and values as guided by the formal code of ethics accepted by the Canadian Purchasing Profession.
- The "purchasing/receiving manager" should be accountable for the implementation and compliance with the requirements of the standards and procedures.
- "Suppliers of goods" and services should be expected to comply with all purchasing and legislative requirements in conjunction with the organization.

4.9.9.2 Procedures

Purchasing is a step-by-step approaches to be followed to ensure all goods and services identified under the scope are covered in order to specify:

- Responsibilities for who is to do what and when
- General and specific job requirements including health and safety-related issues
- Procedure on delivery method

Accident Prevention Elements

- Record-keeping procedures, to include signed and dated documentation
- Evaluation follow-up for purchasing effectiveness and expected outcomes
- To be able to review for continuous improvement

4.9.9.3 Communication and training

All employees should receive general communications with respect to the purchasing activity standards. It can be done through hiring process, rule books, notices, orientation and training, or other effective means and be conducted by the department supervisor.

The training of those responsible for meeting the standards and procedures should include codes of ethics, health and safety training, risk analysis, ergonomic and physical hazards training, jurisdiction standards, and any training associated with goods and services.

4.9.9.4 Evaluation

Purchasing activity performance must be reviewed annually with senior management. Recommendations should be developed and implemented to ensure the continuous improvement of the activity. Effective purchasing procedure includes:

1. **Selecting proper work equipment** which is suitable for its intended use in respect of health and safety considering the severity of the situation:
   - Fatal injuries, falling from a height or being struck by transport: eliminate or safeguard access on top of the tankers, strong tailgate catches on tippers, safe loading bay layout, and safety systems for reversing. Access platforms used on lift trucks to be safe. Entry into the vessels: Free running silos and remotely operated cleaning devices and effective isolation for vessels.
   - Major injuries, slips: equipment such as hoppers and conveyors designed to contain products and avoid spillage. Walkways to be of sufficient slip-resistance.
   - Handling: mechanical aids for stacking and/or loading products into machines for handling, such as drinks containers, wrapping and packing; free-moving trolleys where pushed/pulled. Mechanical parts: elimination or adequate safeguarding of dangerous parts, especially to allow safe blockage clearance, cleaning, and adjustment.
   - Over three-day absence injury, struck by moving/falling objects and use of hand tools, secure storage: Ergonomically designed hand-knives with safe storage for when not in use. Slips: as above, handling: as above.
   - Ill health, musculoskeletal injuries: see handling above.
   - Occupational lung disease: containment or local ventilation of dust due to spices, flour, sugar, or similar food ingredients.
2. **Clearly specify the health and safety**, and hygienic design requirements for the supplier to meet in order to avoid accidents and take following safety measures:

- Conveyors: especially safeguarding at belt roller, belt/fixed parts, and transmission parts which allow safe clearing and blockage clearance. Interlocking guards if daily access is required.
- Band saws: especially easily adjustable blade guards and the use of jigs or holders.
- Lift trucks: especially workplace layout and safe use of equipment including lift truck-mounted working platforms.
- Wrapping machines: especially adequate distance guarding at sealing mechanisms and safeguarding at separating mechanisms, pushers, transmission parts, conveyors, etc.
- Pie and tart machines: especially interlocked guards and mold infills.
- Drink processing machinery: especially cleaning risks and conveyors.
- Dough molders: especially safeguarding access to molding parts via the hopper.
- Horizontal form fills seal machines: especially safeguarding at the forming and sealing dies.
- Dough machines: especially safeguarding access to cutting and feed parts via the hopper.
- Carton production machines: especially safeguarding access to paddles/beaters/ribbons via the discharge on larger cleaning.
- Roll plant: especially safeguarding access to rollers and blade conveyors.
- Slicers: especially safeguarding contact with the blade during processing and feed ram on larger machinery. Catering machines to have circumferential blade guards.
- Vertical form fill seal machines: especially interlocked and fixed guarding at access to the sealing jaws and head.
- Shrink wrapping machines: especially safeguarding at sealers and cutters when cleaning blockages and conveyors.
- Mincer/grinder/mixers: safeguarding access to blades.
- Deriding machine (meat skinner): especially restricting the blade and roller opening to a safe limit and selecting the appropriate machine type.
- Drinks labeling/marketing machines: especially safeguarding access to the piston/rams and forming head.
- Bottling machines: especially cleaning risks, safeguarding at filler heads, parts, conveyors, and access at height.
- Palletizers/de-palletizers: especially safeguarding against entry and safe systems of work for cleaning blockage and maintenance.
- Strapping, banding, and taping machines: especially safeguarding against trapping by moving parts or the band/product during processing.

3. **Check that equipment supplied** to see if it meets your specification and if the supplier has met their legal duties. For best results manufacturers also must:
   - Set up an appropriate product safety organization and procedures

Accident Prevention Elements

- Initiate testing, examination, and proactive research as necessary to supply articles which are as safe as reasonably practicable
- Establish an information system to keep up to date with legal requirements
- Train all staff/workers in the supply chain from designers to sales personnel and agents in the new requirements and product information
- Help purchasers specify and select suitable equipment
- Design and construct equipment to meet the relevant market standards
- Complete the relevant conformity assessment procedures
- Provide information to new or previous purchasers on problems revealed by subsequent experience.

4.9.10 WORK REFUSAL

In Ontario, Canada, a worker can refuse to work if he or she has reason to believe that one or more of the following is true:

- Any machine, equipment, or tool that the worker is using or is told to use is likely to endanger himself or herself or another worker.
- The physical condition of the workplace or workstation is likely to endanger the worker.
- Any machine, equipment, or tool that the worker is using, or the physical condition of the workplace, contravenes the act or regulations and is likely to endanger himself or herself or another worker.

The worker must immediately tell the supervisor or employer that the work is being refused and explain why.

The supervisor or employer must investigate the situation immediately, in the presence of the worker and, if there is such, one of the following people:

- A joint health and safety committee member who represents workers, if there is one. If possible, this should be a certified member, if one is required.
- A worker health and safety representative, in workplaces where there is no joint committee.
- Another worker, who, because of knowledge, experience, and training, has been chosen by the workers to represent them.

The refusing worker must remain in a safe place near the workstation until the investigation is completed. This interval is known as the "first stage" of a work refusal. If the situation is resolved at this point, the worker will return to work.[274]

- The worker must immediately inform the supervisor or employer of any safety concerns and refusal to work.

48　　Occupational Health and Safety in the Food and Beverage Industry

- The worker must stay in a safe place near his/her workstation unless is assigned to a reasonable direction by his/her employer. This could also be subject to the term of any union collective agreement if exist.
- Certain Acts and regulations may specify the worker to be paid on the first stage of work refusal till the situation is resolved. Also, a person acting as a worker representative during work refusal must be paid at regular or the premium rate agreed whichever applicable.
- The supervisor must investigate the situation immediately with the presence of the worker.
- A joint committee member who represents workers if there is one. If possible, this should be a certified member, a health and safety representative, in workplaces where there is no joint committee or another worker, who, because of knowledge, experience, and training, has been chosen by the workers or by the union to represent them. If the situation is resolved, the worker will return to work.

If the investigation was not satisfactory, the complaint can be referred to the board:

- The onus of proof lies on the employer to show that they did not act otherwise.
- Even where the Board finds that there was a cause for the employer's disciplinary action, the Board has the power to substitute a lesser penalty as it considers "just and reasonable" in all the circumstances.[274]

Example: An employee can refuse dangerous work on board an aircraft in operation if he or she has a reasonable cause to believe that[563]:

- The use or operation of a machine or thing on the aircraft constitutes a danger to themselves or another employee.
- A condition exists on the aircraft that constitutes a danger to the employee.
- The performance of an activity on the aircraft by the employee constitutes a danger to the employee or another employee.

There is also a pre-2000 state when the right to refuse work because of dangerous work was extraordinary and applied in more restricted circumstances.

It should be noted that this narrowing of the work refusal right does not remove or limit an employee's right to make representations to his or her employer about health and safety in a general sense. Indeed, there are many other aspects of the code that are specifically directed at providing employees with such opportunities (such as the mandatory establishment of workplace health and safety committees).[461]

Workers have the right to refuse unsafe work. In fact, workers must not carry out (or cause to be carried out) any task that they have reasonable cause to believe would create an undue hazard to the health and safety of any person.[519]

When a worker discovers an unsafe condition or believes that he or she is expected to perform an unsafe act, the worker must immediately report it to the supervisor

Accident Prevention Elements

or employer. The worker may be assigned other work at no loss in pay while the reported unsafe condition is being investigated.

4.9.10.1 Communication/training/evaluation

Management and supervisors should be aware of the refusal to work provisions of the act and be trained in how to carry out the various steps in a refusal situation.

An evaluation of compliance begins with measuring (comparing work refusal activity against the standards and procedures). Information can be obtained from interviews, observations, and work refusal records.

This information should be reviewed and analyzed by a designated person. Deviations from requirements should be reported and corrective action taken.

4.9.11 WORK STOPPAGE

Refusing/stoppage could be due to dangerous situations for employees; however, work stoppage can also be due to collective bargaining contracts which exist in certain countries such as Canada. In that case workers may have a union, where the contract guidelines between the union and the employer can specify details of such stoppage within.[275]

4.9.12 WORKING ALONE

Crime promoters are people who, by contrast, play roles which increase the risk of criminal events, with varying degrees of intentionality and responsibility. They include someone accidentally provoking the (potential) offender; a "friend" encouraging the offender to avenge an insult; or simply someone forgetting to lock their house.

The aim of such prevention is to switch the careless promoter or unprepared worker into the role of careful preventer. The usual focus is on preventers and promoters exercising influence over the probability of criminal events from within the proximal circumstances.[564] But also, preparedness in case of an emergency situation.

According to the Regulation in BC, Canada, to work alone or in isolation "means to work in circumstances where assistance would not be readily available to the worker in case:

(a) If an emergency, or
(b) The worker is injured or in ill health."

To determine whether or not assistance is "readily available," ask the following questions:

- Are other people in the vicinity?
- Are those people aware of your worker's need for assistance?
- Are they willing to provide assistance?
- Are they able to provide assistance in a timely manner?

50 Occupational Health and Safety in the Food and Beverage Industry

Examples of working alone or in isolation are retail (convenience store) employees, taxi drivers, truck and delivery drivers, home care and social service employees, by-law officers and security guards, forestry workers, doing high-hazard work with no regular interaction with other people, warehouse workers in cold rooms or freezers, night cleaners and custodians in private and public buildings, and night-shift employees.

Working alone or in isolation does not include independent realtors and/or owner-operators of trucks. Identify hazards during work inspection by following these basic guidelines:

- Consider all aspects of your business – Think about the location, nature, and circumstances of your business or industry.
- Consider previous accidents – How many incidents have there been in your workplace and what happened? What about incidents at nearby businesses or previous work locations?
- Involve your employees – Ask for their input regarding current problems, concerns, and possible solutions.

Of primary importance is to ensure at a minimum that a system of immediate or predetermined timed communication is in place between the supervisor and the employee and that this system is understood and followed.

Examples: cell phones, two-way radios, on-the-hour check-in phone calls, etc.

Regardless of which system is used, a hazard assessment of the work-alone situation should be conducted by the supervisor and the worker to determine potential risks and establish procedures to mitigate those risks.

It is important to make sure there are guidelines in place for those who have been assigned or will have to work alone in certain situations such as in a production tank remote area of the plant or any other situation be monitored.[479]

4.9.12.1 Workplace inspection

Besides correcting any hazards that you observe from day to day, set aside time for regular workplace safety inspections, and control any hazards that you find during an inspection. It is far better – and less costly – to prevent accidents than to deal with their consequences. Because safety inspections are preventive in nature, they are an important part of your overall health and safety program.[519]

4.9.12.2 When to inspect

You need to inspect your workplace often enough to prevent unsafe working conditions from developing. In food processing, this should be at least once per month. You also need to inspect your workplace when you've added a new process or when there has been an accident. Inspection is an ongoing task because the workplace is always changing.

Accident Prevention Elements

4.9.12.3 Who should be inspected?

Inspections should be conducted by a supervisor and a worker. If possible, the worker health and safety representative (or members of the joint health and safety committee) should be involved.

4.9.12.4 How to inspect?

During an inspection, identify unsafe conditions and acts that may cause injury so you can take corrective measures. Follow these guidelines:

- Use a checklist to ensure that your inspection is thorough and consistent with previous inspections.
- Ask yourself what hazards are associated with the job that you are observing or that would be performed in that work area.
- Observe how workers perform tasks. Do they follow safe work procedures and use personal protective equipment as required?
- Ask workers how they perform their tasks.
- Talk to workers about what they're doing. Ask about safety concerns.
- Record any unsafe actions or conditions that you observe.

While your first inspections may seem slow and difficult, over time inspections will become much easier and ultimately will help make your health and safety program more effective.

4.9.12.5 What to inspect?

There are different ways of approaching safety inspections, depending on the objectives of your health and safety program. For example, you can focus on the most common tasks your workers perform or on a specific issue addressed by your program, such as ergonomics.

Here are some activities and situations that warrant inspection:

- Rarely performed, non-routine, and unusual work, which presents an increased risk because workers may not be familiar with procedures
- Non-production activities such as housekeeping, maintenance, machine setup, etc.
- Sources of high energy such as electricity, steam, compressed gas, flammable liquids, and explosive substances
- Situations that may involve slipping, tripping, or falling hazards; or overhead hazards such as falling objects
- Lifting situations posing a risk of back and muscle injuries
- Repetitive motion tasks such as work involving computers or repetitive, constant uninterrupted motions
- Work involving contact with toxic substances

Check whether safe work procedures are being followed. For example:

- Is equipment locked out during maintenance?
- Are gloves being used for handling garbage?
- Are safe lifting procedures being used?
- Do workers know the procedures for working alone?

4.9.12.6 Inspection topics

Following provide an idea of the type of hazards associated with food processing areas[519]:

- Environment: dust, gases, noise, temperature, ventilation, lighting
- Floors: slipping and tripping hazards, cluttered aisles
- Building windows, doors, floors, stairs, roofs, walls, elevators, exits, docks, etc.
- Containers: scrap bins, disposal receptacles, barrels, carboys, gas cylinders, solvent cans
- Electrical switches, cables, outlets, grounding, extension, breakers, grounding, etc.
- Fire protection: fire extinguishers, hoses, hydrants, sprinkler systems
- Hand tools: wrenches, screwdrivers, power tools, hydraulic tools, explosive actuated tools, pressurized tools
- Hazardous materials: flammables, explosives, acids, corrosives, toxic chemicals
- Materials handling: conveyors, cranes, hoists, hoppers, manual lifting, forklifts
- Pressurized equipment: boilers, vats, tanks, piping, hoses, couplings, valves, etc.
- Production equipment: mills, cutters, drills, presses, lathes, saws
- Support equipment: ladders, scaffolds, platforms, catwalks, staging, aerial lifts
- Powered equipment: engines, electrical motors, compressor equipment
- Storage facilities: racks, bins, shelves, cabinets, closets, yards, floors, lockers, store rooms, mechanical rooms, flammable substances cabinets
- Walkways and roads: aisles, ramps, docks, vehicle ways, catwalks, tunnels
- Personal protective equipment (PPE)
- Hard hats, safety glasses, respirators, gas masks, gloves, harnesses, lifelines
- Protective guards: gear covers, pulleys, belts, screens, workstations, railings, drives, chains
- Devices: valves, emergency devices, warning system limit switches, mirrors, etc.
- Signage, cover plates, lighting systems, interlocks, local exhaust systems
- Controls: startup switches, steering mechanisms, speed controls, etc.
- Lifting devices: handles, eyebolts, lifting lugs, hooks, chains, ropes, slings
- Hygiene and first aid: drinking fountains, washrooms, safety showers, eyewash facilities, toilets, fountains, first aid supplies

Accident Prevention Elements

- Offices: workstations, chairs, computer equipment, ventilation, floors, stairs, equipment, emergency equipment, storage cupboards, filing cabinets.

After the inspection, review the following guidelines:

- Remedy serious hazards or unsafe work practices immediately. For example, if you find that a ladder has a loose or damaged rung, immediately remove the ladder from service and repair or replace it.
- Prioritize other, less serious hazards, and assign someone to remedy each one.
- Follow up on any actions that will take time to complete such as getting a machine.
- Communicate your findings and plans to workers.

As an essential part of a health and safety program, workplaces should be inspected. Inspections are important as they allow you to[276]:

- Listen to the concerns of workers and supervisors
- Gain further understanding of jobs and tasks
- Identify existing and potential hazards
- Determine underlying causes of hazards
- Monitor hazard controls (personal protective equipment, engineering controls, policies, procedures)

Types of workplace hazards include:

- Safety hazards such as inadequate machine guards, unsafe workplace conditions, and unsafe work practices
- Biological hazards caused by organisms such as viruses, bacteria, fungi, an parasites
- Chemical hazards caused by a solid, liquid, vapor, gas, dust, fume, or mist
- Ergonomic hazards caused by anatomical, physiological, and psychological demands on the worker, such as repetitive and forceful movements, vibration, temperature extremes, and awkward postures arising from improper work methods and improperly designed workstations, tools, and equipment
- Physical hazards caused by noise, vibration, energy, weather, heat, cold, electricity, radiation, and pressure

Inspection summary report should include but not limited to[276]:

1. Basic layout plans showing equipment and materials used
2. Process flow
3. Information on chemicals
4. Storage areas
5. Workforce size, shifts, and supervision

6. Workplace rules and regulations
7. Job procedures and safe work practices
8. Manufacturer's specifications
9. Personal protective equipment (PPE)
10. Engineering controls
11. Emergency procedures – fire, first aid, and rescue
12. Accident and investigation reports
13. Worker complaint reports regarding particular hazards in the workplace
14. Recommendations of the health and safety committee
15. Previous inspections
16. Maintenance reports, procedures, and schedules
17. Regulator inspection reports or other external audits (insurance, corporate specialist)
18. Monitoring reports (levels of chemicals, physical or biological hazards)
19. Reports of unusual operating conditions
20. Names of inspection team members and technical experts involved

Type of inspection report could be ongoing, pre-operation, or periodic.

How often inspections are performed will depend on several factors:

- Frequency of planned formal inspections may be set in your legislation.
- Past accident/incident records.
- Number and size of different work operations.
- Type of equipment and work processes – those that are hazardous or potentially hazardous may require more regular inspections.
- Number of shifts – the activity of every shift may vary.
- New processes or machinery.

High-hazard or high-risk areas should receive extra attention.

4.9.12.7 Responsibilities

Development and implementation of workplace inspection must be established by the employer in accordance with the legislative requirements of the state.

Personnel who directly control where the hazards exist reside with those who have direct control over these hazards.

They could be middle management, supervisors, employees, H&S officers, or JHSC representatives. These people should be measured against their compliance with the requirements of the procedure.

Inspection frequency or length should be based on the legal requirements, the need of the company, workplace special consideration, etc. The physical condition of the workplace is recommended to be inspected at least once per month.

4.9.12.8 Procedures

It is a step-by-step method to ensure all areas and practices are covered and should specify and ensure:

Accident Prevention Elements

- Types of inspection (cranes, hoists, lift trucks, etc.)
- Responsibility of inspection by whom and when
- Types of tools to be used
- Record-keeping methods including documentation of dates and signatures
- Recommendations are dealt with
- Follow-up and evaluation of the effectiveness of procedures
- Be reviewed for continuous improvement

4.9.12.9 Communication and training

It is important that everyone affected by workplace inspections be able to have access to the results, time frame, training, etc. The results should be communicated with all effective means within the organization such as posting, meetings, etc.

Inspectors should receive the appropriate training in steps of inspection, hazard identification and control, and reporting procedures.

4.9.12.10 Evaluation

All inspection activities must be reviewed annually with senior management. Recommendations should be developed and implemented to ensure continuous improvement as follows: (1) the person responsible for evaluation measurement, (2) skills needed to carry out the evaluation, (3) method of evaluation, (4) tools required for the evaluation, (5) adequacy of the workplace inspection standards, (6) developing recommendations to address compliance and non-compliance, (7) communicating the results, and (8) monitoring of recommendations.

As the inspection purpose is to eliminate, reduce, and prevent any potential hazards in workplace, it is therefore important that due diligence be performed to prevent future prosecution besides preventing harm to people, property, and environment.

4.9.12.11 Due diligence

In Canada food company owners and managers in particular are increasingly at risk of being taken to criminal court to face huge fines and imprisonment, even if they did not mean to do any harm or did not know they were doing anything wrong or did not know that an employee was committing a prohibited act.[480]

In England in any proceeding for an offence under food hygiene regulation, it shall be a defense for the accused to prove that he took all reasonable precautions and exercised due diligence to avoid the commission of the offence by him or by persons under his control.[481]

4.9.12.12 Field work and external safety considerations

In certain situations, like in academic institutions, the range of activities conducted in the field is as broad as the multiplicity of the interests among the disciplines of teaching and research in the university. These projects can take place over the surface of the planet, its seabed, or in outer space. Managing risks associated with such activities and safety responsibilities begins with the individuals involved in the project. Prior to embarking on any research project, it must be evaluated from a safety perspective, and if it is concluded that the risks are not manageable, then the project

56 Occupational Health and Safety in the Food and Beverage Industry

should not be carried out. Planning and preparation include, but are not limited to, the following as has been referred to in McGill health and safety mission[365]:

- Identifying the hazards associated with the project
- Assessing the risks associated with the project
- Determining the controls required to manage the risk to an acceptable level
- Putting in place a safety plan to address the risks
- Getting departmental/unit clearance
- Identifying the participants and verifying their qualifications and applicable certifications such as a first aider
- Training and equipping the participants
- Assigning/documenting the tasks for all of the participants
- Advising participants of risks and obtaining their informed consent to participate and to agree to the means of controlling the risks
- Assuring of health and immunization status of the participants
- Preparing first aid and emergency supplies
- Identifying and preparing protective equipment and safety gear
- Arrangements for appropriate facilities to conduct the work
- Preparation of any hazardous materials for their safe transportation
- Making pre-arrangements for the ultimate safe and environmentally sound disposal of any wastes generated by the work
- Arranging for appropriate transportation to and from the site
- Arranging for adequate food and accommodations
- Preparing contingency plans for emergency transport, provisions, accommodations, and medical care
- Identifying legislative needs of foreign governments or other jurisdictions
- Ensuring leadership of the project on a continuous basis
- Advising the necessary authorities regarding insurance including medical, travel, general liability, etc.

There have also been following five tips for ensuring a safe environment for workers on the field trip investing in ergonomic equipment such as tools designed to work with the body's natural movements and minimize the risk of musculoskeletal disorders and recognize the dangers of confined spaces. Atmospheric hazards are common in manholes, maintain visibility, use proper ladders carefully, beat the heat by proper hydration and light clothing, and heavy-duty work tents and umbrellas.[458]

4.9.12.13 Safety promotion and recognition

Recognize and encourage people to more effectively promote health and safety procedure which must be recognized and awarded by management in many forms:[7]

- For an individual or group.
- Monthly voucher gift.
- Moral incentive for contribution to a charity for every safety act.
- Safety raffle.

Accident Prevention Elements

- For safety act practices.
- Monthly, quarterly, or annual awards, etc.
- The elements of recognition, improvement, and correction demonstrate the essence of continuous improvement by using the integrated management system.
- The award or trophy designed for such a purpose has to have a meaningful and lasting effect on the provider.
- The leadership team must ensure that information learned by evaluating performance against company standards, rules, and expectations is used for continuous improvement.

4.9.12.14 Commendation is an important element

Some say the first law of selling is "sell yourself first." People "buy you" before they buy your product. Therefore, you need to know[360] enough about safety not only to understand it yourself but also to be able to assist others to understand it, remote safety with individuals, in meetings, through posters, ads, newsletters, magazines, and e-media, and provide incentives, recognition, and awards, by personal examples. Remember to apply the principles of communication, reciprocation, recognition, and repetition.[360]

5 Occupational Hazard Origin and Preventive Measures

INTRODUCTION

HISTORICAL PERSPECTIVE

In 1713, Bernardino Ramazzini (now considered the father of occupational medicine) described the occupational hazards for chemists in his book known as *Diseases of Workers* that chemists certainly deserve the praise, for so devoted are they to the investigation of the obtuse matters and to the enrichment of natural science that they do not hesitate to risk their lives for the good of the public in discovery, invention, and advancement of knowledge. For they must stand by and observe the whole process, enduring the test of fire, the fumes of coal, and or similar conditions.[61]

Every year 1.5 million workers suffer from ill health caused or made worth by work. In the UK food and drink industries, an estimated 29,000 workers (4.8% of the workforce) suffered from occupational work during 2001/02, according to the self-reported work-related illness (SWI) survey for those years.[61]

Some of the major food industry accidents are provided below[744]:

May 2, 1878: The Washburn "A" Mill in Minneapolis was destroyed by a flour dust explosion, killing 18. The mill was rebuilt with updated technology. The explosion led to new safety standards in the milling industry.

January 15, 1919: The Boston Molasses Disaster. A large molasses tank burst and a wave of molasses rushed through the streets at an estimated 35 mph (56 km/h), killing 21 and injuring 150. The event has entered local folklore, and residents claim that on a hot summer day, the area still smells of molasses.

February 6, 1979: The (Roland Mill), located in Bremen, Germany, was destroyed by a flour dust explosion, killing 14 and injuring 17.

September 3, 1991: 1991 Hamlet chicken processing plant fire in Hamlet, N.C in the USA where locked doors trapped workers in a burning processing plant, causing 25 deaths.

September 3, 1998: 1998 Haysville KN grain elevator explosion in Haysville, Kansas.

Dust explosions in a grain storage facility resulted in the deaths of seven people.

DOI: 10.1201/9781003303152-5

February 7, 2008: The 2008 Georgia sugar refinery explosion in Port Wentworth, Georgia, USA. Thirteen people were killed and 42 injured when a dust explosion occurred at a sugar refinery owned by Imperial Sugar.

March 12, 2008: Morin-Heights, Quebec, Canada. A roof collapse in the Gourmet du Village bakery warehouse killed three workers.

In general, the main causes of occupational health are divided into biological, chemical, physical, and psychological hazards which will be discussed.[64]

5.1 Hazard prevention strategy/hygiene

Occupational diseases common to food and beverage industries are normally originated through biological, chemical, and or physical agents which will be discussed below. These could be through ingestion, inhalation, or contact with materials, and they may be biologically, chemically, and/or physically oriented.

The food and drink industry (FDI) has a strong contribution to EU-28's economy as reported by Eurostat (2014). FDI involves more than 4.25 million workers divided into 289,000 companies with a relevant degree of diversification. A research study of more than 6,000 OHS accidents in Europe can help analysts to better address the measures to be adopted in a work environment, in order to prevent occupational accidents and give a concrete contribution to risk management.[756]

A prerequisite for high-quality and safe products is correct hygienic design and maintenance of the production systems. It is recommended hygienic design should be planned so[129]:

1. To avoid contamination by foreign organisms and materials like animals, insects, birds, microorganisms, migrations, construction/equipment parts, lubricants, residues of other products and cleaning agents, etc.
2. To avoid conditions that enhance the growth of microorganisms, like accumulation of dust, water, moisture, and other product.
3. To improve sanitation.

Risks due to poor hygienic designs are mainly caused by incorrect positioning of equipment and utility installations, horizontal surfaces, hollow bodies, dead spaces, bad drainage, insufficient access for cleaning, etc.

Hazard sources could have biological, chemical, physical, or psychological origins.

Intentional (non-accidental) hazard sources are such as food adulteration, tampering, food terrorism, cybercrime, uprising, rioting, or war.

Unintentional (accidental) hazard sources are such as food additives or toxins, natural toxicants, or disasters such as earthquake, volcanic eruption, sinkhole, storm, tornado, flood, etc.

Hazard sources could be biological, chemical, physical, and or psychological in nature.

5.1.1 Water sources of contamination

Chemicals, heavy metals like lead and mercury, and living organisms such as bacteria and viruses can all be threats to a safe water supply. These substances can also contaminate food and drink sources such as unintentional contamination of water as a result of chemical leaks or spills, natural disasters, or other causes have been the main concern than deliberate contamination.[334]

5.1.2 What you can do?

- With the exception of a known accident (such as a chemical spill into the water supply) or an announced terrorist or criminal incident, you probably would not know that you had consumed contaminated water or food unless you developed symptoms. To reduce your risk of consuming contaminated food or water and to be better prepared for public health emergencies affecting the water supply:
- Don't eat food or drink water or any other beverage that looks or smells suspicious. In general, it is not a good idea to eat or drink something when you don't know who has prepared or provided it or where it has come from.
- When shopping, avoid food or beverage items that look like they may have been tampered with – for instance, if the seal is broken or you think that the container or packaging has been opened.
- Remember that most cases of food poisoning, including botulism, happen by accident. Follow guidelines for preparing and cooking food safely, keeping your kitchen clean, and washing your hands and utensils. If you preserve and can foods at home, learn and follow proper canning and freezing techniques to ensure safety. Discard cans or jars with bulging lids or leaks.
- Know where your household's water comes from. Is it from the city water supply? Most public water supplies are carefully monitored and treated to guard against contamination. Does a private well supply your water? Private water supplies are unlikely to be targets of intentional contamination. But they can become contaminated by accident and may not be as closely monitored as city water supplies.[334]
- Consider storing emergency water and food supplies.
- Learn how to purify water. And make sure that you include the supplies for this in your emergency kit. Knowing how to purify water is useful in any situation where you have to rely on untreated water.
- If there is an emergency affecting the water supply:
  - Follow all instructions from local authorities about water purifying (commonly called "boil orders") or using other water sources until authorities notify your community that it is safe to drink from the regular water supply again.
  - Do not strictly ration emergency drinking water supplies. Try not to waste any water, but drink what you need. On average, a person needs about 2 qt (2 L) of water a day. Individual water needs vary depending

on age, health, diet, and climate. Learn the signs of dehydration in children and adults in advance.

- Use the safest water you have first before turning to other water sources.
- If you know or suspect that your skin has come in direct contact with water that has been contaminated by a hazardous chemical or radiation fallout, follow the steps for personal decontamination to get the substance off your body as completely and quickly as possible.[334]

5.2 BIOLOGICAL HAZARDS

A pathogen can cause human or animal disease and can be readily transmitted from one individual to another, directly or indirectly. There are four potential risk groups based on World Health Organization for Risk Group 1–Risk Group 4 based on the risk involved.[355]

Each country classifies the agents in that country by risk group based on the pathogenicity of the organism, modes of transmission, and host range of the organism. These may be influenced by existing levels of immunity, density, and movement of the host population presence of appropriate vectors and standards of environmental hygiene.[336]

Availability of effective preventive measures that may include the followings two steps:

- Prophylaxis by vaccination or antisera; sanitary measures, e.g. food and water hygiene; the control of animal reservoirs or arthropod vectors; the movement of people or animals; and the importation of infected animals or animal products.
- Availability of effective treatment. This includes passive immunization and post-exposure vaccination, antibiotics, and chemotherapeutic agents, taking into consideration the possibility of the emergence of resistant strains. It is important to take prevailing conditions in the geographical area in which the microorganisms are handled into account. Note: Individual governments may decide to prohibit the handling or importation of certain pathogens except for diagnostic purposes.

Biological hazards in general are originated from bacteria, viruses, parasites, and fungi agents through inhalation, ingestion, and contact that cause a number of diseases or be lethal due to pathogenic microorganisms, and chemical agents.[67]

In addition, certain plants some edibles such as mushrooms or fish species could be poisonous for consumption.[130]

The source of it can be workers' health condition, personal hygiene, and or practices as well as other issues discussed in the program previously.

Occupational Hazard Origin and Preventive Measures 63

5.2.1 FOODBORNE ILLNESSES BY PATHOGENIC MICROORGANISMS

The association between the consumption of food and human diseases was recognized very early and it was Hippocrates (460 BC) who reported that there is a strong connection between food consumed and human illness. Foodborne pathogens (e.g. viruses, bacteria, parasites) are biological agents that can cause a foodborne illness event. A foodborne disease outbreak is defined as the occurrence of two or more cases of a similar illness resulting from the ingestion of a common food.[617]

The renewed interest in food production, including food derived from animals, has been prompted by concern about food security as well as challenges of meeting future demands for food sustainability and safety.[131]

Biological sources of disease and contamination could be from many sources including but not limited to *Campylobacter, Salmonella, Clostridium porringers, Giardia lamblia, Staphylococcus* food poisoning, *Toxoplasma gondii, E. coli, Shigella, Yersinia enterocolitis,* Enterotoxigenic *E. coli, Streptococci,* Astrovirus, Rotavirus, *Cryptosporidium parvum, Bacillus cereus, Cyclospora, Brucellosis, Vibrio,* Hepatitis A, *Listeria monocytogenes, Brucella, Salmonella typhi, Trichinella, Vibrio cholera.*[132]

Food safety has long been a focus also in the retail food industry and is increasingly being focused on by consumers and governments worldwide. In some instances, the food purchased by the consumer will require further preparation, while in other cases it will be ready to eat. In all instances, the consumer expectation is that the products they purchase are safe to consume as purchased or that they will be able to prepare the products in such a way that they will be safe to consume. Food safety at retail starts with good procurement practices and ends with good recall procedures that can be quickly implemented should a food safety issue occur.[68] Factors that influence the kind and number of microorganisms present in processed food include:

1. The initial contamination of the raw material and ingredients, and
2. The additional contamination gained within the plant from sources such as manual contact, contact with surfaces like machinery, air-borne, and waterborne contamination, inadequate temperature during process and storage, post-processing contamination, and so on.[69]

Many cases of microbial contamination in particular been reported in recent years in fruits and vegetables have been sources of contamination; from carrots and green beans in 2007 to baby spinach and fresh leaf lettuce in 2006, says Carolyn Yates from McGill tribune.[70]

Fungi and bacteria will infect fresh vegetables during storage. There has been a report that extending the shelf life of refrigerated foods might increase microbial risks in modified atmospheric packaged produce where[71]:

1. Increasing the time in which food remains edible increases the time in which even slow-growing pathogens can produce toxins,
2. Retaining the development of competing spoilage organisms, and

64 Occupational Health and Safety in the Food and Beverage Industry

3. Packaging or respiring produce could alter the atmosphere so that pathogenic growth is stimulated.

Other major factor to be considered has been the high risk of emerging zoonoses, changes in the survival of pathogens, alterations of vector-borne diseases, and parasites in animals which may necessitate the increased use of veterinary drugs, possibly resulting in increased residue levels of veterinary drugs in foods of animal origin.[728–730] This poses both chronic risks to human health and is directly linked to an increase in antimicrobial resistance in human-animal pathogens. Two other factors to be considered are:

1. The application of pesticides, and the subsequent residues in food, is an ongoing concern that is expected to become more prevalent due to climatic changes, with shifts in farming systems and farmers' behavior to adapt to the changing climate, and
2. The increased frequency of inland floods linked to climate change will impact environmental contamination and chemical hazards in foods through the remobilization of river sediments and subsequent contamination of agricultural and pastureland soil.

Foodborne illnesses are caused by one of the following biological agents.[133]

- **Mycotoxins** produced by microorganism such as aflatoxins in some leafy vegetables and patulin in fruit juices.
- **Bacteria** such as *E. coli, Salmonella, Clostridium botulinum* as well as molds can cause food poisoning and of course not all microorganisms are dangerous. In general they could have an inert, beneficial, spoilage, or pathogenic effect.[134] They are often a point of contamination in shredded vegetables in salad bars and with fresh fruit and vegetable product.
- **Viruses** are protein-wrapped genetic materials and are transmitted to human hosts by food or food contact surfaces such as hepatitis A and Norovirus.[133]
- **Parasites** need a host to survive such as Trichinella Spiralis in game animals or Cyclospora in imported fruits and vegetables including fresh basil and raspberries, that are consumed raw or lightly cooked from where Cyclospora are common.

Foodborne pathogens are causing a great number of diseases with significant effects on human health and economy. Previously the most ten common parasites in food sources have been reported by FAO–WHO (2014) as follows[363]:

1. *Taenia solium* (pork tapeworm) common in pork meat.
2. *Echinococcus granuloses* (dog tapeworm) common in fresh produce.
3. *Echinococcus multilocular* is (a type of tapeworm) in fresh produce.
4. *Toxoplasma gondii* (protozoa) common in red and game meat.

Occupational Hazard Origin and Preventive Measures

5. *Cryptosporidium* spp (protozoa) common in fresh produce.
6. *Entamoeba histolytica* (protozoa) common in fresh produce.
7. *Trichinella spiralis* (pork worm) common in pork meat.
8. Opisthorchiidae (family of flat worm) common in fresh – water fish.
9. *Ascaris* spp (small intestinal round worm) common in fresh – water fish.
10. *Trypanosome cruzi* (protozoa) in fruit juices.

Most recent references to common pathogenic bacteria such as (*Bacillus cereus, Campylobacter jejuni, Clostridium botulinum, Clostridium perfringens, Comebacker sakazakii, Escherichia coli, Listeria monocytogenes, Salmonella* spp., *Shigella* spp., *Staphylococcus aureus, Vibrio* spp. and *Yersinia enterocolitica*), viruses (Hepatitis A and Noroviruses), and parasites (*Cyclospora cayetanensis, Toxoplasma gondii,* and *Trichinella spiralis*)[617] and to this list US Health Official have added also Norovirus and Vibrio Vulniticus.[618]

A 2020 report from WHO has referred to the following facts[619]:

- Access to sufficient amounts of safe and nutritious food is key to sustaining life and promoting good health.
- Unsafe food containing harmful bacteria, viruses, parasites, or chemical substances, causes more than 200 diseases ranging from diarrhea to cancers.
- An estimated 600 million – almost 1 in 10 people in the world – fall ill after eating contaminated food and 420,000 die every year, resulting in the loss of 33 million healthy life years (DALYs).
- US$110 billion is lost each year in productivity and medical expenses resulting from unsafe food in low- and middle-income countries.
- Children under five years of age carry 40% of the foodborne disease burden, with 125,000 deaths every year.
- Diarrheal diseases are the most common illnesses resulting from the consumption of contaminated food, causing 550 million people to fall ill and 230,000 deaths every year.
- Food safety, nutrition, and food security are inextricably linked. Unsafe food creates a vicious cycle of disease and malnutrition, particularly affecting infants, young children, elderly, and the sick.
- Foodborne diseases impede socioeconomic development by straining healthcare systems and harming national economies, tourism, and trade.
- Food supply chains now cross multiple national borders. Good collaboration between governments, producers, and consumers helps ensure food safety.

Fresh fruits and vegetables have been identified as a significant source of pathogenic contaminants. As a result, there has been a wealth of research on identifying and controlling hazards at all stages in the supply chain.[73]

Food safety has been an increasingly important public health issue. Governments all over the world are intensifying their efforts to improve food safety. These efforts are in response to an increasing number of food safety problems and rising consumer concerns.

In Canada, each year 11–13 million Canadians suffer from foodborne diseases with symptoms such as cramps, nausea, vomiting, diarrhea, fever, and death.[136,340]

Making food safe is an effort, involving the suppliers, the production plant or factory, and all of us to protect the beneficiary's health. The main risks are contamination by microorganisms, toxins, foreign matters, chemical contaminants, excess moisture content, degradation of the nutritional value, or pest infestation.[135]

There are more than 250 known foodborne diseases. The majority are infectious and are caused by bacteria, viruses, and parasites. Other foodborne diseases are essentially poisonings caused by toxins and chemicals contaminating the food. All foodborne microbes and toxins enter the body through the gastrointestinal tract and often cause the first symptoms there. Nausea, vomiting, abdominal cramps, and diarrhea are frequent in foodborne diseases.[337]

5.2.1.1   Magnitude of foodborne illness

Foodborne diseases are a widespread and growing public health problem, both in developed and developing countries. Even in America Food safety is an important public health priority. Foodborne illness (sometimes called "foodborne disease", "foodborne infection", or "food poisoning") is a common, costly – yet preventable – public health problem. The Centers for Disease Control (CDC) estimates that each year roughly 1 in 6 Americans (or 48 million people) gets sick, 128,000 are hospitalized, and 3,000 die of foodborne diseases.[338]

Foodborne diseases are an important cause of morbidity and mortality worldwide.

Diarrheal diseases alone – a considerable proportion of which is foodborne – kill 1.9 million children globally every year.[339]

The high prevalence of diarrhea diseases in many developing countries suggests major underlying food and water safety problems.

Foodborne illness may be caused by bacteria, viruses, or fungi. Pathogens, the microorganisms that cause foodborne illness, thrive in specific environments. Some are aerobic requiring oxygen and others are anaerobic and grow in absence of oxygen. Desirable moisture levels, "danger zone" temperature ranges (4–60 C), and pH levels are three most important factors affecting microbial growth. The food that which best support microbial growth are those with a neutral pH, including milk and milk products, poultry, meat, fish, seafood, and some vegetables such as potatoes and corn.[67]

5.2.1.2   Prevention

Prevention can be achieved through personal hygiene, adequate cooking time and temperature, and avoiding cross-contamination during processing, preparing, storage, or serving food. If the source of foodborne illness is due to the consumption of contaminated drinking water with organisms such as *Cryptosporidium parvum* then the water must be boiled for at least one minute prior to consumption.[364] It is

Occupational Hazard Origin and Preventive Measures 67

important to keep hot food hot and cold food cold which will let the food stay out of "danger zone" and microbial growth.[67]

5.2.1.3 Integrated Food Safety System (IFSS)

The integrated food safety system is a vision of joining government food safety resources and authorities at all levels into a more unified and coordinated system. This chapter will explore the impact of integration on the food protection system, the history and evolution of food safety laws, and the national efforts to create an IFSS.[483]

5.2.2 HAZARDS ASSOCIATED WITH LIVE OR DEAD ANIMALS

In the food processing industry the following factors are considered good manufacturing practices (GMP) to avoid hazardous food processing:

(1) Control Program, (2) Training (3) Operational Controls, (4) Environmental Controls, (5) Hazard Analysis Critical Control Point (HACCP) as discussed in the previous chapter that include personal hygiene practices, during shipping, receiving, handling, and storage, as well as sanitation practices, equipment maintenance, pest control, recall and traceability, construction design, and water supply safety.[77]

5.2.3 FOOD SAFETY PRIORITIES

Food safety questions may arise in many ways for consumers and could be such as[204]:

- Must peanut butter be refrigerated after it's opened? That depends on the type. Regular peanut butter does not need refrigeration, though that might help retain its flavor and slow the process of separation.
- If I cut those tiny molds off the cheddar, will it be ok? Maybe not, because spoilage and sometimes toxins could go beyond the visible mold. The US Department of Agriculture says that if you cut off at least one inch around and below the mold spot, hard cheese can still be safely consumed.
- I noticed an old can of peach fruit with no expiration date. Is it safe to consume from the can? You can store high-acid foods such as peaches and tomatoes for up to 18 months; low-acid foods such as meat and vegetables for 2–5 years. But never store cans above the stove, under the sink, in a damp room, or where they could be exposed to temperature extremes.
- How long can I keep herbs and spices? Ground spices can be kept for 2–3 years; whole spices for 3–4 years; herbs for 1–3 years. Air, light, moisture, and heat are enemies. Therefore, store spices and herbs tightly sealed in a cool and dry place. Red spices retain their color better and are best protected from insects if refrigerated if not used frequently.
- How long do refrigerated eggs last? Federally graded fresh eggs in their shells are safe for 3–5 weeks after the expiration date on the carton; raw yolks and whites for 2–4 days; hard-cooked eggs for one week.

Recent events worldwide have shown that we do not consistently control the agents that cause foodborne illness. Directly or indirectly, domestic, and wild animals are the source of zoonotic pathogens, which are often carried asymptomatically and shed by animals. It is generally accepted that zoonotic pathogens can be concentrated in food animals if manure is used improperly.[78] Dr Holley points to some key measures to improve the Canadian food safety system as follows:

- The Canadian food safety system is reactive, and its development has not kept pace with new knowledge used by other countries to protect consumers.
- Food-borne illness surveillance is a passive patchwork of regional systems that feed poor-quality data on illness outbreaks or diseases.
- These surveillance systems ignore the large sporadic illness component, which is more useful in discovering the major sources of illness in developing policy.
- Food operations are now inspected at three government levels to different standards; operation of the multi-government agency interface for both illness surveillance and food inspection should be seamless.
- We must address changes in pathogens (identity or antimicrobial resistance) and foods as they evolve through insightful policy based on sound Canadian data.

In another review of Canadian Food Safety Priorities[79] reveals the importance of the following criteria:

1. Canadian Food Inspection Agency: further improvement for increased hiring of qualified inspectors, as well as more uniform training is in demand.
2. Product Traceability: although it will be more expensive but more detailed food production history will be required for better recall and traceability in the market.
3. Validation Models: are needed to help processors select and optimize intervention ingredients and process options to establish product shelf life.
4. Mandatory National Food Borne Illness Surveillance Program: A more uniform and mandatory national program involving various health-professionals coordination will be required during emergency foodborne illnesses or outbreaks.
5. Retail and Consumer Food Safety Training: Cost control measures have reduced ongoing food safety training and improvement both in industrial and government sectors, also a better health and wellness program should be considered at school.
6. Compatibility of Provincial with Federal Inspection, Standards, and Food Legislations.
7. Crisis management: progress has been made during the crisis in food chain intervention but there is room to improve in terms of jurisdictions during a crisis.

Occupational Hazard Origin and Preventive Measures

5.2.4 FOOD ALLERGY, SENSITIVITY, AND FOOD INTOLERANCE

Food allergy and allergens, in general, have become increasingly important to the food industry because they can represent a serious health hazard to consumers. Food Processors are therefore obliged to prevent food from being contaminated with these allergens and be labeled properly as it could be fatal for failing to do so at times.

Health Canada has defined the foods that represent 95% of allergic reactions in Canada. These are termed "priority allergens" are peanuts, tree nuts (almonds, Brazil nuts, cashews, hazelnuts, macadamia nuts, pecans, pine nuts, pistachio nuts, and walnuts), sesame seeds, milk, eggs, seafood (fish, crustaceans, and shellfish), soy, wheat, sulfites, and mustard.[703]

Recent studies provide some input in comparative structural studies of peanut allergens demonstrating the utility of bio-informatic tools, resources, and strategies to better understand the different allergenicity of peanut allergens with their closely related yet less allergenic counterparts from other legume species.[139]

When a person presents with a food-related health complaint, the history is paramount and dictates the differential diagnosis, appropriate testing, and treatment. The physical examination may add some additional information if the patient happens to be seen acutely at the time of a reaction. There are five crucial elements to the history[147]:

- The suspect(s): What food(s) does the patient believe caused the reaction(s)
- Timing: How long from exposure to symptom onset?
- The nature of the reaction: What are the symptoms?
- Reproducibility: Has it happened more than once? Does it happen every time?
- What treatment was administered and what was the response?

Many foods have immunological cross-reactivity with other foods, and prior exposure may have been to such a food sharing similar proteins. Thus, patients may appear to have an allergic reaction on their first exposure to a food, but in fact they have had prior relevant allergen exposure.

This prior exposure was sufficient to cause sensitization (i.e., Urticarial/anaphylactic) antibody production but not a clinical reaction. Once sensitized, a subsequent exposure to a larger amount causes an allergic reaction.[147]

Food intolerance is a digestive system response where a person can't digest or break down food. While food allergy is an immune system response to an ingredient in food such as protein which creates a defense system (antibodies) to fight it.[141]

Effects can range from common digestive system problems including abdominal pain and spasms, diarrhea, constipation, and flatulence, to headaches, skin rash and eczema – all due to the body's inability to thoroughly digest food caused by the lack of proper amounts of certain enzymes.

Histamine is the first product worldwide that decreases histamine levels that cause food intolerance by replenishing the body's digestive enzyme Diamine Oxidase

(DAO). This clinically shown dietary supplement regulates histamine levels in the body, unlike antihistamines, which only block the histamine.[137]

Likewise, food intolerance is a physiological response which appears gradually, happens when one consumes a lot of the food, when consumed often, and is not life threatening.[358]

Food intolerance has also been expressed as a negative reaction to a food, beverage, additives, or compound found in foods that produces symptoms in one or more body organs and systems, but it is not a true food allergy. Some foods that can cause intolerance had been reported to be dairy products, salicylates in fruits and vegetables, caffeine, tea, beer, honey, wine, banana, and avocado. Food poisoning, lactose intolerance, gluten and nut intolerance, and carbohydrate intolerance are examples of such sensitivity to food.[142,143]

Other terms that are occasionally used to describe types of food intolerance include food toxicity or food poisoning, idiosyncratic reactions, and pharmacologic reactions to foods. Food toxicity may be the result of natural or acquired toxins in some foods or the result of microorganisms or parasitic contamination of natural or processed foods.

Finally, a pharmacologic reaction occurs to some foods containing chemicals (e.g., caffeine), and some food additives (e.g., food colors) have drug-like effects.[356]

The immune system misinterprets a chemical component of a food as harmful and releases histamine and other chemicals to combat it which will result in allergy symptoms mentioned earlier. The Food and Drug Cosmetic Act requires a complete listing of food ingredients on food labels.[91]

During the processing of food, it is not uncommon for formulation errors or oversights to occur. These errors could just end up being a labeling issue, or they could become a serious health threat. There could be cross-contamination, inclusion of undeclared components in the raw materials, unlabeled recipe change, or the use of the wrong recipe or ingredient. There are no inherent grounds for assuming that genetically modified (GM) foods are more or less allergenic than traditional foods.

Chemical sensitivity is reported when exposure to airborne chemicals such as cigarette smoke, engine exhaust, perfumes, household detergents, and solvents causes symptoms but food allergy is an abnormal response of the immune system to an otherwise harmless food.[565]

Food intolerances include metabolic food disorders, such as lactose intolerance, phenylketonuria, and favism, as well as direct adverse reactions to naturally occurring or industrially added chemicals in food.[139]

Allergen refers to any protein from any of the following foods, or any modified protein that includes any protein fraction derived from any of the following food often such as almonds, Brazil nuts, cashews, hazelnuts, macadamia nuts, pecans, pine nuts, pistachios, or walnuts; peanuts; sesame seeds; wheat or triticale; eggs; milk; soybeans; crustaceans; shellfish; fish; mustard seeds, etc. [341]

5.2.4.1 Food allergy prevention

Essential steps in developing an effective allergen management program include[144]:

Occupational Hazard Origin and Preventive Measures 71

- Allergen assessment: prepare a master list of all ingredients used, identify if any of these food ingredients contain food allergens and finally indicate those allergens on your master list for special measures to be taken.
- Supplier information: obtain product specification from your supplier for ingredients being received, verify supplier has a documented allergen control program, and finally obtain a letter from your supplier that he guarantees ingredients are free of undeclared allergens.
- Shipping, receiving, handling, and storage: inspect the shipment for spills or damage, verify ingredient is exactly what you ordered, store allergenic food ingredients separately from nonallergenic ones and on the bottom of racks and identify ingredients with allergenic ingredients with a color code.
- Production and plant scheduling: plan the production of long runs of allergen-free products to minimize changes in production lines, products containing allergens can be produced on specific days and possibly be done the last in the production line.
- Evaluate product flow during the production cycle in conveyors to prevent their accidental contamination.
- Store and isolate packaging materials to avoid allergen cross-contamination.
- Cleaning and sanitizing: test your cleaning procedures to ensure no residual allergens are present before production.
- Employee training: avoid consumer threats as well as financial consequences of a product recall by providing adequate training.
- Labeling: confirm label accuracy against the product's formulation, the CFIA enforces Canada's labeling laws of all foods and use common food terms for allergens such as milk instead of casein.
- Documentation: keep records to show your facility is taking adequate measures to control possible allergens cross-contamination and label verification.

5.2.5 FOOD SAFETY OF BIOTECHNOLOGY-DERIVED PRODUCTS

Since 1973 the safety aspect of genetically engineered food has been the subject of considerable debate in many countries.

In 1983, three reports from the University of Ghent, the University of Washington and the Monsanto Company showed that the Ti plasmid of *Agrobacterium tumefaciens* could be used to transfer foreign DNA into a plant genome, thus producing the first GM plants.[487]

A novel food is defined as any food that has not been used previously to any significant degree for human consumption. A similar definition exists for a novel process but determining the safety of food derived from genetically modified food (GMO) is more difficult than that of single food constituents or defined chemical mixtures that will be used in food.[146]

It started with little fanfare over a decade ago, when genetically modified foods quietly appeared on American grocery shelves. But in the years since, GM foods have sparked a global controversy.[484]

The Canadian federal government defines biotechnology as "the application of science and engineering to the direct or indirect use of living organisms or parts or products of living organisms in their natural or modified form".[145]

In fact, genetic engineering debate can risk potential health, environment, and agronomic calamities which are done by simply removing the desired gene from its source and inserting it into the recipient genome (an organism's DNA).[443]

On the other hand, others argue that GM plants for food and feed could be very promising as in the 21st century, overpopulation and exhaustion of resources will increase.[446]

5.2.6 SKIN DISEASE/DERMATITIS

Dermatitis is one of the leading causes of ill health and affects people working in many industrial sectors in particular catering and food processing food industry.

In the UK, nearly 3,000 cases of contacts dermatitis are reported by occupational physicians, dermatologists and or other health professionals.[80]

One study identified Irritant Contact Dermatitis (ICD), which has been related to and provoked by work materials or workflows, has been a frequent cause of Occupational Skin Disease (OSD). [485]

It is common for exposure to occur to more than one irritant and more than one allergen at any one time. Such exposures may give rise to a cumulative irritant and cumulative allergic response. An irritant contact dermatitis may also develop first, rendering the skin more susceptible to penetration by sensitizers. It is also possible that an original allergic contact dermatitis might be later sustained by an irritant.[343]

Contact dermatitis is inflammation of the skin, caused by contact with a range of substances such as detergents, toiletries, chemicals, and even natural products like food and water over a prolonged period. It can affect all parts of the body, but it is most common to see the hands affected. There are three main types of contact dermatitis[344]:

1. Irritant contact dermatitis; is caused by things that dry out and damage the skin such as detergents, solvents, oils, and prolonged or frequent contact with water.
2. Allergic contact dermatitis occurs when someone becomes allergic to something that comes into contact with his or her skin. The allergic reaction can show up hours or days after contact. Common causes include chemicals in cement, hair products, epoxy resins, and some foods.
3. Contact urticarial is a different kind of allergy and occurs within minutes of the material touching the skin. Things like plants, foods, and rubber latex gloves.

5.2.6.1 High-risk occupations

General professions at risk are healthcare workers, hairdressers, beauticians, printers, those in cleaning, construction, and workers using metalworking fluids. But remember it can occur in just about every workplace. Professions at risk in the food

Occupational Hazard Origin and Preventive Measures   73

industry include agriculture workers, bakers, confectioners, meat processors and butchers, food service workers and caterers, distillers and bartenders, beverage and soft drink workers as well as ingredient, spice, and additive workers which will be discussed later.[148]

5.2.6.2 Prevention

Health and Safety Executive (HSE) in the UK recommends the following for the prevention of dermatitis[80]:

1. Substitute a more hazardous material with a safer alternative,
2. Automate the process,
3. Enclose the process as much as possible,
4. Use mechanical handling,
5. Use equipment for handling,
6. Don't use hands as tools,
7. Use a safe working distance,
8. Train workers how to protect their skin,
9. Remind workers to wash any contamination from their skin promptly,
10. Tell them about the importance of thorough drying after washing,
11. Provide soft cotton or paper towels,
12. Supply moisturizing pre-work and after-work creams,
13. Provide appropriate protective clothing/gloves and equipment,
14. Make sure gloves are made of suitable materials,
15. Select gloves that are the right size and right for the task involved,
16. Use and store gloves correctly,
17. Replace gloves when necessary,
18. Consider job rotation, and
19. Have regular skin checks for early signs of dermatitis for treatment.

5.2.7 RESPIRATORY ILLNESSES

Asthma and rhinitis are two common respiratory illnesses due to inhalation of dust from hazardous substances from non-food such as building material dust or common food substances such as flour, wheat, spices, or food additives. Food additives are chemicals added to foods to keep them fresh, enhance color, flavor, texture, aroma, and shelf life and those also working in milling, malting, baking, fish processing and coopering, etc. are at particular risk due to exposure to such food or additives dust.[149]

5.2.8 HEARING PROBLEMS

Hearing can play an important role when the safety and security of a worker are concerned. A person may not feel comfortable living alone or working in an area where fire, smoke, or burglar alarms are inaudible.[150]

Toxic chemicals such as manganese which is often found in welding fumes can attack the cochlea, the ear's spiral-shaped cavity and sensing organ that picks up

74 Occupational Health and Safety in the Food and Beverage Industry

sound. Other ototoxic solvents such as styrene, toluene, and xylene also affect the auditory vestibular nerve, which transmits or balances information to the brain.

These substances can harm the auditory cortex, the part of the brain that processes sound but still for many chemicals, scientists can't say conclusively that exposure can cause hearing loss. The most common hearing protection wear is defending ears with suitable-fit ear plugs.[449]

5.2.8.1 Preventive management

1. Monitor the sickness absence due to such problems and see if they have been in contact with such substances,
2. Investigate worker complaints with safety representatives and employees,
3. Consider wearing personal protective equipment,
4. Consider replacing the type of protective wear,
5. Modify process or formulation if possible,
6. Consider job rotation, and
7. Provide proper training.

Providing training on health and safety issues on all the above occupational disease is very important in letting workers know how to minimize the risk, monitor if symptoms develop, reporting and seek medical attention. Let them know the importance of the Material Safety Data Sheet (MSDS) available as a worker's right to know in order to wear protective wear or equipment to prevent contact or inhalation of harmful substances. One must be aware that certain intermediary substances are produced during the processing of food and beverages that could be harmful to health.[155]

5.2.9 FOOD HYGIENE SAFETY AND SANITATION CONCERN

Although city food inspectors and other public health officials are involved with controlling the safety of food and beverages for public consumption area, they may not adequately oversee food handling activities going on within many public institutions such as universities, schools, hospitals, health care and correctional facilities, etc. Therefore, if there is going to be an event that involves the preparation and/or consumption of food products by an employee, they may need to get a "Temporary Permit" for such events.

There are generally following steps involved for sanitary food handling, preparation, and serving which include: permits, booth, menu, cooking; reheating; cooling and storage; transportation; hand washing techniques; health and hygiene; food born handling; dishwashing; ice; wiping cloth; and finally insect control and proper waste disposal discussed previously.[152]

As of November 2009, Quebec[154] like other health establishments in the rest of Canada and North America hold establishments in charge such as food service operators in universities, hospitals, and hotels or else responsible for having trained/certified personnel for applying principles of food safety in such institutions for public

Occupational Hazard Origin and Preventive Measures 75

safety as it involves purchasing, receiving, storing, preparing, selling, or serving food products to the public which must cover the following topics[153]:

1. The disease and symptoms that are carried or transmitted by food.
2. Points in the flow of food where hazards can be prevented or reduced and how the procedures meet the regulatory requirements.
3. The relationship between personal hygiene and disease spread, especially during cross-contamination, hand contact with ready-to-eat-foods, and hand washing.
4. How to keep injured or ill employees away from food and contact surfaces.
5. The length of time that potentially hazardous foods may remain at temperatures where disease-causing microorganisms can grow.
6. Safe cooking temperatures and times for the safe refrigerated storage, hot holding, cooling, and reheating of potentially hazardous foods.
7. Correct procedures for sanitizing utensils and food contact surfaces equipment.
8. The types toxic materials used, how safely store, dispense, use, and dispose.
9. The need for equipment that is sufficient in number and capacity, properly designed, constructed, located, installed, operated, maintained, and cleaned.
10. The source of the operation's water supply to be kept sanitary and clean.
11. How the operation complies with the modern food safety principal standard known as Hazard Analysis Critical Control Point required.
12. The rights, responsibilities, and authorities the local legislation assigns to employees, managers/operators, and the local health department.

In many institutions "local Food Permit Policy" guidelines must be adhered to besides regulatory public health agencies' requirements to prevent infections or other diseases among workers.[155]

5.2.9.1 Musculoskeletal injury

Every workplace has hidden ergonomic problems that need to be addressed. It is also important to look into various production lines and machinery involved in evaluating possible musculoskeletal injuries in workplace.[524] Health and safety at work summary statistic in Great Britain indicated that musculoskeletal injuries, illnesses, and working days lost in the food and drink industry have been at 41%, which is the third highest among all industry sectors after metallic manufacturing (1.7%) and transport manufacturing products.[722]

This represented 3,200 injuries arising from five different causes as has been reported by HSE[156] as follows:

85 % happened while the load was being handled manually.
60 % happened due to body overload from heavy weight.
48.5 % happened by lifting and lowering loads.

76 Occupational Health and Safety in the Food and Beverage Industry

The type of work can be divided as follows during such accidents:

1. Stacking/un-stacking containers such as boxes, crates sacks.
2. Pushing wheeled racks such as oven racks and trolleys or food products.
3. Cutting, boning, joining, trussing, and evisceration such as in meat and poultry processing sectors.
4. Packing products such as cheese, confectionary, and biscuits.
5. Handling drinks containers such as the delivery of casks, kegs, and crates.

5.2.9.1.1 How do we know if there is a problem?

All sorts of work may involve invisible musculoskeletal injuries and which can lead to further harm. This may require a multitude of several factors to properly address the problem such as posture and position of work done, frequency, part of the body affected, length of time exposed, type of work involved, etc.[486]

These unaddressed hearing problems will have financial losses for both the employer and employer such as sickness absence, high staff turnover, loss of production, etc. Compensation cases may also affect your medical and company insurance premium.[156]

5.2.9.1.2 Preventive management

1. Identify tasks that present a serious risk of acute injury (e.g. lifting) or chronic injury (e.g. repetitive upper body work).
2. Evaluate to identify factors in detail that lead to the risk.
3. Consider engineering or ergonomics designs and mechanization (e.g. powdered trucks, conveyors, vacuum filters, bulk handling, or automation).
4. If mechanization is not possible then, consider other options such as reducing loads of boxes to below 25 kg.
5. Provide training on proper ergonomics and handling, rotation of jobs and workers, and consider medical surveillance.
6. Consult trade unions and employees as well as health and safety professional advice.

5.2.9.2 Working with animals

Working with animals can pose the following risks[88]:

1. Transmission of zoonoses (transfer of diseases from animal to human). The transmission of such diseases such as Campylobacteriosis has been reported through carcass contamination in sheep slaughterhouses in UK.[87]
2. Sensitization (the development of allergies). The risk could come from tissues, body fluids, hair, dander and cage, mites, and animal accessories.

5.2.9.2.1 Precautions

1. Exercise aerosol control during cleaning operations in animal housing,
2. Maintain animal housing areas in a hygienic state,
3. Practice good personal hygiene,

Occupational Hazard Origin and Preventive Measures

4. Use automatic cage washers instead of manual cleaning operations,
5. Obtain veterinary surveillance for animals,
6. Practice appropriate handling and restraint techniques to prevent accidental bites and scratches,
7. Seek immediate medical attention for bites and scratches.

5.2.9.3 Transient emissions

Transient emissions are commonly observed in multifloor buildings housing multiple laboratories and activities.[61] Due to their intermittent nature and the complexity of building mechanical systems, transient emissions can be difficult to track down. To find the exact sources typically requires detective work. In this regard, lab personnel can play a key role. Steps to follow if experiencing transient emissions are as follow[61]:

1. Report all incidents to the ventilation department and health and safety officer,
2. Evacuate the areas affected at the first sign of symptoms discomfort,
3. Determine what areas of the building are affected,
4. Pour water into all laboratory drains in affected areas (sinks, floor drain, etc.),
   Keep a logbook of all incidents to record,
   a. time and duration of incident,
   b. description of odor characteristics,
   c. symptoms experienced,
   d. areas affected,
   e. wind conditions,
   f. direction at the time of the incident,
   g. use the troubleshooting guide to attempt in identifying possible pathways, and
   h. arrange to have an air quality monitoring analysis performed by an occupational hygienist.

5.2.9.4 Burn

Burns are among the most common household injuries, especially in children. Depending on the cause and degree of injury, most people can recover from burns without serious health consequences. More serious burns require immediate emergency medical care to prevent complications and death. Burn levels could be 1st, 2nd, or 3rd-degree damage based on the depth and damage done and its origin could be thermal (steam, heat, cold, electrical), chemical, or radiation exposure.[159]

The type of burn is not based on its cause. Scalding, for example, can cause all three burns, depending on how hot the liquid is and how long the skin makes contact. Chemical and electrical burns warrant immediate medical attention because they can affect the inside of the body, even if skin damage is minor.[159]

78 Occupational Health and Safety in the Food and Beverage Industry

Burn to certain sensitive parts of the body increases, risk of infection with symptoms such as dry skin, pain, redness, edema (swelling), blisters, waxy and white skin as well as damage to muscles and blood vessels. For details on treating and providing first aid based on the type and level could also view the following reference.[487]

See thermal hazards for further details. In another chapter we will look into particular, occupational hazard and injury management in food manufacturing more specifically.

5.3 CHEMICAL HAZARD

Each year about 6,000[359] new chemicals are created and many of these find their way into the workplace. These are in addition to many already being in use over the years which are health threatening. It is important to differentiate between toxic and hazardous. Toxic refers to the ability of a material to produce harm to a living organism while hazardous substance refers to the probability of causing harm.[359]

There are more than 1,000 pesticides used around the world to ensure food is not damaged or destroyed by pests. Each pesticide has different properties and toxicological effects.

Many of the older, cheaper (off-patent) pesticides, such as Lindane and dichlorodiphenyltrichloroethane (DDT), can remain for years in soil and water.

These chemicals have been banned by countries that signed the 2001 Stockholm Convention – an international treaty that aims to eliminate or restrict the production and use of persistent organic pollutants. WHO and FAO have jointly developed an International Code of Conduct on Pesticide Management voluntary framework published in 2014. One example could be chlordecone used in banana fields to repel insects that have been responsible for prostate cancer[768] and or chlorpyrifos, a potent neurotoxicant member of the organophosphate insecticide family. Chlorpyrifos can harm workers, communities, and the environment but is not generally detected on peeled bananas. Children are especially sensitive to chlorpyrifos toxicity.[769,770]

Chemical hazards exist and could harm people through air, water, and land and could be present as vapor, gas, mist, smoke, liquid, or solid from farm to fork.

Despite an increased appreciation of chemical hazards, production continues to grow. Our creative pursuit generally outpaces our efforts to fully understand these new chemical substances being poured into market.[160]

Wexler reports that major efforts by the Environmental Protection Agency (EPA) in the current chemical management in the USA should adopt the following recommendation:

1. Chemicals should be reviewed against risk-based safety standards based on sound science that reports human health and environment.
2. Manufacturers should provide EPA with necessary information to conclude that new and existing chemicals are safe.
3. EPA should have clear authority to take risk management actions when chemicals do not meet the proposed safety standards with flexibility to take

Occupational Hazard Origin and Preventive Measures

into account sensitive subpopulations, cost, social benefits, and other relevant considerations.

4. Manufacturers and US EPA should assess and act on priority chemicals, both existing and in a timely manner.

5. Green chemistry should be encouraged and provisions assuming Transparency and Public Access to Information be strengthened.

6. US EPA should be given a sustained source of funding for implementation.

Polyaromatic compounds such as metabolites, alkylated polyaromatic, mixed halogenated compounds, or heterocyclic compounds originated from both man-made and natural sources. Many of these compounds have been studied as a result of food contamination incidence or occupational exposure.[161]

Shopping for food was easy when most foods came from the farm. Now factory-produced foods with numerous chemical and food additive names where most people are not even able to pronounce the chemical name properly on a label would like to know about their hazards and property. Where some such as tocopherol (Vitamin E), ascorbic acid (Vitamin C), calcium or sodium propionate, alginate, carrageenan, citric acid, etc. may be safe but others such as artificial food coloring, acesulfame-K, artificial sweetener cyclamate, nitrite, etc. could be hazardous. [106]

The hazards associated with industry can be through poisoning, fire/explosion, and burns. On the other hand, there are pathways for consumers to get exposed to chemicals which could be present as pesticides or their metabolite, biopesticides, additives, color, environmental contaminants, animal drug, etc. where they can be transferred by air, water, and or land into food chain[98] and with many linked to environmental sources.[489]

Chemicals are also naturally present in many food plants while some have healing properties while others have toxic effects and in particular above certain limits. Naturally occurring toxicants are like aflatoxin from growing *Aspergillus flavus* on corn or peanut, *parahydrazinobenzoic* acid in common mushrooms, or alkaloids in potatoes.[105]

The presence of high amounts of nitrates or nitrites in fruits and particularly vegetables and or nitrosamine has been associated with "methemoglobinemia" or known as "baby blue" and or nitrosamine which forms during food processing such as frying or smoking at high concentrations may cause health problems.[100]

Chemical hazards responsible for such illnesses were chlorine, ammonia, processing bromelain, smoke and pesticides have been reported in the food industry as well as diacetyl and its substitute 2, 3 Pentanedione used in a wide variety of food and beverage flavorings.[158]

All Solanaceae plants, which include tomatoes, potatoes, and eggplants, contain natural toxins called solanines and chaconne (which are glycoalkaloids). To minimize harm[704]:

1. Do not assume that if something is "natural" it is automatically safe,

2. Throw away bruised, damaged, or discolored food, and in particular moldy foods,

80   Occupational Health and Safety in the Food and Beverage Industry

3. Throw away any food that does not smell or taste fresh or has an unusual taste, and
4. Eat mushrooms or other wild plants that have been identified as non-poisonous.[704]

5.3.1 Natural Toxicants/Prevention

Food contamination can be from toxigenic (poisonous) bacteria producing toxins or naturally occurring toxicants in food.[66] Some natural toxins in food plants are tabulated in Table 5.1.[703]

TABLE 5.1
Sample of Food Product and Contamination Sources

Food commodity	Toxin
Ackee fruit	Hypoglycin
Cassava root, bamboo shoots, stone fruit	Cyanogenic glycoside
Fiddlehead	Unidentified
Green beans, red kidney beans, white kidney beans	Lectin
Wild mushrooms	Amanitin's, gyromitrin, muscarine, phallotoxins
Parsnip	Furocoumarins
Rhubarb	Oxalic acid
Cabbage, cauliflower, broccoli, mustard, turnip	Goitrogens

5.3.2 Chemical Hazard Prevention

Always ask yourself "What would happen if…". Answering such questions requires you to understand the hazard associated with chemicals, equipment, and machine involved. The reactivity, flammability, corrosively, and toxicity of the chemical being used will dictate the precautions you must take. In laboratory scale safety American Chemical Society[163] for performing safe chemical works recommend acquiring practice and habit of accident prevention, using appropriate engineering control, using smaller quantity and PPE, and anticipate consequences.

Both mixing improper combination of chemicals but also incompatible storage of chemical could cause accidents.[166]

Many chemical substances could be harmless such as salt (sodium chloride), corrosive such as many caustics or acidic materials, toxic and poisoning and others flammable and could cause fires, explosion and burns, Some preventive recommended measures are[16]:

- Substitute harmful with harmless or less harmless substances or processes.
- Take effective steps to prevent the liberation of harmful substances using PPE to prevent inhalation, absorption, and ingestion of hazardous substances.

Occupational Hazard Origin and Preventive Measures

- Take effective measures for the protection of workers from harmful radiation.
- Avoid smoking, drinking, or food in such areas by using proper facilities.
- Practice personal hygiene.
- Have emergency measures possibilities in case of an accident such as a first-aid kit, spill response resources, fire prevention equipment, etc.
- Make sure you have the updated Material Safety Data Sheet on site.
- Hazardous chemicals should be properly labeled, segregated, and stored.
- Avoid storing chemicals above eye level and store heavier items on lower shelves.
- Use edge guard as well as trays and absorbent in case of leaks, breakage, etc.
- Provide training for handling, storage, and other emergency measures.
- Have accident/incident and occupational disease forms for reporting.
- Have regular training on required OHS practices and monitor workers.
- Have regular inspections and follow-ups for evaluation for reported accidents, equipment/machine evaluation, maintenance, and performance.
- Formal and regular procedures to ensure that sufficient full-time personnel are trained in the proper use of emergency equipment and procedures.
- Proper disposal guidelines for hazardous chemical waste.
- Provide proper communication and training for handling, storage, and other emergency measures such as first aid and medical emergencies, etc.
- Consult updated MSDS or GHS for further and detailed information.
- Keep an updated inventory of chemicals at all times.
- Use recommended equipment, instrumentation, and machines in handling chemicals.
- Organization and proper housekeeping is an accident-prevention tool at all times.

5.3.3 Workplace Hazardous Material Information System (WHMIS 2015)

Anyone who purchases, stores, uses, or resells hazardous chemicals must comply with regulations called Workplace Hazardous Material Information System which came into effect on October 31, 1988.[103,164,165,169]

WHMIS key elements are

1. Supplier's labels (name of product, supplier, hazard symbol, risk phrases, precautionary measures, and first-aid information);
2. Workplace labels (name of product, handling precautions, what to do when exposed, no use abbreviation and formula);

3. Material Safety Data Sheet (product information, hazardous ingredients, physical properties, preventive measures and fire, explosion, reactivity, and toxicological data),[167,104] however, over time other information such as composition, fire-fighting, accidental release, handling and storage, chemical properties, stability and reactivity, ecological information, disposal consideration, and transport and regulatory information have been added;[168]
4. Training: All employees must receive training by employer and comply with WHMIS Act.[169] The training is usually valid for a period of three years as is the expiry for any MSDS for a chemical product.

5.3.3.1 Controlled products

Controlled products have been designated the following classes within WHMIS;

- Class A: Compressed gases;
- Class B: Flammable material;
- Class C: Oxidizing material;
- Class D_1: Poisonous and Infectious material causing immediate and serious toxic effects;
- Class D_2: Poisonous and infectious material causing other toxic effects;
- Class D_3: Poisonous and infectious material with bio-hazardous infectious effects; Class
- E: Corrosive material; Class F: Dangerously reactive material.[167,170]

5.3.4 FUTURE TRANSITION TO GLOBALLY HARMONIZED SYSTEM (GHS)/WHMIS AFTER GHS

WHMIS as described above has been a national standard to provide information about hazardous materials used in Canadian workplaces,[342,345,346] however, to expand its scope GHS has been developed internationally. GHS is a system that defines and classifies the hazards of chemical products and communicates OHS information on labels and material safety data sheets (called Safety Data Sheets, or SDSs).[347]

As a result, a number of countries and organizations have developed laws and regulations regarding the use of these chemicals through a unified labeling system called SDS.

GHS in general provides,[566] protection of public health and environment, is a recognized framework for countries without existing system, reduces the need to test and evaluate chemicals, facilitate international trade for such chemicals whose hazards have been previously identified, and provides better classification and wider contents.[567]

The UN Sub-Committee of Experts on GHS Classification and Labeling of Chemicals has been established to address the implementation and maintenance of the GHS.[260]

Occupational Hazard Origin and Preventive Measures

In comparison to MSDS, SDS provide better symbols and pictograms including health hazard, flame, exclamation, gas cylinder, corrosion, exploding bomb, flame over circle, environment, and skull/crossbones for better hazard communication.[490]

The key elements in the GHS vocabulary are SDS, Hazard Group, Category, Hazard Statement, Precautionary Statement, Signal words, and Pictograms.[654]

The GHS hazard groups and building blocks are divided into three groups.[654]

Physical hazards, health hazards, and environmental hazards as described below:

1. The **Physical** hazard groups are explosives, flammable gases, aerosols oxidizing gases, gases under pressure, flammable liquids, flammable solids, self-reacted substances and mixtures, pyrophoric liquids, pyrophgoric solids, self-heating substances and mixtures, substances and mixtures which, in contact with water, emit flammable gases, oxidizing liquids, oxidizing solids, organic peroxides, and corrosives to metals.[654]
2. The **Health** hazard groups are acute toxicity, skin corrosion/irritation, serious eye damage/eye irritation, respiratory or skin sensitization, germ cell mutagenicity, carcinogenicity, reproductive toxicity, specific target organ toxicity – single exposure specific target organ toxicity – repeated exposure, and aspiration hazard.
3. The **Environmental** hazard groups are hazardous to the aquatic environment (acute and chronic) and hazardous to the ozone layer.

The adoption of GHS has not changed the current roles and responsibilities of suppliers, employers, and workers in WHMIS 2015 or WHMIS after GHS.

Suppliers, importers, and producers' duties continue to include, classifying hazardous products, preparing labels and SDS, and providing these elements to customers.[655]

The labels required on these products by their respective governing legislation are accepted as WHMIS labels in the workplace. MSDSs are not required for these products; however, the worker education of WHMIS does apply.[348]

GHS style labels and safety data sheets are already been established in workplaces and USA and other countries have already adopted GHS into their hazard classification and communication laws for workplaces and include health, physical as well as ecological hazards.[349] It has taken more than a decade internationally to develop GHS through United Nations.[350]

5.3.5 Fire safety

Fire is a chemical reaction that requires fuel, heat, and oxygen and gives off light and heat.[367]

Most fires involve combustible solids, although in many sectors of industry, liquid and gaseous fuels are also to be found.[495]

The difference between a flammable and combustible liquid is in the degree of fire hazard and this is related to the flash point of the liquid which is the temperature at which the liquid will evaporate into the air and from an ignitable mixture. Therefore, lower flashpoint liquids have a greater fire and explosion hazard.[494,180]

The heat sources reaching ignition temperature could be an open flame, the sun, hot surfaces, sparks, friction, chemical action, electrical energy, compressed gases, and arcs.

The oxygen source must be approximately 16% (normal air contains 21%) and some fuel materials contain sufficient oxygen within their makeup to support burning.[368]

There are many types of causes for fire including both fire and explosion due to the incompatibility of mixing chemical or their waste.[112]

As for fire strategy and building design protecting occupants of a building is the primary objective in the design of a fire protection system. As an atrium is an open design linking many floors, a fire started on one floor can lead to smoke spreading to other parts of the building via atrium.[491]

A large number of other toxic and irritant gas species also contribute to the hazard from fire gases to a lesser extent. The yields of most of these species will depend on both the material properties (especially the elemental composition) and the fire conditions.[492]

The key factors in understanding the explosions of gases and vapors have to do with the flammable range of the material; that is the percent of the mixture by volume in air which is flammable by mixture with air.[372]

Every year numerous fires cause huge losses in the food industry. In 1990 a deadly fryer fire at Poultry Processing Plant at Imperial Foods in Hamlet, NC, USA,[109] prompted industry-wide interest in upgrading the hydraulic fluids in food processing.[171]

The Chernobyl Nuclear Plant explosion resulted in 31 deaths and released radioactive materials into farms and the surrounding environment, which also resulted in many other losses in 1986.[172] In another major disaster, 14 workers lost their lives and 34 were injured and burned in the deadliest sugar dust explosion at the Imperial Sugar Refinery Plant in Georgia, USA, on February 7, 2008. Please see the Appendix providing information on various case studies including this disaster.[73,738–741]

Companies that handle, transfer, and store flammable liquids should contact manufacturers to determine if these liquids can accumulate dangerous levels of static electricity and if they can form explosive vapor-air mixtures inside storage tanks. If so, extra precautions – beyond normal bonding and grounding – may be necessary.[493]

5.3.5.1 Fire classes

The National Fire Protection Association (NFPA) categorizes fires as follows:[369]

Class A: Class A fires (designation symbol is a green triangle) involve ordinary combustible materials like paper, wood and fabrics, rubber. Most of the times, this type of fire is effectively quenched by water or insulated by other suitable chemical agents.

Class B: Class B fires (designation symbol is a red square) mostly involve flammable liquids (like gasoline, oils, greases, tars, paints, etc.) and flammable gases. Dry chemicals and carbon dioxide are typically used to extinguish these fires

Class C: Class C fires (designation symbol is a blue circle) involve live electrical equipment like motors, generators, and other appliances. For safety reasons, non-conducting extinguishing agents such as dry chemicals or carbon dioxide are usually used to put out these fires.

Class D: Class D fires (designation symbol is a yellow decagon) involve combustible metals such as magnesium, sodium, lithium potassium, etc. Sodium carbonate, graphite, bicarbonate, sodium chloride, and salt-based chemicals extinguish these fires.

Class K: Class K fires are fires in cooking appliances that involve combustible cooking media (vegetable, animal oils, or fats).

The relevant graphics and letter designations that accompany these classes are specified by NFPA 10, the standard for portable fire extinguishers.[369]

5.3.5.2 Fire extinguisher classes

There are five classes of fire extinguishing that could be summarized as follows[115,174]:

1. Carbon Dioxide: It is used for Class B and C and dissipates so quickly that hot fuel may ignite.
2. A_B_C Dry Chemical: It is used for Class A, B, and C and is most versatile, but leaves milder corrosive powder which must be cleaned.
3. Water or Mist: It is used only for Class A and could be very dangerous being used around the due environment with electrical equipment and water-reactive chemicals.
4. Class D Dry Chemical: It is used for Class D and designed for reactive metal fires.
5. Class K: It is used for cooking oil or fat on a fire. These types of fire extinguishers contain "Potassium Acetate", low PH, and are best for food preparation facilities.

There are also water fire extinguishers designed for practices as can be seen on the bench counter of the laboratory below:

5.3.5.3 Type and size of fire extinguishers

Portable fire extinguishers differ by the extinguishing agent they expel onto a fire. There are a wide variety of choices available based on the type of fire they will be fighting.[174] Portable extinguishers are also rated for the size of fire they can handle. This rating is expressed as a number from:

- 1 to 40 for Class A fires (Class A fire).
- 1 to 640 for Class B (Class B fire) fires.
- The "C" on the label indicates only that the unit can has been tested for use on electrical fires (Class C fires).
- Extinguishers for Class D (Class D fires) must match the type of metal burning.

5.3.5.4 Inspection and maintenance of fire extinguishers

Fire extinguishers, like all other fire protection systems, need monthly periodic inspections (NFPA 10 Section 7.2.1.2) and annual maintenance (NFPA 10 Section 7.3.1.1.1) to ensure that they are in serviceable condition. The monthly periodic inspection checklist must consist of at least the following items:

(1) location in designated space;
(2) no obstructions to access or visibility;
(3) pressure gauge in the operable range; and
(4) no obvious physical damage or conditions such as leakage or corrosion to prevent its operation. In addition, a dry chemical extinguisher with a gauge indicating a low-pressure level must be serviced before it can be used again.[497]

5.3.5.5 Fire prevention management

Planning an effective method of combating fires requires[112]:

- Minimizing quantities of flammable liquids kept in the lab, not to exceed the maximum container sizes for flammable liquids in the lab,
- Use and store flammable liquids and gases only in well – ventilated area, conduct work involving the likely release of flammable vapors in a lab fume hood,
- Keep containers closed, use NFPA or UL (Underwriters Laboratories) approved flammable liquid storage cabinets for flammable liquids, use only approved,
- "Explosion Safe" refrigerator when storing and preferably use a flammable storage cabinet when storing flammable liquids or solvent rooms,
- Keep flammable away from heat, sparks, flames, electrical motors, and direct sunlight, bond, and ground large metal containers of flammable liquids in storage, and when dispensing, bond both containers to each other to avoid the build-up of static charges,
- Keep combustible material such as sodium, potassium, and phosphorus away from humidity and room temperature as they rapidly burn at room temperature, use portable safety cans,
- Whenever possible, for storing, dispensing, and transporting, clean up spills of flammable liquids promptly to minimize the surface area of the spilled liquid and to avoid the vapor concentration risk exceeding the lower flammable limit,
- Apply these precautions to waste liquids since they would still be fire hazards, invest in flame – retarded products in your facilities,
- Maintain and use proper fire extinguishing equipment, provide training to all employees within a fire prevention policy,
- Never pour or let flammable liquids in drains,

Occupational Hazard Origin and Preventive Measures 87

- Refill gasoline, propane, and diesel-fueled motor vehicles at authorized areas only and remember the motor should be off, avoid spills of flammable when transferring and use proper tools and in a ventilated area, respect and follow safety signs, do not use portable electric or spark-producing tools where flammable vapor, dust, or liquid is involved,
- Have and consult update MSDS/SDS on site at all times for workers, make sure emergency exits are not blocked and visible, make sure fire extinguishing equipment is not blocked, and are functional and maintained regularly,
- Have a "Fire Prevention Safety Committee" and fire prevention management as part of your organization "Fire Protection Measure" in place, provide training and inspection on fire hazards and prevention, properly indicate exits for emergencies, have firefighting equipment for proper classes available and record accident, evaluate and improve for safety practices because of the constant and catastrophic threat to the work environment, fire loss control management is among the critical aspects of any business. Elements of such a Fire Loss Control Program are[368]: hazard inventory, written fire plan, training program, inspection program, fire drills practices, fixed responsibility and accountability, and management involvement.

5.3.5.6 Emergency fire procedures

Emergency fire procedure steps involve[370]:

- Alert all others nearby (shout fire),
- Call Fire Department in your area,
- Evacuate the lab and immediately close the door,
- Pull the nearest Fire Alarm (break the cover glass if needed),
- Exit the building quickly but calmly, by way of the designated escape route, If you cannot escape because of smoke, heat or other reason, return to your lab, close the door and place wet towels over openings around the door,
- Let the Fire and/or Security Department know where you are, use a fire extinguisher to assist you in escaping or on small fires which you can safely extinguish and if your cloth catches fire, "drop and roll on the floor", use a fire blanket to wrap yourself, or use the emergency shower if applies.

5.3.5.6.1 Emergency evacuation and duties

The following are descriptions of the duties of the emergency evacuation team. The monitors will be called on in the event of a fire or other emergency that would require the swift orderly evacuation of all building occupants to ensure their safety. As a rough guide, each exit leading from a floor area should be provided with an exit monitor and each monitor should have two searchers working with him or her.[370]

Emergency trained personnel are the emergency warden/building director, floor sweeper, exit monitor, and building exit monitor.

5.3.5.6.2   *Emergency evacuation procedures for disabled persons*

A special procedure should be established to assist those with a permanent or temporary physical challenge so that they may exit a building safely during an emergency.

5.3.5.7   Fire prevention special hazard

5.3.5.7.1   *Hot work policy*

Hot work is defined as any welding, cutting, grinding, or any other activity involving open flames, sparks, or other ignition sources which may cause smoke, fire, or which may trigger detection systems. Hot work is like a hidden hazard which must be carefully examined before its implementation as such, therefore, it is important to have a Hot Work Permit policy in place for employees and contractors.[176]

5.3.5.8   Combustible dust and explosion hazards

Combustible dust sources could be metal, plastic, wood, and many others. Explosive dust in the food industry includes flour, custard powder, instant coffee, sugar, dried milk, potato powder, and soup powder. The process that provokes such explosion could include flour and provender milling, sugar grinding, spray drying of milk and instant coffee, conveyance/storage of whole grains, and finely divided materials.[113]

Combustible dust explosions reported by the US Department of Labor Occupational Safety and Health Administration are as follows[498]:

Agriculture and Food Products dust originate from products such as egg white, milk powder, soy flour, corn rice, meal, and wheat starches, beet sugar, tapioca, sugar milk, sugar beet, whey, flour, apple, beetroot, carrageenan, carrot, cocoa powder, coconut shell dust, coffee dust, cottonseed, garlic powder, gluten, grass dust, hops (malted), lemon peel dust and pulp, locust bean gum, malt, oat flour, oat grain dust, olive pellets, onion and parsley powder, dehydrated parsley, peach, peanut meal and skins, peat, potato flour and starch, raw yucca seed dust, rice dust, rice flour, rice starch, rye flour, semolina, soybean dust, spice dust or powder, sunflower seed dust, tea, tobacco blend, tomato powder dust, walnut dust, wheat flour, wheat grain dust, and starch and xanthan gum.

The US Chemical Safety and Hazard Investigation Board website reported that even organic material in general such as plastic, corn starch, pharmaceuticals and metal powdered such as aluminum can produce an explosion.[178]

Other issues as reported dust explosions could be[175]:

- Hazardous levels of dust accumulation in the workplace due to poor housekeeping practices.
- Electrical equipment and powered industrial trucks are not approved for locations handling combustible dusts.
- Dust collectors were located inside buildings without proper explosion protection systems, such as explosion venting or explosion suppression systems.
- Deflagration isolation systems were not provided to prevent deflagration propagation from dust handling equipment to other parts of the plant.

Occupational Hazard Origin and Preventive Measures 89

Additional elements to form a combustible dust explosion will be the dispersion of dust particles in sufficient concentration to fall within the explosive and confinement of the dust in an environment besides fuel, oxygen, and fuel. The addition of the latter two elements to the fire triangle creates (Explosion Pentagon). Once the dust cloud is ignited within a confined or semi-confined vessel, area, or building very rapidly could lead to an explosion.

One must also look into primary ingredients, intermediary and finished products as sources of dust explosion products and combustible dust could include as reported.[177]

- Organic materials such as plastics, wood, coal, solid fuels, starch, flour, grain.
- Un-oxidized metals such as aluminum, silicon.
- Other oxidizable materials such as zinc stearate, sulfur, iron sulfides (pyrites).

5.3.5.8.1 *Dust identification analysis and prevention measures*
Any such prevention measure should include but not limited to identify:

- Combustible materials when finely divided.
- Processing steps that require, consume, or manufacture combustible dusts.
- Open and hidden areas where combustible dusts may accumulate.
- Manner and means where combustible dusts may be dispersed in air.
- Presence of potential ignition sources.
- Use appropriate electrical equipment and wiring system.
- Eliminate static electricity by methods such as grounding.
- Avoid or control smoking, open flame, and sparks.
- Use separator devices to remove foreign materials capable of ignition.
- Isolate heating systems from dust.
- Proper use and type of industrial trucks, cartridge-activated tools, etc.
- Proper maintenance of all equipment mentioned above.
- Avoid the use of aluminum tools, grounding metallic tools and care with spiral wound reinforcing wires also need to be considered.[179]

5.3.5.8.2 *Preventive management in dust explosion*
Topics include facility and systems design and construction, identification of combustible dust flash fire or explosion hazard areas, process equipment protection, fugitive dust control and housekeeping, ignition source identification and control, fire protection, training and procedures, and inspection and maintenance.[371]

5.3.5.8.3 *Other preventive measures are:*
- Minimize dust escape through processes such as a ventilation system.
- Use approved filter dust collection systems where motor and electrical equipment is not in the path of the air drawn through the vacuum cleaner.

90 Occupational Health and Safety in the Food and Beverage Industry

- Provide access to open and hidden areas for inspection and apply nondust generating cloud and non-igniting source methods for dust elimination.
- Use non-ignition equipment such as a vacuum cleaner for dust elimination.
- Locate relief valves away from dust-hazard areas.
- Develop and implement a hazardous dust inspection, testing, housekeeping and control program with established frequency for this purpose.
- Avoid using aluminum tools.
- Avoid picking up wet, burning, or hot material by a vacuum cleaner.
- Empty the dust collection bag frequently to avoid overloading the vacuum cleaner.
- Change disposable filters regularly and non-disposable ones at fixed interval times rather than relying on the discretion of the operator.
- Make sure the operator is trained to refit the filters correctly as the filters are the only devices preventing the food particle dust like flour contacting the sparking inside the motor.

The following measures have also been recommended by the US Department of Labor Occupational Safety and Health Administration[498]:

1. The facility has an emergency action plan.
2. Dust collectors are not located inside buildings. (some exceptions)
3. Rooms, buildings, or other enclosures (dust collectors) have explosion relief venting distributed over the exterior wall of buildings and enclosures.
4. Explosion venting is directed to a safe location away from employees.
5. The facility has isolation devices to prevent deflagration propagation between pieces of equipment connected by ductwork.
6. The dust collector systems have spark detection and explosion/deflagration suppression systems.
7. Emergency exit routes are maintained properly.

5.3.5.9 Chemical storage safety

Release of chemicals into the environment and within the facilities becomes an increasing popular topic in the news media. This might be at a much smaller scale for laboratories and limited chemical storage facilities but a pool of liquid from a leaking drum may enter a sewer drain, or spilled chemicals that have been stored outside the building may migrate across property boundaries into the ground but for the most part release of lab chemicals occur inside building.[181]

Surface runoff has to be considered with equipment layout. If surface runoff from one area goes directly to another area, the feature of separation is then not accomplished [499].

5.3.5.9.1 Chemical storage options

Storage facilities exist at various locations around the world; however, the types and quantities of such chemicals are not the topic of our discussion here and which may be regarded for disposal or longer storage period.[500]

Occupational Hazard Origin and Preventive Measures

Depending upon the quantity and class, flammable and combustible liquids may be stored in cabinets, storage areas, rooms and/or a location indoor or outdoor. The proper flammable storage cabinet is shown in Figure 5.1.

5.3.5.9.2 Accident prevention management

Following are precautions that need to be considered when storing chemicals[181]:

1. Store chemicals in a secure location, accessible to authorized personnel.
2. Avoid storing fire-causing hazardous materials in the "Chemical Store - Room".
3. Minimize quantities of chemicals kept when possible.
4. Keep glass containers off the floor and protected from collisions.
5. Store chemicals away from direct heat sources.
6. Store hazardous liquids in compartments or devices capable of containing spills.
7. Store large and heavier containers on lower shelves to minimize the extent of splash or spillage in the event of container fall or breakage.
8. Install rust-proof and chemical-resistant shelving for longer protection.
9. Install shelf edge protector to safeguard against the accidental fall.

FIGURE 5.1 Proper flammable storage cabinet in a food science laboratory. Figure 5.1 provides a vented storage yellow cabinet for flammable chemical in a food science laboratory. The cabinet is lockable and labeled for sorted chemicals. The ventilation tube is connected to the bottom left of cabinet to exhaust hazardous vapors.

10. Cover the storage areas when and where possible with absorbing material to minimize both friction and spillage when the accident occurs.
11. Use only sturdy shelves with a load capacity well in excess of the weight of the chemicals placed upon them.
12. Check the proper assembly of shelves (assembled to the wall) and cabinet doors by verifying all clamps, supports, braces, and shelf brackets are correctly positioned and make sure they are not overcrowded.
13. Follow weight limit on shelf and volume content permitted in solvent cabinet and other types of chemicals storage containers are adhered to.
14. Segregate chemicals physically by reactive class and flammability.
15. Make sure proper ventilation, electrical, fire extinguishing system, fire alarm, and other safety features are regularly inspected and functional.
16. Ensure that PPE as well as those in case of an accident such as first aid, spill response kits, fire extinguishers, etc. are functional and accessible.
17. Examine chemical storage areas weekly to replace faded or loose labels on containers; and arranged for disposal surplus, expired, and unwanted chemicals.
18. Make sure carts, bottle carriers, or pails are in good physical condition.
19. Metal containers should be free of rust and chemical leaks and have labels.
20. Container shape must be normal, free from any sign of pressure building up.
21. Have spill control trays or basins to dispense or transfer liquids available.
22. Never store chemicals in hoods, alphabetically, in the fridge, above eye levels, one bottle supporting another one, out of the way, or with no inventory control.
23. Proper solvent cabinet connection should be direct not through hoods.

5.3.5.10 Laboratory safety

Health and safety are everyone's responsibility. Every laboratory worker must dress and behave appropriately, be aware of emergency procedures, and observe all laboratory safety rules. In most recent references in this regard, students or employees must also be trained such as avoiding unauthorized experiments, food, drink, and cosmetic application, no smoking, and all other regulation for PPE that may apply for the proper use and familiarity in the lab environment.[183]

5.3.5.11 Chemical inventory management, segregation, and storage

Segregation schemes and compatibility charts are generally more complicated than required for most laboratories and require substantial knowledge of chemistry to be used effectively. Compatibility charts can contain as many as 50 categories based on functional groups, many more than one would be likely to find in a typical food science laboratory.[373]

However, there are some relatively simple ways to segregate incompatible materials that can be managed by most laboratory workers. Managing chemical inventory requires[501]:

1. A realistic appraisal and selection of chemicals that are necessary versus those that are desirable but not required to conduct a viable experiment.
2. An understanding of the incompatibility of chemicals.
3. A qualitative risk assessment of the available space and the location of the chemical containers therein.
4. A reasonable method of identifying chemicals for disposal, and
5. A proper disposal strategy, which includes labeling and monitoring.

The appropriate storage of chemical is shown in Figure 5.2.

The incompatibility of storing the chemicals or chemicals that cannot be kept together are tabulated in alphabetical order in Tables 5.2A to 5.2D.

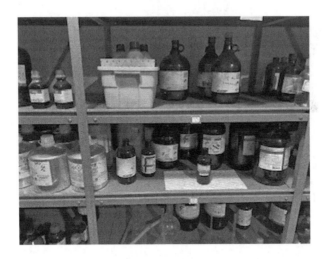

FIGURE 5.2 Example of solvent/flammable storage room. Figure 5.2 provides solvent and flammable storage room equipped with metallic edge guarded shelves and absorbent material (light green sheet) in case of spill in a ventilated fireproof room. The shelves contain different flammable solvents and are secured to solid fixture (cemented walls). On the top-shelf a blue chemical resistance container is placed for non-stable containers.

94   Occupational Health and Safety in the Food and Beverage Industry

TABLE 5.2A
Chemicals That Are Incompatible and Should Be Stored in Separate Spots

Chemical	Incompatible chemicals
Acetic acid	Chromic acid, nitric acid, perchloric acid, peroxides, permanganates, and other oxidizers
Acetone	Concentrated nitric and sulfuric acid mixtures, and strong bases
Acetylene	Chlorine, bromine, copper, fluorine, silver, mercury
Alkali metals	Water, carbon tetrachloride or other chlorinated hydrocarbons, carbon dioxide, halogens
Ammonia, anhydrous	Mercury, chlorine, calcium hypochlorite, iodine, bromine, hydrofluoric acid
Ammonium nitrate	Acids, metal powders, flammable liquids, chlorates, nitrites, sulfur, finely divided organic or combustible materials
Aniline	Nitric acid, hydrogen peroxide
Arsenic materials	Any reducing agent
Azides	Acids
Bromine	Same as chlorine
Calcium oxide	Water
Carbon (activated)	Calcium hypochlorite, all oxidizing agents
Carbon tetrachloride	Sodium
Chlorates	Ammonium salts, acids, metal powders, sulfur, finely divided organic or combustible materials

TABLE 5.2B
Chemicals That Are Incompatible-2

Chemical	Incompatible chemicals
Chromic acid and chromium trioxide	Acetic acid, naphthalene, camphor, glycerol, glycerin, turpentine, alcohol, flammable liquids in general
Chlorine	Ammonia, acetylene, butadiene, butane, methane, propane (or other petroleum gases), hydrogen, sodium carbide, turpentine, benzene, finely divided metals
Chlorine dioxide	Ammonia, methane, phosphine, hydrogen sulfide
Copper	Acetylene, hydrogen peroxide
Cumene hydroperoxide	Acids, organic or inorganic
Cyanides	Acids
Flammable liquids	Ammonium nitrate, chromic acid, hydrogen peroxide, nitric acid, sodium peroxide, halogens
Hydrocarbons	Fluorine, chlorine, bromine, chromic acid, sodium peroxide
Hydrocyanic acid	Acids
Hydrofluoric acid	Ammonia, aqueous or anhydrous, bases, and silica
Hydrogen peroxide	Copper, chromium, iron, most metals or their salts, alcohols, acetone, organic materials, aniline, nitromethane, flammable liquids
Hydrogen sulfide	Fuming nitric acid, other acids, oxidizing gases, acetylene, ammonia (aqueous or anhydrous), hydrogen
Hypochlorites	Acids, activated carbon
Iodine	Acetylene, ammonia (aqueous or anhydrous), hydrogen
Mercury	Acetylene, fulminic acid, ammonia

Occupational Hazard Origin and Preventive Measures

TABLE 5.2C
Chemicals That Are Incompatible-3

Chemical	Incompatible chemicals
Nitrates	Sulfuric acid
Nitric acid (concentrated)	Acetic acid, aniline, chromic acid, hydrocyanic acid, hydrogen sulfide, flammable liquids, flammable gases, copper, brass, any heavy metals
Nitrites	Acids
Nitroparaffins	Inorganic bases, amines
Oxalic acid	Silver, mercury
Oxygen	Oils, grease, hydrogen; flammable liquids, solids, or gases
Perchloric acid	Acetic anhydride, bismuth, and its alloys, alcohol, paper, wood, grease, and oils
Peroxides, organic	Acids (organic or mineral), avoid friction, store cold
Phosphorus (white)	Air, oxygen, alkalis, reducing agents
Potassium	Carbon tetrachloride, carbon dioxide, water
Potassium chlorate and perchlorate	Sulfuric and other acids, alkali metals, magnesium, and calcium
Potassium permanganate	Glycerin, ethylene glycol, benzaldehyde, sulfuric acid
Selenides	Reducing agents
Silver	Acetylene, oxalic acid, tartaric acid, ammonium compounds, fulminic acid

TABLE 5.2D
Chemicals That Are Incompatible-4

Chemical	Incompatible chemicals
Sodium	Carbon tetrachloride, carbon dioxide, water
Sodium nitrite	Ammonium nitrate and other ammonium salts
Sodium peroxide	Ethyl or methyl alcohol, glacial acetic acid, acetic anhydride, benzaldehyde, carbon disulfide, glycerin, ethylene glycol, ethyl acetate, methyl acetate, furfural
Sulfides	Acids
Sulfuric acid	Potassium chlorate, potassium perchlorate, potassium permanganate (or compounds with similar light metals, such as sodium, lithium, etc.)
Tellurides	Reducing agents

5.3.5.12   Special chemical hazards

Typical laboratory may come cross substances that must be handled with special care due to their light, humidity, physical shock, toxicity, or thermal effect such as benzoyl peroxide, picric acid, toxic effects like mercury, ammonium nitrate[505] or other functional groups that pose special consideration such as aside, ozonide, nitro, nitroso, diazo, halogen – substituted amine ether, perchloric acid, and peroxides are

96   Occupational Health and Safety in the Food and Beverage Industry

also explosion hazard.[185] It is important to get training and have MSDS or SDS available and verified in advance with making sure of precautionary recommendations in place.

5.3.5.12.1   Chemical spills

In general, hazardous chemical spills or releases could end up in water, land, or atmosphere and cause pollution, fire, or health hazards to human, animal, plant, and the environment[187] or cause an explosion.[188]

Despite the best efforts of researchers to practice safe science in the laboratory, accidents resulting in the release of chemicals will occur.[189]

The Bhopal 1984, India gas tragedy which released methyl isothiocyanate and other toxins into the air and groundwater did cause numerous injuries, suffering, and death.[196,197] In the same year Environment Canada published a summary of such chemical hazards and identified many such substances.[194]

In May 2010 oil spill into the Gulf of Mexico deep water in USA which endangered and caused a lot of environmental as well as economic damages for years to come.[195]

5.3.5.12.2   Industrial cleaning sanitation

Food processing and food service operations must have an effective system in place to ensure adequate and appropriate cleaning of the facility and equipment. An effective program must cover[376]:

1. Areas and equipment,
2. The designated food handler responsible for sanitation and cleaning,
3. The chemicals and or cleaning products, including concentration and process,
4. The procedures used, including temperature and time involved,
5. Frequency of cleaning and sanitizing,
6. Inspection and monitoring of equipment and machines,
7. Records to be completed after cleaning and sanitizing,
8. Cross-contamination points such as tools, boards, counters, etc.,
9. Safety guards must be removed before sanitizing,
10. MSDS/SDS must be accessible at all times,
11. Pesticides such as insecticides, rodenticides, fungicides, herbicides, bactericides through improper handling or storage or pest management, etc.,[192,193]
12. Lubricants, oils, ink, etc. from machineries,
13. Other chemical source materials being used such as additives such as nitrate in processed meat, food color, etc., and or natural toxicants such as alkaloids in potatoes or aflatoxin in peanut or pistachio, etc.,
14. Detergents, polishes, caustics, cleaning, and drying agents,[190]
15. Hazards associated with cosmetic and personal care products,
16. Toxic metals leaching through food packaging and containers such as galvanized (zinc-coated) or lead-based products such as lead-glazed ceramics in food processing.[191]

Occupational Hazard Origin and Preventive Measures 97

5.3.5.12.3 *Slips, trips, and falls due to contamination*

Slipping on the floor, stairs, and/or other working areas can be due to the presence of chemical, food, water, debris, and or other contaminated material sources.

Each year nine million injuries are reported due to slips, trips, and falls. Body balancing acts are controlled by several parts including the inner ear, vision, muscle, and nerves in the spinal cord. Biomechanics researchers at Virginia Tech in the USA are using a special slip simulator to study ways to reduce worker injuries associated with tripping and falling.[503]

According to the US Bureau of Labor Statistics, the incidence rate of lost-workday injuries from slips, trips, and falls (STFs) second most common cause of lost-workday injuries in hospitals. An analysis of workers' compensation injury claims from acute-care hospitals showed that the lower extremities (knees, ankles, feet) were the body parts most commonly injured.[504,199,615,616]

HSE has reported slips and trip injuries have one of the highest injury rate among all industries compared to food processing and beverage manufacturing. The slips prevention methods can at least cut these injuries by 50%. Most slips (90%) occur when the floor is wet with water or contaminated with food products. Most trips (75%) are caused by obstructions or uneven surfaces.[198]

5.3.5.12.3.1 Managing slips

1. Eliminate contamination of the floor from water, food ingredients/products, and/or other chemical or sanitary cleaning agents used.
2. Prevent contamination of walkways, especially around equipment and machine, working tables, drainage channels, cleaning incoming footwear.
3. Reduce contamination effects by its immediate treatment of spills, spoilages, and use good ventilation and dehumidification for drying.
4. Ensure sufficient roughness on the floor with less than 20 microns roughness (Rtm) for water-wet floors, to 45 microns for dairy and milk, and 70 microns for olive oil-type industries. This type of roughness does not prevent hygienic cleaning.
5. Make sure the sanitary product you use for both machines and surfaces is both effective and not slippery at the end.
6. Provide appropriate footwear to prevent slips for example microcellular urethane with a good pattern on the softer sole of the least slippery footwear on wet floors.
7. Pushing/pulling and/or carrying loads can increase the slip risk and should be eliminated or reduced where possible.
8. Eliminate holes, uneven, or slope on surfaces.
9. Provide training to employees and inspection of facilities.
10. Eliminate materials or objects likely to cause tripping and slips.

Besides the above practices also[504]:

1. Keep proper housekeeping to include how frequently the floor has to be cleaned or maintained, where spill control kits are stored, proper signage, and cleaning methods in the working area to eliminate hazardous slips.

2. Mats should be large enough so that several footsteps will take place on the mat; if there is water around or beyond the mat, it means that the mat is not large enough and/or is saturated and needs to be replaced.
3. Secure mats from moving and make sure they have slip-resistant backing. Remind staff to lay mats in the correct position daily and use visual cues such as tape on the floor if necessary.
4. Make sure that drip pans of ice machines and food carts are properly maintained so that water does not spill onto the floor.
5. Restrict access to spilled or wet areas.
6. Remove all signs once the floor has been cleaned or maintained.
7. Replace smooth flooring materials in areas normally exposed to water, grease, and/or particulate matter with rougher surfaced flooring when renovating or replacing healthcare flooring.
8. Make sure elevators are leveled properly so that elevator floors line up evenly with hallway floors.
9. Pay particular attention to where there are hazards associated with slips, trips, and falls which could be indoor or outdoor such as walkways, floors, around ground drains, etc.
10. Repair and patch possible accident cases such as floor cracks, holes, carpeting, tiles, or debris if in the way.
11. Keep records and plan accordingly for the type of injuries and causes for prevention.

The National Institute of Safety and Health (NISH): You can **slip** when you lose your footing, you can **trip** when you catch your foot on or in something, and you **fall** when you come down suddenly. Spills, ice, snow, rain, loose mats, rugs, and stepladders are some of the common causes of slips, trips, and falls. In addition, poor lighting and clutter can cause injuries such as sprains, strains, bruises, bumps, fractures, scratches, and cuts. The followings are some safety tips recommended by the National Institute of Safety and Health (NISH)[615]:

1. If you see something you might slip or trip on, tell your supervisor right away.
2. Clean up spills and anything slippery. Check with your supervisor about how to use cleaning products. Don't use cleaners that could make the floor slippery.
3. Clear walkways, stairs, and lobbies of anything that might be a tripping hazard, such as cords, wires, empty boxes, and clutter.
4. Make sure that floor mats lay flat rather than wrinkled or bunched.
5. Use handrails when you walk up and down steps.
6. Before using any ladder or stepladder, make sure it opens fully.
7. Check that ladder extensions are fully locked and that the ladder legs are stable on a flat, non-slippery surface.
8. Clean off slippery material on the rungs, steps, or feet of a ladder before using.
9. Don't go over the load limit noted on the ladder.

Occupational Hazard Origin and Preventive Measures

5.3.5.12.4 Cross-contamination of food

With the ever-increasing awareness and importance of producing foods that clearly label if a product contains known allergens, either as a deliberate ingredient or as a possible contaminant, allergen risk assessments and management must be introduced to the food process. Food cross-contamination hazards can spread through[505]:

1. Hands that touch the raw food and the touch cooked or ready-to-eat- foods.
2. Various types of utensils and tools that may be handling food or ingredients with.
3. When food contact surfaces that touch raw food and are not cleaned and sanitized and they touch ready-to-eat-foods.
4. Cleaning cloths and sponges that touches raw food and equipment used for food processing or preparing; are not cleaned and sanitized and are then used on surfaces or equipment and utensils, etc. for ready-to-eat-foods.
5. Raw or contaminated foods that touch or drip fluids on cooked or ready – to-eat-foods[153].
6. Adding raw or contaminated ingredients to food that receives no further cooking.

5.3.5.12.5 Hazardous waste management

A material is usually defined as waste when it is determined that the material should no longer be used for various reasons such as being contaminated, expired, safety reasons, etc.

In the laboratory setting the worker has often been the one who determined when an unwanted material is to be declared a waste; more recently, federal and state regulations have expanded our understanding of what constitutes a waste.[185]

Human exposure to hazardous substances occurred when prehistoric humans inhaled noxious volcanic gases or carbon monoxide from inadequately vented fires in caves. Over time, to protect workers, new legislations have come through in order to provide right-to-know provisions for their protection.[200]

In recent years in the USA especially with the strong interest of the legal profession in following the various activities and regulations promulgated by such agencies as OSHA, National Institute for Occupational Safety and Health (NIOSH), the Food and Drug Administration (FDA), Consumer Product Safety Commission (CPSC), Department of Energy and Defense new attitude has developed in the public part, employee, and employer[23] in particular when dealing with pollution reduction and sustainable practices.[201,202]

Besides chemical waste special consideration must also be given to other hazardous waste categories such as biomedical, radioactive, and other wastes.[203]

5.4 PHYSICAL HAZARDS

Physical hazards may have different sources such as volcanic eruption, earthquake, flood, landslide, asteroid impact, avalanche, aviation hazard, sink hole, and so on which have its effect on agricultural productivity.[506]

100 Occupational Health and Safety in the Food and Beverage Industry

Physical hazards in food and beverage manufacturing involve various hazards from the presence of foreign matters to noise, slip and falls, electricity, machineries, vibration, radiation, pressure, confined spaces, and many others which will be discussed.

5.4.1 PHYSICAL HAZARDS IN FOOD AND DRINK MANUFACTURING

Stones in lentils or beans may damage a tooth. Years ago, home cooks checked lentils for stones to protect their families from any harm. The number of occurrences of foreign materials in food is now becoming less frequent, and consumers are less used to checking primary foodstuffs for stones, sharp metal pieces, or bones, as the food industry continuously improves the safety and quality of their products.

However, this trend also decreases public acceptance of such occurrences to a level that approaches "zero tolerance". Any incidence of foreign material harms the consumer, undermines confidence in the brand, and generates headlines. The name of the producer and the grocery chain that sold the article are widely reported.[162]

Physical hazards include glass, plastic, stones, metal, and bones. The introduction of physical hazards has been characterized as inadvertent contamination from growing, harvesting, processing, and handling intentional sabotage or tampering, and chance contamination during distribution and storage.[507]

In Canada, physical hazards in food and drinks have been defined as either foreign material unintentionally introduced to food products (e.g., metal fragments in minced meat) or naturally occurring objects (e.g., bones in fish) that are a threat to the consumer. A physical hazard can enter a food product at any stage of production. Hard or sharp objects are potential physical hazards and can cause injury and cuts to the mouth and throat, intestine and teeth, or gum.[210]

5.4.2 COMMON PHYSICAL HAZARDS IN FOOD AND DRINK

Foreign objects in food products could be divided into several categories[379]:

1. Physical matters such as glass, jewelry, metal shaving, stones, wood, torn sifter, bone chip, BB Shots/needles, insulation materials, personal effects, insects, rodents, birds, or their excreta or, hard or sharp parts from broken light bulb or from food like shells from various nut products, etc. In general, the type of foreign objects in food affects the technology needed to detect it could include[162]:
   Glass: Sharp glass contamination often occurs during filling processes in glass containers if a container is accidentally broken. Another source, but less frequent, are light bulbs broken during building maintenance.
   Metal: Sharp metal objects may include screws and equipment splinters, blades, broken veterinary needles, fragments, and clippings of prior processing procedures.
   Plastics: Soft and hard plastics may come from packaging material of an intermediary production phase.
   Wood: Wood splinters may have their origin at the farm or may come from handling wooden pallets.

Occupational Hazard Origin and Preventive Measures

Stones: Small stones are common in crops like peas or beans contaminated during harvest.

2. **Chemical matters** such as flaking paint, plastic, pesticides, lubricants, sanitary products, etc.

3. **Biological matters** originate from three sources:
   - **Insects** (internally infested ingredients, widespread infestation in over boards above sensitive or exposed ingredients or product zones, equipment infestations where product adulteration is likely, houseflies and fruit flies in excessive numbers with little control provided, and any other roach activity on or in product zone.
   - **Rodents** (visual presence of live rodents their excreta or gnawing on raw materials and decomposed rodents).
   - **Birds** (resident avian in processing areas or warehouse/nesting and bird excreta on product zones, raw materials, or finished product).

Canadian Food Safety Code of Practice refers to pests such as birds, insects, and rodents and their hair or excreta are not only unpleasant but they can also cause choke and external or internal injuries and may carry millions of microorganisms that cause foodborne illnesses and food contamination by simply crawling on food product. In addition, they can cause internal or external body injury such as skin or mouth cut or tooth break.[134] The injury rate depends on size, person's age or condition, type of product for its hardness or sharpness, and other physical characteristics. They could be originated from processing lines or food service facilities for workers. These are the primary concern in a food production line.[34]

5.4.3 CLASSIFICATION OF PHYSICAL HAZARDS

Some companies have even developed a series of software applications and systems which address the unique needs of the food and beverage industry and work to streamline the processes, procedures, regulatory standards, and unique challenges that companies in the industry face.[24]

5.4.4 DEVELOPING AN EFFECTIVE PHYSICAL HAZARDS PLAN

To develop an effective physical hazard identification program, processors need to collect detailed information for every step of every food process in the facility. Information on potential physical hazards can be obtained by closely observing each process during all phases of its operation. The Canadian Food Inspection Agency has developed a Reference Database for Hazard Identification that contains valuable reference information for food processors.[205]

According to the CFIA three classes of physical hazards are found on the basis of likelihood Table 5.3 provide such a classification[378]:

TABLE 5.3
Classification of Physical Hazard by CFIA

Class of physical hazard	Hazard ranking
Class 1(High Likelihood)	For high risk little or no control measures established. Major and critical infractions occur.
Class 2 (Moderate Likelihood)	For medium risk some control established, but gaps and inconsistencies occur.
Class 3 (Low Risk)	The agency also rates the likelihood of occurrence based on the level of control that a food processor has to eliminate the risk: A. **Low risk** (good control measures established), but minor infractions occur. B. **Medium risk** (some control measures established, but gaps or inconsistencies occur and C. **High risk** (little or no control established. Major and critical infractions occur. Every food process has its own specific and potential hazards. Evaluation of the type of product, the intended market for the product, and other factors needed to be considered to determine the risk category for a possible physical hazard).

5.4.5 Current regulations around the world

The Food and Drug Administration (FDA) regulations define the food adulteration by hard or sharp objects as follows: [162]

A physical hazard in food is any extraneous object or foreign matter that may cause illness or injury to the consumer. FDA considers a product adulterated if it contains a hard or sharp foreign object that measures 7 mm to 25 mm in length and is ready to eat or requires only minimal preparation steps that would not eliminate, invalidate, or neutralize the hazard prior to consumption.

Canadian regulations support the FDA concept based on the 7 mm to 25 mm size criterion.[162] In EU regulations all food products sold in the EU must comply with regulations on the hygiene of foodstuffs, which demand, above all, that these products must be safe.

To achieve such a level of safety, food processors, packagers, and distributors should use the HACCP concept, complemented with a quality control system, such as GMPs or ISO 9000.[162]

5.4.6 Prevention of physical hazards in food and drinks

Table 5.4 indicates some possible physical contaminants that may be found in meat processing operations as the eighth most common food categories implicated as foreign objects.[205]

Occupational Hazard Origin and Preventive Measures

103

TABLE 5.4
Possible Physical Contaminants in Meat Processing Operations

Food category	Number of complaints	Percent
Bakery	272	10.2
Soft drinks	228	8.4
Vegetables	226	8.3
Infant foods	187	6.9
Fruits	183	6.7
Cereal	180	6.6
Fishery	145	5.3
Chocolate and cocoa products	132	4.8

5.4.7 PREVENTIVE MANAGEMENT

1. When pests are found discard all contaminated food, clean, and sanitize the immediate and surrounding areas to prevent further contamination, destroy nesting and breeding places and seal them off and involve a licensed pest operator in eradicating pests and rodents.
2. Have a good warehouse monitoring system and use a "first-in-first-out" method for food stock, particularly for cereal and grains.
3. Keep stored items on shelves or raised skids at least 15 cm above the floor.
4. Remove garbage and keep doors and windows shut or use screens.
5. Inspection of raw materials and food ingredients for contaminants such as stones in cereals that were not found during receiving.
6. Handle food according to standard procedures recommended such as Good Manufacturing Practices (GMP) with steps to prevent falling of artificial fingernails, hair, or jewelry etc. into the food products.
7. Eliminate potential sources of physical hazards in processing and storage areas such as using protective acrylic bulbs.
8. Install an effective detection and elimination system for physical hazards such as metal detectors, magnets, X-ray machines and food radar system in the processing line.
9. Establish an effective maintenance HACCP program for the equipment and machinery in use to prevent their old parts from falling into the food products.
10. Provide proper training to workers and evaluation at each level of food processing.[204] Other steps may include[205]:
11. Inspecting raw materials and food ingredients for field contaminants that were not found during the initial receiving process.

104   Occupational Health and Safety in the Food and Beverage Industry

12. Follow good storage practices and evaluate potential risks in storage areas (e.g., sources of breakable glass such as light bulbs, staples from cartons, etc.) and use protective acrylic bulbs or lamp covers.
13. Develop specifications and controls for all ingredients and components, including raw materials and packaging materials. Specifications should contain standards for evaluating the acceptability of ingredients or packaging materials (e.g., recycled cardboard used for packaging sometimes contains traces of metals that can be detected by metal detectors). A limit for metal detection should be established to avoid false positive detection of metal in food.
14. Set up an effective detection and elimination system for physical hazards in your facility (e.g., metal detectors or magnets to detect metal fragments in the production line, filters, or screens to remove foreign objects at the receiving point).

General prevention steps include supplier approval program pest management, glass and brittle plastic control, cleaning and sanitation program, building management, preventive maintenance program, pallet management, equipment and utensil programs, personnel and personal hygiene, and production standard operating procedures.[380]

5.4.8 SLIPS, TRIPS, AND FALLS

Most common falls could be from a ladder, roof, or other types of equipment and construction building. Almost one-third of all fatal accidents in the construction industry in the USA are due to falls.[508] If working at height on a vehicle is unavoidable, you must think about preventing slips and trips as part of the combination of things you do to prevent falls. There are several factors for slip, trip, and fall at work[211]:

A. Flooring,
B. Footwear,
C. Contamination,
D. Environment, and
E. Working condition

Figure 5.3 illustrates contributing factors and elements causing slip, trip, and fall accidents.[198]

A. **Flooring**
1. The floor in a workplace must be suitable for the type of work activity that will be taking place on it.
2. Where a floor can't be kept dry, people should be able to walk on the floor without fear of a slip despite any contamination that may be on it, so it should have sufficient roughness.

Occupational Hazard Origin and Preventive Measures

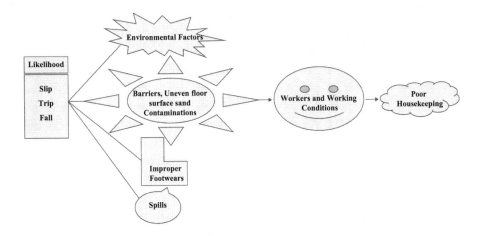

FIGURE 5.3 Illustration of likelihood factors and elements for slip-trip and fall. The diagram illustrates various potential factors and elements which can cause slip-trip and fall in a food production facility.

3. The floor must be cleaned correctly to ensure that it does not become slippery or keeps its slip-resistance properties (if a nonslip floor).
4. The floor must be fitted correctly to ensure that there are no trip hazards to ensure that nonslip coatings are correctly applied.
5. The floor must be maintained in good order to ensure that there are no trip hazards e.g., holes, uneven surfaces, curled up carpet edges.
6. Ramps, raised platforms, and other changes of level should be avoided, if they can't then stairs should have, (a) high visibility and be nonslip on the step edges, (b) a suitable handrail, (c) steps of equal height, and (d) steps of equal width.

B. **Footwear**

One out of every four-foot injuries sustained in the workplace is caused by workers not wearing the proper protective footwear, according to National Safety Council (NSC). Preventable foot injuries in the workplace include crushed or broken feet, cuts or severed feet or toes, and sprained or twisted ankles and recommend[448]:

1. Walk in new footwear to ensure it is comfortable.
2. Boots should have ample toe room (toes should be about 12.5 mm from the front).
3. Make allowances for extra socks or special arch support when buying boots.
4. Boots should fit snugly around the heel and ankle when laced.
5. Lace up boots fully. High cut boots provide support against ankle injury.

6. Use a protective coating to make footwear water resistant.
7. Inspect footwear regularly for damage.
8. Repair or replace worn or defective footwear.
9. Electric shock resistance of footwear is greatly reduced by wet conditions and with wear.

Footwear can play an important part in preventing slips and trips.[199]

1. Where you can't control footwear e.g. pedestrians using a shopping center thoroughfare. It is vitally important to ensure that smooth floors are kept clean and dry.
2. For work situations where you have some control over footwear, but where floors are mainly clean and dry, a sensible footwear policy can help reduce risks. For slips and trips sensible means: flat, with a sensible heel, with the sole and heel made in a softer material that provides some grip.
3. In work situations where floors can't be kept dry or clean e.g. food preparation, the right footwear will be especially important, so a slip-resistant shoe may be required. If an employer introduces a slip-resistant shoe policy, the footwear will be considered to be personal protective equipment and will be subject to the requirements of the Personal Protective Equipment Regulations e.g. will have to be provided to employees free of charge.
4. Choosing the most suitable slip-resistant footwear for a particular environment/work activity can be difficult. Descriptions of slip-resistance given in suppliers brochures range from "improving the grip performance" to "excellent multi-directional slip-resistance", but often do not describe the work environments for which footwear are, or are not, suitable.
5. Footwear, which claims "slip-resistance", may not perform well in your work environment. So how can you make the best choice?
6. Undertake a footwear trial before buying sufficient stock for your entire workforce.
7. Footwear can perform differently in different situations. For example, footwear that performs well in wet conditions might not be suitable where there are food spillages.
8. A good tread pattern is essential on fluid-contaminated surfaces. The pattern is characterized by, among other things, leading edges in all directions to sweep away lubricant leaving dry contact under cleats.
9. Sole tread patterns should not become clogged with any waste or debris on the floor. If they do, then that design of sole is unsuitable for the situation.
10. Sole material type and hardness are key factors; caution is needed in making generalizations and testing is always recommended.
11. When choosing footwear take into account factors such as comfort, durability, and any additional safety features required, such as steel mid-sole. The final choice may have to be a compromise.

Occupational Hazard Origin and Preventive Measures   107

C. **Contamination** due to mainly spills, rainwater, diesel, oil, mud, food product, or ice getting on to the surface.
D. **Environmental factors** such as poor lighting or glare, working in high winds, wet weather, low visibility, surface condensation, and loud noise.
E. **Worker's performance** due to tiredness, sickness, or lack of training, and **working conditions** such as poor housekeeping.

5.4.9 IDENTIFYING AND ASSESSING FALL HAZARDS

The following ten factors have been reported to be the main causes or contributing factors for slips, trips, and falls[199,206]:

1. Floor contamination,
2. Poor drainage, pipes, and drains,
3. Indoor walking surface irregularities,
4. Outdoor walking surface irregularities,
5. Weather condition (ice and snow),
6. Inadequate lighting,
7. Stairs and handrails,
8. Step stools and ladders,
9. Tripping hazards: clutter, including loose cords, hoses, wires, etc., and
10. Improper use of floor mats and runners

Slips, trips, and falls are a major source of injury and death in industry. About two-thirds of all falls take place on the same level. A simple trip or slip claim can result in significant costs cost reduction.

5.4.10 FALLS FROM HEIGHT AND PRINCIPLES OF WORKING AT HEIGHT

OSHA regulates that different employees may possess different levels of responsibility for fall protection that could include[549]:

1. When an employee is on a walking/working surface that has unprotected side or edge.
2. When an employee is constructing a leading edge.
3. When an employee may fall through a hole in the walkway/working surface.
4. When an employee is working on the face of formwork or reinforcing steel.
5. When employees are on ramp, runways, and other walkways.
6. When employees are working at the edge of excavation, well, pit, or shaft.
7. When employees are working above dangerous equipment fewer than 4–6 feet.
8. When an employee is performing overhand bricklaying and related work.
9. When an employee is performing roofing work on slow-sloped roof.
10. When an employee is working on steep roof.

108 Occupational Health and Safety in the Food and Beverage Industry

11. When an employee is engaging on precast concrete erection (with certain exceptions).
12. When an employee is engaged in residential construction (with certain exceptions).

The most common area of the vehicle for people to fall from is the load area, followed by the cab access steps and fifth wheel catwalk – you don't have to fall far to land hard. If you are in control of the work, you must think in terms of a hierarchy of controls[211]:

1. Plan to avoid work at heights where you can.
2. Where you can't, make sure you use work equipment to prevent falls, a. First choice – vehicle-based systems, 2. Second choice – on-site systems.
3. Where the risk of a fall can't be eliminated, use work equipment to minimize the distance and consequences of a fall.
4. Always consider measures that protect everyone at risk (e.g., platforms and guardrails) before measures that only protect the individual (e.g., safety harness).

5.4.11 FALL HAZARDS ASSESSMENTS

1. Working conditions like poor lighting, slippery surfaces, and poor housekeeping,
2. Protective devices like guardrails are missing,
3. Ladders and scaffolds in poor shape,
4. Personal protective equipment is not available, not used or misused,
5. Worn or improper footwear used,
6. Poor work practices like unclear job procedures, lack of training or workers rushing to meet deadlines,
7. Access platform on bulk tank trucks.

5.4.12 PREVENTIVE MEASURES; FALLS FROM HEIGHT

1. Recognize that some people have acrophobia, a medical term for the fear from heights. A person as such will be at particular risk if assigned to work in elevated environments.
2. Slips and trip hazards must be recognized and removed.
3. Clean and tidy workplace conditions result in lower accident rates.
4. Adequate lighting should be provided for all areas.
5. Appropriate footwear should be worn for the type of process.
6. Handrails should be added to prevent loss of balance.
7. Loose or damaged floor coverings should be dealt with immediately.
8. Ensure all stairs are fitted with handrails and where appropriate, anti-slip tread in good condition.
9. All elevated work surfaces should be well illuminated.

Occupational Hazard Origin and Preventive Measures **109**

10. All fall related incidents should be investigated for future fall reduction or prevention.
11. Adequate training on technics and PPE or other tools used should be provided.
12. Prevent the need for access to height; this has to be designed or provided for work, for example in plant/machinery or road tankers for sampling, control operating points can be located at ground level where possible. Or cleaning of plant machinery might be carried out from ground level using a foam jet cleaner or cleaning might be reduced by better eliminating fumes, dust, etc.
13. Safe access must be provided or designed where access to height is necessary.
    - For example, a steps and platform with handrails are required.
    - This will allow contractor or portable equipment that are required to be there.
14. Fall caused by slipping; are common and all steps must be taken to ensure lack of contamination by water, grease, or other substances as discussed above.
15. Falls from stairs; is common and often is due again to stairs being contaminated with slippery material, water, oil, food product, etc.
16. Fall from FLT forks are also due to standing on pallets mounted on forks.

5.4.13 LADDER HAZARD PREVENTION MANAGEMENT

Each year there are more than 164,000 emergency room-treated injuries in the USA relating to ladder as reported by the U.S. Consumer Product Safety Commission. Following are general guidelines recommended for reduction or prevention of fall from height[206,207]:

1. Consider a personnel-lift cage attached onto a forklift instead of a ladder for work at elevated sites indoors.
2. Use a vertical lifeline that runs the entire length of a fixed ladder.
3. Documented, periodic inspection of ladders and stepladders should be done.
4. Ladder footing profile and condition should be inspected prior to each use.
5. Unsafe ladders and stepladders must be taken out of service until repaired.
6. All ladders and stepladders should be fitted with non-slip feet.
7. Fall protection must be worn when performing work that has a fall distance of three meters or more.
8. Fiber glass, not metal or wood, ladders or stepladders should be used near or vicinity electrical installations, power lines or electrical equipment.
9. Ensure that workers shoes are clean, dry, and with well-defined heels.
10. Place warning signs in area where work is being performed.
11. Make sure ground or floor has level, solid flooring.
12. Secure the top of the ladder when practical and if not ensure a second person to foot the ladder base at all times.

13. Never use a painted wooden ladder or stepladder since cracked rungs are not clearly visible.
14. Ensure that ladders are of the correct height to avoid reaching or stretching.
15. Do not stand ladder on snow, ice, or other slippery contaminated floors.
16. Fully lock or open any door near a ladder in use.
17. Open step ladder fully.
18. Maintain three points contact when climbing or descending a ladder; always face the ladder when climbing or descending the ladder.
19. Portable ladders should be certified for their type of intended use in workplace.
20. Never use the top three rungs of a straight, single, or extension ladder as a step.
21. Place the ladder at a tilt of 75-degree angle.
22. Make sure ladder weight is supporting does not exceed its maximum load rating.
23. There should only be one person on the ladder at one time.
24. Use a ladder with proper length for the job. Proper length is a minimum of three feet extending over the roofing or working surface.
25. All metal ladders should have a slip-resistant feet.
26. Be sure that all locks on extension ladders are properly engaged. The, ground support under the ladder should be level and firm. Large flat wooden braced under the ladder can level a ladder on uneven or soft ground. A good practice is to have a helper hold the bottom of the ladder.
27. Do not place ladder in front of possible accidental hazard such as in front of the front door that is not locked, blocked, or guarded.
28. Do not use a ladder for any other purpose other intended for.
29. Never leave a raised ladder unattended.
30. Get training of ladder and safety at height, and
31. Follow use and safety instruction labels on ladders by manufacturers.

Use ladder inspection form for regular inspection and maintenance.

5.4.14 Scaffold safety

A lot can go when working on scaffolding. Provincial and federal occupational safety laws outline specific duties and responsibilities for employers, supervisors, and workers. Knowing the rules and regulations is in everyone's best interest for more detailed information refer to local and state authority guidelines. Michelle Morra a freelance journalist based in Toronto; Canada points out to ten important factors in scaffold safety.[214]

5.4.15 Workplace transport

In agriculture and fields as well as within food manufacturing facilities variety of transport equipment are used. Operators of vehicles for the transport of equipment

Occupational Hazard Origin and Preventive Measures   **111**

and those assisting them can be struck by moving, falling, or shifting equipment while loading, unloading, or transporting it. They can be crushed while hutching equipment to a prime mover. They can fall while securing the equipment or be injured while installing the means to secure it or injured when vehicle or the equipment collides with object during transport, which can also result in harm to others as well as the property.[550]

About 40% of fatal accidents in food and drink processing are due to transport related accidents. Some 270 people each year were struck by forklift trucks (FLTs) or other vehicles with serious injuries. Overall struck by vehicle were 31%, by FLT trucks 26%, fall from truck 22%, trapped between vehicle and wall 6%, trapped by overturning 6% and between two vehicles 5%.[212]

It is therefore important to have fall prevention as an integral part of transport loading and unloading as well as select the right kind of vehicle to reduce and eliminate such accidents.[211]

Workplace transport in the food and drink industries are the major cause of serious injuries. The food and drink sectors use a wide varieties of vehicles such as FLTs, tankers, grain lorries, and flatbed vehicles which present a range of risks.

The priorities have been established to be[252]:

Pedestrian safety and pedestrian vehicle segregation, vehicle reversing and falls from vehicles, therefore, specific hazards need to be addressed such as: Overturning of tipping lorries and trailers, tailgate safety on bulk delivery vehicles, forklift trucks falling from loading bays, employee training, noise level within driver's cab and outside the vehicle during operations should be reduced so far as reasonably practicable to protect the hearing of both the driver and the other workers nearby and abiding by the regulation and guidelines.

5.4.16 FORKLIFT TRUCK

Primary equipment used are forklift truck with internal combustion engine; electric forklift truck; cables; fork legs extenders; hoisting ropes; pallets; baskets and buckets.

Workplace hazards with forklift normally exist in warehouses; heavy industry; metal, wood, food and drink, electronics, textile, and similar industries, loading or unloading in docks, airports and other industries in which mechanical lifting equipment is required.[208]

A worker who drives an industrial truck equipped with lifting devices, and performs.

In summary hazards associated with forklift trucks can be:

A. **Physical hazards**
   1. Psychological stress associated with increased risk of accidents involving other vehicles and suddenly appearing pedestrians.
   2. Psychological problems with coworkers (e.g., caused by their requests of a ride).

112 Occupational Health and Safety in the Food and Beverage Industry

3. Exposure to excessive noise levels (especially when operating diesel trucks or working inside closed structures), with resulting hearing impairment.
4. Exposure to whole-body vibration caused by rigid construction of truck (particularly wheels), inadequate shock-absorbing properties of operator's seat or prolonged driving on rough grounds.
5. Exposure to harsh climatic conditions (heat, cold, rain, winds, etc.) while working outdoors.

B. **Chemical hazards**
1. Allergic skin reactions as a result of contact with fuel and/or solvents.
2. Eye injury due to splashes of corrosive materials.
3. Intoxication by exhaust gases, especially asphyxiation by CO resulting from incomplete combustion of fossil fuel, which are emitted from the exhaust pipe and their concentration is rising rapidly inside closed and inadequately ventilated structures
4. Exposure to nitrogen oxides (NOX) emitted together with the exhaust gases inside relatively closed and inadequately ventilated structures.

C. **Biological hazards**
   Exposure to aerosols containing microorganisms, fungi, etc. present in the air with the dust as result of truck movement.

D. **Ergonomic, psychological, and organizational hazards**
1. Cumulative trauma disorders of hands and arms pains resulting from their overexertion while driving a non-laden truck presenting higher resistance to steering.
2. Low-back pain, muscle contraction and other disorders caused by prolonged seating (in a rigid and often awkward posture) in an unergonomic seat.
3. Neck pains as a result of frequent back-turning of head and neck stretching during reverse driving and while transporting bulky load obstructing operator's vision.
4. Vision problems (eyestrain, eye burn, other kinds of irritation, double vision, etc.) due to prolonged work under condition of poor lighting, abrupt changes in visual environment, blinding effect of other vehicles headlights and floodlights, etc.
5. Psychological stress associated with increased risk of accidents involving other vehicles and suddenly appearing pedestrians.
6. Psychological problem with workers (e.g., request for ride, etc.)

5.4.17 STRUCK BY OBJECTS

It has been reported that injury by moving object being from a knife or an object accounts for 15% of injuries in food and drink industries.

Occupational Hazard Origin and Preventive Measures

A third of these injuries are due to falling objects from above (such as storage rack), a quarter are from hand tools (e.g., knives) and by third largest cause being hit by moving pallet trucks, etc.[213]

5.4.17.1 Preventive management; struck by objects

5.4.17.1.1 Falling objects

1. Storage above ground should be stable and not fall by any movement.
2. Store heavier items on lower shelves without extending off the shelf.
3. Consider safe stacking, handling, and their movement to prevent fall.
4. Tall self-standing objects such as gas cylinder or objects leaning against the walls are either stable if knocked or secured properly.
5. Make sure the shelves themselves are solid and can stand the weight over time there could be corrosion, rust, and so on.

Proper and safe handling of gas cylinders is shown in Figure 5.4.

FIGURE 5.4 Proper safe handling of various gas cylinders. Figure 5.4 Shows 9 full capped gas cylinders secured to the cemented fixture by a metal chain tightened above the middle parts of cylinders. In the lower right part of picture 3 small, contained gas cylinders are shown.

5.4.17.1.2 Hand tools

1. Sharp object such as knife causes greatest injuries and should be stored safely.
2. When hand knives are in use, knife resistant protective clothing and gloves must be worn as determined by risk assessment.
3. Hand tools should be maintained in good condition to prevent accident.

5.4.17.1.3 Moving objects

1. Moving pallet trucks, racks, trolleys, etc. need to be used in designated routes to prevent pedestrians.
2. The pushing/pulling person should have good visibility on the designated routes.
3. Assessing risk is required for specific hazards in a particular area such as presence of hoist hooks, kegs, or items being ejected from machines.
4. Make sure to provide proper training on all above criteria as may be required.

5.4.18 FOOD PROCESSING MACHINERY, TOOLS, AND EQUIPMENT SAFETY

The shifting from hunting and gathering societies to agrarian societies allowed people to stay in one place and grow their food. Until the industrial revolution, most people had to grow their own food, using their own labor. Farmers spend many hours operating machinery in the field to perform tillage, planting, cultivation, and harvesting.[215]

Automatic process control in food, beverage, pharmaceutical, nutraceutical, cosmetic and chemical process industries have advanced tremendously in recent years and performing tasks which were formerly handled by men.[23]

5.4.18.1 Main causes of injury

Statistic and analysis of accident investigated by HSE demonstrated over a four-year period that these accident percentages of injuries are due to the following types of machinery,[217] conveyors, forklift trucks, band saws, and other food and beverage machinery such as thermoform machines, pie and tart machines, palletizers/de-palletizers, strapping/banding/tapping machines and mincing/grinding/mixing machines, patty formers, cartooning machines, vertical thermoform machines, drinks labeling/marketing machines, shrink wrapping machines and drink process machinery and drink crating/decorating machines, stamping/punching/franking machines, and drink canning machinery.

Also 90% of accidents occur during normal operations of production time.[220]

5.4.18.2 General machine hazard safety

1. When purchase new machineries select suitable machine intended for its work.
2. Specify health and safety as well as hygienic design requirements such as noise level, etc. Make sure the equipment supplied meets your specification and supplier has met their legal duties and standards required in your

Occupational Hazard Origin and Preventive Measures   **115**

country. For example The European Standardization Committee (CEN) has produced a number of standards which set out safety requirements for certain types of food machineries.

3. Provide training to employees for machine or tool use, repair, and maintenance.
4. Inspect tools before use and know the correct method for handling tools.
5. Avoid using defective tools.
6. Keep tools clean and store properly to prevent damage due by accidental fall.

5.4.18.3   Food processing machine safety prevention management

The hazard elimination including machine hazards associated with any processing system requires[215,510]:

1. The identification and assessment of all hazards associated with the final food product.
2. The identification of the steps or stages within food production at which these hazards may be controlled, reduced, or eliminated: the critical control points (CCPs).
3. The implementation of monitoring procedures at these CCPs.

5.4.18.4   Siting of fixed equipment

1. Adequate space around machine allowing operator to work without distraction.
2. Floors in good condition and kept clean to prevent slipping.
3. Machine on a secure base to prevent movement during use.
4. Suitable and sufficient lighting, such that the operator is not distracted by shadows or glare; awareness that fluorescent lights can make moving equipment appear stationary.
5. Suitable positioning of appropriate warning notices.

5.4.18.5   Electrical safety

1. Procedures to maintain the safety of electrical equipment, including user checks formal visual inspection and combined inspection and testing where necessary.
2. Use of sensitive (30 mA max.) residual current circuit breaker (also known as an earth leakage circuit breaker) fitted to pressure washing units and steam cleaners. Also recommended for all electrical equipment in wet or hazardous areas.

5.4.18.6   Maintenance (including cleaning)

1. All cutting edges/blades to be maintained in a sharp state,
2. All electrical equipment isolated prior to any maintenance/cleaning,
3. Use of blade carriers when removing blades for cleaning and sharpening,
4. Regular checks of mechanical and electrical interlocks on equipment service contract carried out,

116   Occupational Health and Safety in the Food and Beverage Industry

5. Provision and maintenance of suitable guards/safety devices,
6. Provision of suitable protective clothing including gloves, aprons, goggles, etc. when using cleaning and sanitary chemicals.

5.4.18.7   Protective clothing

1. Chain mail gloves, arm protectors, aprons to protect against cutting and piercing injuries in operation such as boning work when using band saws,
2. Gloves, aprons, goggles for use when using or preparing cleaning or other chemicals, or handling hazardous materials, e.g. hot oil or fat,
3. Hats/protective clothing in a catering environment; flameproof, and close woven to repel splashes of hot oil and fat liquids,
4. Non-slip safety footwear to protect against slipping and impact damage to feet from falling, heavy, or sharp objects.

Commonly used food processing equipment and machine include but not limited to:

- Knives (hand held),
- Worm type mincing machines,
- Microwave ovens,
- Rotary knife bowl type chopping machines,
- Dough breaks and mixers,
- Food mixing machines including attachments for mincing, slicing, chopping, chipping, etc.,
- Wrapping and packing machines,
- Circular knife slicing machines,
- Chipping machines whether power operated or not,
- Gas/electric cookers and ovens,
- Cold stores/rooms, and
- Prescribed dangerous machines as reported by HSE in UK.[222]

To identify hazard, carry out risk assessments where required, ensure proper maintenance, and use of all equipment through training and supervision, introduce safe systems of work for all activities involving the use of equipment or machinery and ensure that all staffs are aware of hazards and are either informed, instructed, or trained in risk control procedures.

5.4.18.8   Specific equipment/machine/tools hazard

Risk assessment is the data gathering component, while risk control is the application of the risk assessment evaluation.[509]

Conveyors are perhaps the simplest of automation, transferring parts, assemblies, or loose material from one place to another so as to avoid manual handling.[220] Where 90% of these injuries occur on belt conveyors and 90% of these injuries involve well-known hazards such as in running nips, transmission parts, and trapping points between moving and fixed parts with 90% of accidents during normal operations of production time.[218]

Band saws are used for cutting meat and poultry products mostly in grocery meat department, catering, food service and or other similar working areas which a

Occupational Hazard Origin and Preventive Measures

worker performs tasks that put them at risk for injuries due to cutting tools, machinery, high-force or repetitive motions, heavy boxes, slippery floors, and other potentially hazardous conditions.[221]

5.4.19 Hazard prevention

1. Guard all parts of the blade except at operation guarding, fully enclose the pulley mechanism, and make sure the saw has a tension control for adjustment.
2. Maintain appropriate blades that are sharp with appropriate size blade for the job.
3. Set the guard as close as to the stock as possible without affecting movement.
4. Purchasing can be important factor to be taken into consideration for potential hazard when a band saw is being purchased.

5.4.20 Commercial mixers

It is recommended to plan a thorough testing program with a reliable and experienced equipment manufacturer safety features before even committing to a specific type of mixer system.[223] Mixers are very powerful machines and the major hazards are getting caught in and/or crushed by moving parts. Access to these parts can cause crushing, laceration, and/or fracture.

5.4.20.1 Identification and assessing mixer hazards

Rotating attachments in mixers may be present a significant hazard during mixer operation. These attachments such as a whisk, dough hook, and paddle or flat beater can cut, entangle, stab, or abrade fingers, hands, and arms. Occupational health and safety code requires an employer to provide safeguards if a worker can come into contact with moving parts of machinery.[224]

1. Identifying what and where hazards are likely to be present.
2. Worksite assessments to determine exposure levels and who is exposed.
3. Reviewing injury records to determine the frequency, severity, and location.
4. Worksite observations to determine how workers are exposed, is PPE being used?
5. Reviewing what standards and control measures are in place.

5.4.20.2 Mixers hazard prevention management

5.4.20.2.1 Guarding

Two acceptable approaches to guarding are:

1. Interlock guards that is preferable choice and may include a barrier guard or presence sensing device (infrared light, curtain, infrared sensor, etc.) and must have two features[224]:

   Fail/safe that means it must stop operating for whatever reason to prevent the mixer from operation when the guard is in its "Open" position.

118   Occupational Health and Safety in the Food and Beverage Industry

2. Barrier guards used in combination with an administrative procedure describing written safe work practices. A barrier guard must:
A. Prevent body parts including hair and items such as clothing or jewelry from coming into contact with the rotating shaft and attachment,
B. Be solidly constructed and securely mounted,
C. Prevent access to the rotating shaft and attachment from all accessible locations,
D. Do not allow a build-up of product as this might affect the effectiveness of the guard, and
E. Do not cause an injury in itself.

Administrative procedure alone reminds workers to stay clear of a rotating shaft and attachment is insufficient to provide a satisfactory level of worker protection.[224]
Main preventive measure should include:

1. Providing a guard to prevent access to the moving parts such as whisks, paddles, hooks, and other blending accessories.
2. These guards are to be interlocked to prevent operation of the equipment when the guard is opened for pouring or cleaning or when the bowl is lowered.
3. A raised guard that does not stop the machine is an unguarded mixer.
4. The interlock system is part of the guard, not a substitute for a lock-out system which is also required by the Regulations.

5.4.20.2.2   Lock-out system

When cleaning, oiling, adjusting, repairing, or performing any type of maintenance, the following precautions must be taken:

1. Control switches or other mechanisms MUST be locked out.
2. Other effective precautions necessary to prevent any starting shall be taken.
3. You may need an electrician to install a new control box.

Training for lock procedure, how to use it safely during normal operations, how to use the guard, precautions to be taken when cleaning, adjusting, or during maintenance and workplace-specific hazards and provide procedures and safety information of the machine supplier.

5.4.20.2.3   Knife hazards

Many knives cut and muscle strain injuries occur in the meat and food industries. Injuries include[226]:

1. Cuts to the non-knife hand or arm (most common).
2. Cuts to the hand holding the knife which occur when the hand slips off the handle.
3. Cuts which occur with a reverse grip and pulling back toward the body.

Occupational Hazard Origin and Preventive Measures

4. Cuts to another person, inadvertently, where people are too close together when working.
5. Sprains or strains (e.g. from the extra effort required to use knives that are not sufficiently sharp). These can be reduced by using a well-designed/sharp knife.

5.4.20.2.4 Identifying and assessing hazards

1. Identifying how extensive knife use is.
2. Identifying who uses knives, to what extent, and where.
3. Reviewing injury records to determine the frequency, severity, and location of knife-related accidents.
4. Worksite observations to determine how knives are being used, postures of users, types of grips used, lighting in the area, is PPE being used.
5. Reviewing what standards and control measures are in place.

5.4.20.2.5 Legal requirements

Employers have responsibilities under the Occupational Health and Safety Act to ensure a safe working environment without risks to health, so far as is reasonably practicable.[226]

5.4.20.2.6 Knife hazard elimination and reduction

You must ensure that your employees are kept safe from harm so far as is reasonably practicable. You must assess the risk of your employees being cut by knives and take reasonable precautions.[227]

45. 1. Train employees in the safe use of knives and safe working practices when sharpening them.
2. Use a knife suitable for the task and for the food you are cutting consider the type of food, cut performance, length, angle, and sharpness of blade as well as shape, size, and texture of the handle required for the job.
3. Keep knives sharp.
4. Cut on a stable surface.
5. Handle knives carefully when washing up.
6. Carry a knife with the blade pointing downwards.
7. Store knives securely after use, such as in a scabbard or container.
8. Use protective equipment as required. For deboning, it is recommended that a suitable protective glove is worn on the non-knife hand, and a chain-mail or similar apron is worn.

5.4.20.2.7 Knife, other safety practices

1. Maintain enough room between people so that the person using a knife won't be bumped or inadvertently slip and cut someone else.
2. Cutting surfaces (tables, boards) should be maintained in clean, smooth condition.
3. A protective glove, apron, finger stall, and arm guard should be worn.

120   Occupational Health and Safety in the Food and Beverage Industry

4. Wherever practicable, cut away from the body. Protective aprons should be worn during work where it is necessary to pull the knife with the point toward the body, such as in boning cuts.
5. Hand hooks should be provided as a gripping aid. These hooks can also be used to assist in moving cuts of meat from one operation to the next. A knife should not be used to transfer meat by piercing and levering.
6. Cutting should be done at waist height using a purpose-built surface.
7. Keep work surfaces and floors clean and tidy to avoid a slip or trip.
8. Appropriate "non-slip" safety footwear should be worn.
9. Adequate lighting needs to ensure good visibility to eliminate accident.
10. Provide training in proper use of knives.
11. Knife handles should be guarded to prevent hand slipping onto blade.
12. Use the proper knife for the task.
13. Cut away from the body.
14. Only one knife should be carried at a time, with the blade pointed down and close to the side.
15. Allow sufficient room to work safely without endangering nearby workers (two individuals should not work on the same piece at the same time).
16. Knives should be cleaned regularly and as required.
17. Replace worn down knives.
18. When not in use, knives should be stored in a guarded scabbard.
19. All cuts should be immediately reported and first aid administered.
20. Workers should concentrate on the job while working with a knife.
21. One should never try to catch a falling knife.

5.4.20.3   Prevention management; transport hazards

1. Have safe traffic routs and preferable one way for transport.
2. Stack, secure pallets safely; do not exceed maximum weight or height.
3. Install overturning protection.
4. Wear hearing protection appropriate for the noise levels and type of noise consult the supplier or an expert.
5. Protect hands with chemical-resistant gloves; if impractical, use a barrier cream.
6. Wear appropriate eye protection; consult a safety supervisor or a supplier.
7. Do not operate diesel or gasoline-powered forklift truck in confined or inadequately ventilated spaces; use an electrically operated truck.
8. Use a respirator if truck motion raises much dust from the floor.
9. Install an ergonomically designed driver's seat.
10. Have pedestrian and vehicle on separate tracks/marked inside and outside plant.
11. Make sure pedestrians and vehicle have separate doors into buildings with barriers where required.
12. Eliminate reversing totally or as much as possible.
13. Make sure there is visibility and devices such as a camera, mirror where required.

Occupational Hazard Origin and Preventive Measures 121

14. Make sure there is no need for a person to direct operations.
15. Raising workers on fork may be replaced by cages constructed for their lifting.
16. Eliminate top access, avoid sampling or access to certain valves to ground level.
17. Provide training for all operators.
18. Load should not pass over any worker close to ground floor as possible.
19. Never leave a lift truck while not present and no lifting or lowering workers.
20. Follow all traffic signs for road safety being indoor or outdoor.
21. Transport in exterior also requires further attention to weather condition, other motor vehicles, cargo stability, length of driving, etc. for safety.
22. Wheel should be blocked before any unloading or loading, etc.
23. If working under the truck, motor should be off, wheel should be secured and protective wear must be worn.
24. Cargo must have proper sign for their content and be secured.
25. Driver should have proper training not only for driving but also in emergency situation, such as fire, spills, movement of cargo, accident, etc.

5.4.20.2.8 Refrigerator/freezer safety protocols

Use of refrigerators and freezers in science, engineering, and/or production facilities could have potential hazards[228,229]:

1. Biological contamination, exposure to bio-toxins, and bacteria in the food supply are one problem.
2. Chemical hazards involving fire, explosions, and chemical toxin.
3. Physical decontamination before disposal.

Types of refrigerators/freezers could be household or industrial for food and drink, lab safe, or explosion-proof type for hazardous chemicals to be used as has been designed for.

5.4.20.2.9 Purchasing policy and safety

Accident proneness was part of a third important strategy that then appeared: separating particularly vulnerable and dangerous people from risky situations. At the end of the century, however, this third alternative diminished dramatically when physical or engineering means overwhelmingly came to dominate safety, obscuring even the continuing programs to educate people to behave so as to avoid accidents.[511]

The hazards identified should take into consideration the intended end use and when purchasing material.[225]

As identifying food ingredient quality is of prime importance so too are purchasing safe equipment and machinery and other instrumentation operation of critical importance to reduce or eliminate accidents in food production processing areas.[225]

Such policies will have a profound affect in controlling amputation and such hazards must be identified in advance that will affect designing and purchasing of required machines. US Department of Labor has developed examples of such pertinent questions in order to assist when such hazards are of importance.[512]

Employers must ensure that all work equipment they provide conforms with the essential requirements of the relevant community law at the point of purchase and has a Declaration of Conformity and that simple safety checks have been made before bringing it into use – purchasers can save much time and trouble, should the product later cause harm or be otherwise found unsafe.[513] Also view previous chapter for further information on purchasing policy.

5.4.20.2.10 Machine guarding/hazard and safety

As mentioned in the food processing machinery section, safety guard is an important aspect of security measures in working condition. Beyond that, there was at first no systematic way except education and discipline to reduce the grim, scandalous accident rate. And it is true that still in the late 20th century, safety education and safety rules and laws continued to constitute a major resource by which societies attempted to reduce injury and other damage.[551]

In 2014, a large UK food manufacturer had to pay a fine of £800,000 after a serious industrial accident. An engineer was trapped by the machinery while examining a conveyor belt and suffered major injuries and ongoing nerve damage.[753]

Accidents such as this are widely reported, but many people are unaware of the number of hazardous areas found in food and beverage processing plants. Here Darcy Simonis, industry network lead for ABB's food and beverage segment, explains the safety procedures that must be developed in these processing plants.

In milking parlors, exposed platform rollers must be guarded to avoid clothing or employees becoming trapped. Hazards are present throughout the plant, from the handling of the raw material, to production where industrial ovens can often reach very high temperatures to the final packaging of the product ready for transportation.

Across the globe, there are a variety of different regulations for food processing plants. In particular, North America and Europe have strict regulations for safety in these potentially dangerous environments.[753]

In the 1970s, the increase in heavy machinery such as the creation of the steel press led to increased safety guards. Since then, many safety-conscious companies undertake a risk analysis in the initial stages of machine development. In the case of decanters, it is not possible to remove the risk, but it is possible to mitigate the risk to an acceptable level by putting safety guards such as enclosures or emergency stops into place.[753]

To minimize machine guarding hazards, use appropriate engineering and work practice controls including operator training. Such hazards include:

1. Operators making contact with or being pulled into the chipper.
2. Hearing loss.
3. Face, eye, head, or hand injuries.

The safe practices that could prevent such accident happen, are as follow:

Occupational Hazard Origin and Preventive Measures

1. Never reach into a chipper while it is operating.
2. Do not wear loose-fitting clothing around a chipper.
3. Always follow the manufacturer's guidelines and safety instructions.
4. Use earplugs, safety glasses, hard hats, and gloves.
5. Workers should be trained on the safe operation of chipper machines. Always supervise new worker using a chipper to ensure that they work safely and never endanger themselves or others.
6. Protect yourself from contacting operating chipper components by guarding the infeed and discharge ports and preventing the opening of the access covers or doors until the drum or disc completely stops.
7. Prevent detached trailer chippers from rolling or sliding on slopes by chocking the trailer wheels.
8. Maintain a safe distance (i.e., two tree or log lengths) between chipper operations and other tree work or workers. When servicing and/or maintaining chipping equipment (i.e., "unjamming") use a lockout system to ensure that the equipment is de-energized.[551]

Most machines today have a safety guarding to prevent accident and therefore machine guarding is an important part of Occupational Health and Safety Act of 1970 in the USA to protect operator from hazard associated with machine operations. Machine guarding specifically prevent injury from following sources[514]:

1. Direct contact with moving parts,
2. Splashing of chipped or hot metal as well as hot chemicals,
3. Mechanical and electrical failures,
4. Human failures, resulting from curiosity, distraction, fatigue, worry, anger, illness, violence, or deliberate actions.

Each year number of accidents claims life or limb of many due to moving parts with lack or improper machine guarding being reported and some preventive measures some in Canada and the USA.[230,381,620,621]

5.4.20.2.11 Identifying and assessing machine hazards

1. The point of operation: that point where work is performed on the material, such as cutting, shaping, boring, or forming of stock.
2. Power transmission apparatus: all components of the mechanical system that transmit energy to the part of the machine performing the work. These components include flywheels, pulleys, belts, connecting rods, couplings, cams, spindles, chains, cranks, and gears.
3. Other moving parts: all parts of the machine that move while the machine is working. These can include reciprocating, rotating, and transverse moving parts, as well as feed mechanisms and auxiliary parts of the machine.

To prevent hazards while servicing machines, each machine or piece of equipment should be safeguarded during the conduct of servicing or maintenance by:

124   Occupational Health and Safety in the Food and Beverage Industry

1. Notifying all affected employees (usually machine or equipment operators),
2. Stopping the machine,
3. Isolating the machine or piece of equipment from its energy source,
4. Locking out or tagging out the energy source,
5. Relieving any stored or residual energy; and verifying that the machine or equipment is isolated from the energy source.

Machine safeguarding results in the following[231]:

1. Prevents employee contact with the hazard area during machine operation,
2. Avoids creating additional hazards,
3. Is secure, tamper-resistant, and durable,
4. Avoids interfering with normal operation of the machine,
5. Allows for safe lubrication and maintenance,

Various types of **machine guards** available for **machine guarding**[622]:

1. Fixed guards,
2. Fixed limited access guards,
3. Fixed adjustable access guard,
4. Interlock guards,
5. Automatic guard,
6. Safety by Machine Controls,
7. Safety by Precautions and Maintenance,
8. Criteria for Machine Guard Selection, and
9. Built-in Safety Devices.

5.4.21 LOCKOUT (ENERGY CONTROL)

The fundamental goal of any safety program is to ensure that workers are not exposed to sources of energy such as high-voltage electricity, high-temperature fluids, toxic chemicals, moving parts, or falls from heights. Therefore, before working on a piece of equipment, the associated sources of high energy must be identified and secured. In the case of a pump, e.g., the following energy sources are probably present[517]:

1. Rotating energy. The driver, drive shaft, and impeller all turn. It is important that they be secured from inadvertent movement (even if the motor has been de-energized) before anyone works on the pump.
2. Electrical energy. If the pump has an electrically driven motor, the electricity supply to it must be properly isolated.
3. Heat energy. If the pump is driven by a steam turbine, or if there is steam tracing around it, it is important to ensure that the steam – and the associated steam condensate system – are properly isolated.
4. Chemical energy. If the pump normally handles hazardous chemicals that are toxic or a health hazard, it has to be properly cleared of them.

Occupational Hazard Origin and Preventive Measures 125

FIGURE 5.5 Example of a lockout station system. Figure 5.5 Illustrates a standard locks & tags station (lockable) as part of employer's responsibility to protect employees from hazardous energy sources on machines and equipment during service and maintenance.

5. Flammable/explosive energy. If the pump handles hydrocarbons, or other materials that could ignite, need to be cleared, using an inert gas such as nitrogen.
6. Potential energy. If the pump is not located at grade, it may be possible for a person to fall off it (and if it is at grade there may be a pit below it).
7. Workers injured on the job from exposure to hazardous energy lose an average of 24 workdays for recuperation.[232]

A standard locks and tags station (lockable) as part of employer's responsibility to protect employees from hazardous energy sources on machines and equipment during service and maintenance is shown in Figure 5.5.

5.4.21.1 Multiple workers/contractors and lockout/tagout
When multiple workers must simultaneously service a machine, each should be assigned a separate lock and key. If work is not completed before the end of a shift and new workers are to continue the work, the new arrivals should apply their locks before the departing ones remove theirs. Each worker must move away from the machine before it is re-energized.[515]

5.4.21.2 Lockout/tagout exercise
When a machine needs repairs or other maintenance, you want to be sure the equipment stays off while you work on it to prevent accidental operation by others.[516]

5.4.21.3 Hoists

Workers can be seriously injured or even die as a result of hazards involving material handling. Inspectors will target hazards involving material handling and they will check that employers, supervisors and workers are complying with requirements under the Occupational Health and Safety Act (OHSA) and its regulations.[233.]

5.4.21.4 Hazard's priorities

Examples of material handling hazards include:

1. Lifting loads improperly or carrying loads that are either too large or too heavy resulting in strains and sprains,
2. Being struck by or caught between materials or being caught in pinch points resulting in fractures and bruises,
3. Materials that fall or collapse when they have been improperly stored or when ties or other securing devices have been incorrectly cut or unfastened resulting in cuts, bruises and crushing type injuries,
4. Contact with moving equipment, vehicles, lifting devices, and/or their unsecured loads that fall or collapse resulting in critical or fatal injuries,
5. Same level falls and falls from heights when attempting to move, place, store, or access materials in an unsafe manner resulting in critical or fatal injuries,

The risk of back injury and muscular strains (musculoskeletal injuries/disorders) from lifting and moving heavy or bulky items of stock is common in workplaces that handle materials on a regular basis.

5.4.21.5 Glassware, sharp, and bottling hazards

Glass is a component of many industries such as pharmaceutical, cosmetic, food and beverage industry, etc. Food industry makes a considerable use of glass as a packaging medium for liquid, paste, and or solid form food or beverages where broken glass hazard management has to be considered as well as those use in laboratories.[234]

5.4.21.6 Hazard prevention

Reduce the risk of injury by following these guidelines:

1. Wear cut-resistant gloves when using sharp instruments.
2. Use the right tool for the job, and make sure it is sharp.
3. Always cut away from your body.
4. Carry tools in a sheath or holster.

5.4.21.7 Handling sharp objects

Handling sharp objects such as broken bottles or glass requires cut-resistant gloves when using cutting tools and sharp objects in laboratories or other manufacturing

Occupational Hazard Origin and Preventive Measures 127

facilities. When handling glassware are under stress, such as inserting tubing through a rubber stopper or freeing a "frozen" glass joint.[382,383]

5.4.21.8 Glass/bottling hazards prevention management in processing areas

Bottling equipment is associated with glass breakage. Use mechanical barriers, operating and maintenance procedures to minimize hazards.

Mechanical barriers include filler guards in the counter pressure section of the bottle filler, filling station guard, and conveyor covers.

Proper illumination at points where bottles are handled is important to minimize broken bottle hazards. Steps for prevention and or reduction or injuries are:

1. Instruct glass handlers in the importance and necessity of those safe procedures that apply to their particular operation.
2. Have employees wear PPE.
3. Store glass containers in cases, cartons, or racks.
4. When glass containers are broken during the packing process, stop the machinery and lock it out before attempting to remove the broken glass from the machine.
5. Instruct all glass handlers to obtain first aid for injuries immediately.
6. Open pressurizes or vacuum-packed glass containers before discarding to avoid unexpected explosions or implosions.
7. Instruct employees not to throw glass broken or not into open containers. This will prevent the danger of flying broken glass. Instruct employees to use only clearly marked and labeled shatter and leak proof receptacles.
8. Where possible use industrial vacuum cleaners for cleaning up broken glass. If this is not possible, use a brush and dustpan. Place warning signs for broken glass.
9. When exposure top broken glass is possible, wear gloves, sleeves, if the arms are exposed and heavy use soled, slip-resistant safety footwear. Use safety eye or face protection even when other means such as equipment shields are provided.
10. Use bottles that are free of defects such as blisters, surface cracks, chips, or a frosty appearance.
11. Train all line employees to spot a defective bottle and in breakage handling.
12. Use shatter proof lights such as shatter shield.

5.4.21.9 Fragile devices

Each organization and manufacturing site might have certain fragile devices such as sight glasses or cast-iron piping, etc. established procedures must be in place to minimize the hazards associated with flexible joints and fragile devices where it could be.

5.4.21.10 Radiation/and food irradiation hazards

Radioactivity was not banned in consumer products until 1938. This is four years after death of radioactive discoverer, Madame Curie in 1934 due to leukemia.[568]

128   Occupational Health and Safety in the Food and Beverage Industry

Various radiation emitting devices are capable of emitting ionizing radiation when activated. This definition applies to devices such as X-ray machines, whose sole function is the generation of ionizing radiation. These hazards can be separated in two classes as follow as ionizing radiation (manmade source of radiation) and non-ionizing radiation.[235] If there are too many or too few neutrons, the nucleus is unstable, and the atom is said to be radioactive.[236]

5.4.21.11   The food irradiation process and hazard

5.4.21.11.1   Introduction

After World War II, scientists looked for new ways to improve life utilizing applications of nuclear energy. One of these applications is that of food irradiation. This process is used to kill dangerous diseases associated with foods.[237]

Food irradiation causes chemical changes producing so called "Radiolytic Products", these products are also naturally present in other non-irradiated food at levels that are many times greater than the levels generated by irradiation.[239]

In 2011 Japan's nuclear disaster contaminated enormous amount of food such as milk, spinach, and other food products due to presence of Cesium–137 which has a longer-term half-life.[240]

5.4.21.11.2   Food irradiation process

Irradiation is used to extend the life period of food. This is advantageous in areas where food refrigeration is not available. The purpose of food irradiation is to clean food in the very best way possible. The radiation penetrates all parts of the food, killing harmful bacteria that cause diseases, such as salmonella in seafood, trichinosis in pork, and cholera from fish. Food irradiation is the process of exposing food to ionizing radiation to destroy and check the multiplication microorganisms, bacteria, viruses, or insects that might be present in the food. Further applications include sprout inhibition, delay of ripening, and improvement of re-hydration. Cobalt-60 is the most commonly used radionuclide for food irradiation. For irradiating the foods the quantity of dosage is very much important.[775] It has been referred that it is a type of adulteration since it causes disruption of internal metabolism of cells, DNA cleavage, formation of free radicals, disrupts chemical bonds, high capital costs, possible development of resistant MO, inadequate analytical procedures to detect irradiation in food and public resistance.[775]

5.4.21.11.3   The effect of irradiation on food

Ionizing radiation also breaks some of the chemical bonds within the food itself. The effects of chemical changes in foods are varied. Some are desirable, others are not. Examples:

- In some food that turn mushy, such as lettuce, changes happen in their structure as they may be too fragile to withstand the irradiation,
- On the other hand ripening and maturation are slowed in certain fruits and vegetables, which lengthens their shelf life,

Occupational Hazard Origin and Preventive Measures

- At times there will be reduction or destruction of some nutrients, such as vitamins A, B, and E, which reduces their nutritional value,
- There could be formation of compounds that were not originally present and therefore will require the strict control of radiation levels, such as free radicals, which can recombine with other ions.[775]

Others believe although food irradiation deals with very dangerous materials, it does not seem to harm the food in any way.[238]

5.4.21.11.4 Workers' protection and hazard management in food irradiation facilities

1. Reduce direct radiation through the use of regular glasses.
2. Use goggles or full-face shield to reduce exposure to UV.
3. Cover exposed body parts to ensure adequate protection from UV radiation.
4. UV radiation may be reflected by some surfaces, such as stainless steel.
5. Radiation is controlled at the source by shielding equipment that emits radiation.
6. Control along the path is done by increasing the distance from source.
7. Using protective wear such as lead apron, gloves, and goggles.
8. In some cases the location area of irradiation can be isolated such as using protective shielding in welding.
9. A dosimeter can be worn for control of workers being exposed to radiation.
10. Train employee for proper handling, processing, protective equipment, and emergency procedures.

5.5 NOISE HAZARD/HEARING PROBLEM

Regulations applicable to noise levels within an industrial facility are intended to protect employees from noise-induced hearing loss. The regulations are usually based on an employee noise exposure limits such as 85 dB(A) averaged over 8 h.

The regulations normally allow for exposure to higher noise levels for reduced time periods and mandate the use of hearing protection if necessary. This allows the employees' averaged noise exposure level to remain below the safe limits.[521]

Typically, these regulations set a series of action levels based on employee noise exposure levels as follows:

1. First action level – daily personal noise exposure exceeds 85 dB(A)
   - Employer is responsible for initiating a hearing conservation program and having noise assessment made by a competent person.
   - Employer is required to reduce the risk of hearing damage from exposure to noise to the lowest level reasonably practicable.
   - Appropriate hearing protection should be available to the employee on request.
2. Second action level – daily personal noise exposure of 90 dB(A)
   - Appropriate hearing protection must be made available to all persons exposed.
   - Employer must reduce noise as far as reasonably practicable.

130 Occupational Health and Safety in the Food and Beverage Industry

Other requirements include assessment records and audiograms, designated ear protection zones, training on maintenance and use of equipment, and provision of information to employees and information to employees.

Above permitted-level noise and sound levels could be the source of injury and be considered an occupational hazard in industry. A low-level sound as low as an air condition (30 dBA) is considered normal; however shouting at a close distance (100 dBA) can be considered hazardous, particularly based on the length of time one is exposed to.[243]

Tables 5.5–5.10 compare various noise and sound levels measured with point of reference in decibels (dBA) at various locations.

TABLE 5.5
Noise level of various machines at home

Machinery	Noise level (dBA)*
Refrigerators	50
Electric toothbrush	50–60
Washing machine, Air conditioner	50–75
Electric shaver	50–80
Coffee percolators	55
Dishwasher,	55–70
Sewing machines	60
Vacuum cleaner	60–85
Hair dryer	60–95
Alarm clock	65–80
TV audio	70
Coffee grinder	70–80
Garbage disposal	70–95
Flush toilet	75–85
Doorbells, whistling kettles	80
Food mixer or processor, blenders	80–90
Squeaky toy held close to the ear	110
Noisy squeeze toys	135

* The level of noise is measured in decibels (dB), after adjusting the decibel to human hearing the noise level is described in decibels A (dBA).

Occupational Hazard Origin and Preventive Measures

TABLE 5.6
Various Machines' Noise Level versus Sources of Sound or Noise at Work

Machinery	Noise level (dBA)*
Quiet office, Library	40
Large offices	50
Power lawn mowers	55–95
Manual machines, tools	80
Handsaws	85
Tractors	90
Subways	90–115
Electric drills	95
Factory machineries, woodworking class	100
Snow blowers	105
Power saws, leaf blowers	110
Chains saw, hammer on nail, pneumatic drills, heavy machine, jet planes (at ramp), ambulance sirens	120
Chainsaw	125
Jackhammer, power drill, air raids, percussion section at symphony	130
Airplanes taking off	140
Artillery fire	150
Rockets launching from pad	500

* The level of noise is measured in decibels (dB), after adjusting the decibel to human hearing the noise level is described in decibels A (dBA).

TABLE 5.7
Various Noise Levels at Recreation Centers

Machinery	Noise level (dBA)*
Quiet residential area	40
freeway traffic	70
heavy traffic, noisy restaurant	85
trucks, shouted conversation	90
motorcycles	95–110
snowmobiles, school dance, boom box	100
Discos, busy video arcades, symphony concerts, car horns	110
rock concert	110–120
personal cassette players on high	112
football game (stadium)	117
band concerts	120
auto stereos (factory installed)	125
stock car races	130
bicycles horns	143
firecrackers	150
cap guns, balloon pops	156–157
fireworks (at three feet), rifles, handguns, shotguns	162–170

* The level of noise is measured in decibels (dB), after adjusting the decibel to human hearing the noise level is described in decibels A (dBA).

TABLE 5.8A
Points of Reference Measured in Decibels (dB) or Decibels A (dBA)* for Various Machineries

Machinery	Noise level (dBA)*
Refrigerators	50
Electric toothbrush	50–60
Washing machine, Air conditioner	50–75
Electric shaver	50–80
Coffee percolators	55
Dishwasher,	55–70
Sewing machines	60
Vacuum cleaner	60–85
Hair dryer	60–95
Alarm clock	65–80
TV audio	70
Coffee grinder	70–80
Garbage disposal	70–95
Flush toilet	75–85
Doorbells, whistling kettles	80
Food mixer or processor, blenders	80–90
Squeaky toy held close to the ear	110
Noisy squeeze toys	135

* The level of noise is measured in decibels (dB), after adjusting the decibel to human hearing the noise level is described in decibels A (dBA).

TABLE 5.8B
Points of Reference Measured in Decibels (dB) or Decibels A (dBA)* for Other Noise Examples

Noise example	Noise level (dB, dBA)*
The softest sound a person can hear with normal hearing	0
Normal breathing	10
Whispering at five feet	20
Soft whispers	30
Rainfalls	50
Normal conversations	60
Shouting in ear	110
Thunders	120

* The level of noise is measured in decibels (dB), after adjusting the decibel to human hearing the noise level is described in decibels A (dBA).

Occupational Hazard Origin and Preventive Measures

TABLE 5.9

Typical Noises in Different Food Processing Sectors

Food processing sector	Noise level (dBA)*
Drinks	Bottling halls, bottling filling/labeling, decarting, and washing (85–95); Casking/kegging (85–100); Cooperage machine (above 95)
Meat	Animal in lairage (80–110), Powered saw (up to 100); Blast-freezers/chillers (85–107); Bowl–choppers (Above 90); Packaging machinery (85–95)
Milling	Mill areas (85–95); Hammer mills (95–100); Grinders (85–95); Seed graders (90); Bagging machinery (85–95)
Bakery	Dough-mixing room and Baking plant (85); de-panning (90); Bread Slicing (85–95); Fruit washing (92)
Dairy	Production areas (85–95); Homogenizers and Bottling lines (95–100); Blast- chillers (87–95); Pneumatics (85–95)
Confectionary	Hopper feed (95); Mold-shaker (90–95); Wrap-bagging (85–95); High boiling (85)

* The level of noise is measured in decibels (dB), after adjusting the decibel to human hearing the noise level is described in decibels A (dBA).

TABLE 5.10

Type of Operation/Machinery and Noise Level

Type of machinery	Noise level (dBA)*
Glass bottling	85–100
Product impact on hoppers	90–100
Wrapping, cutting wrap, bagging, etc.	85–95
Choppers	95 or more
Pneumatic noise and compressed Air	85–95
Milling	85–100
Saws/cutting machinery	85–107
Blast chillers	85–107
Packaging machinery	85–95
Trolley racks up to	107

* The level of noise is measured in decibels (dB), after adjusting the decibel to human hearing the noise level is described in decibels A (dBA).

5.5.1 Hearing problem and noise level

Improving the hearing level and preventing noise will actually improve not only your employee's health and safety but also increase the production and efficiency. American Chemical Society refers to normal noise level to be below 140 dBA and recommend earmuffs rather than earplugs for protection.[244]

134   Occupational Health and Safety in the Food and Beverage Industry

Generally, earmuffs and earplugs have a greater attenuation factor than earplugs. Some people have experienced discomfort at sound levels below 90 dBA.

5.5.2 FOOD AND BEVERAGE INDUSTRY NOISE HAZARDS

High impact industries are those that cannot be located near non-industrial uses due to their site emissions such as smoke or noise.

Bottling or canning facilities (200 tons of production or more) or meat processing including rendering (1,000–50,000 tone/year) are examples as such and following table can provide a comparable noise level for various activities.[241,242]

Table 5.10 demonstrates noise level measured at processing machinery and operation involved[245]:

Noise can be defined as levels of sound associated with risk of adverse health effects:

- Auditory effects such as:
    - D. Temporary hearing loss due to exposure to a high level of noise,
    - E. Acoustic trauma due to a brief very loud noise,
    - F. Permanent hearing loss due to exposure to continuous exposure to noise,
    - G. Tinnitus (ringing or buzzing in the ear),
- Possible non-auditory effects such as:
    - D. A. annoyance, irritability, headache, or depression,
    - E. Potential high blood pressure.

5.5.3 IDENTIFYING AND ASSESSING NOISE HAZARDS

1. Walk-through survey to determine potential noise problem.
2. Indicators:

D. People have to raise their voice to talk to someone at one meter away,
E. At the end of a shift, workers have to increase the volume of their radio or TV to a level loud for others.

3. After working for a few years at a workplace, workers have difficulty communicating in a crowd where there are other sounds or many voices.
4. Noise measurement data from studies in similar situations are helpful.
5. Determine actual noise measurement: field survey measurements with a sound level meter, Type II (noise mapping).
6. Personal noise exposure with a noise dosimeter.

5.5.4 NOISE PREVENTIVE MANAGEMENT AND CONTROL

This can be achieved through proper Hearing Conservation Programs as explained below.

Occupational Hazard Origin and Preventive Measures

5.5.5 HEARING CONSERVATION PROGRAM (HCP)

A HCP is a planned, coordinated course of action designed to prevent hearing loss. Effective programs do more than prevent hearing loss. They can:

1. Improve employee morale,
2. Create a sense of well-being,
3. Improve production,
4. Reduce incidence of stress-related disease,
5. Provide base-line measurements for future reference.

HCP include the following control measures: engineering, administrative, PPE, training, and audiometric testing

5.5.5.1 Engineering
- Replace noisy equipment parts, install flexible mounts,
- Enclose noise source or isolate worker from noise source,
- Install noise barriers or sound absorption material,
- Use muffler or silencers on equipment.

5.5.5.2 Administrative
- Plan layout so that noisy areas isolated,
- Change work schedules or rotate workers exposed to noise,
- Specify maximum noise levels on a new machinery (this is best handled through purchasing procedure),
- Consider visual warning sings as well as auditory ones.

Training, appropriate PPE, and periodically audiometric testing are important occupational health and safety consideration to minimize and or prevent noise hazard.

5.5.6 STUNNING ELECTRICAL EQUIPMENT

All electrical stunning equipment is potentially dangerous to staff. It should only be used by properly trained, skilled individuals. Particular care should be taken of the equipment, with regular checks and maintenance carried out by a qualified electrician. All electrical stunning equipment should work by the use of an isolated circuit, in which current flows preferentially between the two electrodes. If a person comes between the electrodes there is a danger of a fatal electric shock.[248]

5.5.6.1 Hazard prevention
1. Use safety switches or triggers so that the current only flows when the switch is held down by the operator.
2. Do not tape switches down so that the electrodes are permanently live.

136 Occupational Health and Safety in the Food and Beverage Industry

3. Have a pre-set timer which regulates the duration of the current flow. With some models of equipment the voltage returns to a low level between each stunning episode. This low voltage is used for immediately sensing the resistance between the electrodes and once detected as within pre-set limits, the stunner switches instantly to the higher stunning voltage which is applied for a pre-set duration.
4. House the control box in a separate area which is always kept dry.
5. Inform operators, via warning lights of the state of the equipment, ready, stun on, stun complete, etc.

5.5.6.2 Cleaning and storage

1. Electrodes should be cleaned regularly to ensure that there is minimal resistance.
2. Clean with a wire brush, a powered wire wheel, or place the electrodes in a cleaning station, after every 20–25 animals.
3. When not in use, tongs should be stored in a dry environment with the electrodes protected from damage.
4. Between stunning operations and/or when filling pens, tongs should be seated on a mounted wall-bracket, or in a cleaning station.

5.5.6.3 How to use work equipment safely

Employers and those who have control over work equipment (e.g. those hiring out work equipment) have responsibilities for equipment provided for use at work.

Work equipment is almost any equipment used by a worker while at work including[249]:

1. machines such as circular saws, drilling machines, photocopiers, mowing machines, tractors, dumper trucks, and power presses,
2. hand tools such as screwdrivers, knives, hand saws and meat cleavers also portable and electric or air powered tools,
3. lifting equipment such as lift trucks, elevating work platforms, vehicle hoists,
4. lifting slings and bath lifts,
5. other equipment such as ladders and water pressure cleaners.

5.5.6.4 Vibration

Vibration transmitted through the seat or feet (known as whole-body vibration or WBV)

Drivers of some mobile machines, including certain tractors, forklift trucks, and quarrying or earth-moving machinery, may be exposed to WBV and shocks, which are associated with back pain. Other work factors, such as posture and heavy lifting, are also known to contribute to back problems for drivers, however further study is needed into the impact of WBV.[255]

Vibration could affect:

Occupational Hazard Origin and Preventive Measures

1. Hand–arm vibration syndrome (HAVS). This is often noticed in worker who uses handheld vibrating tools. It is a chronic disorder that can lead to permanent disability, with its effect lasting from few months to several years.
2. Carpal Tunnel Syndrome. It has been linked to frequent use of smaller handheld tools such as a computer mouth.
3. Whole-body vibration. Workers near vibrating tables and lift trucks, etc. are subject to this chronic whole-body vibration transmitted through the seat or floor of their workstation.[267]

5.5.6.5 Pressure vessels and pipelines

Pressure vessel systems could be a source of occupational hazard, and some examples are listed below[256]:

1. boilers and steam heating systems,
2. pressurized process plant and piping,
3. compressed air systems (fixed and portable),
4. pressure cookers, autoclaves, and retorts,
5. heat exchangers and refrigeration plant,
6. valves, steam traps and filters,
7. pipework and hoses, and
8. pressure gauges and level indicators.

Certain hazard could be associated with the following factors:

1. poor equipment and/or system design,
2. poor maintenance of equipment,
3. an unsafe system of work,
4. operator error, poor training/supervision,
5. poor installation, and
6. inadequate repairs or modifications.

The main causes and some examples of hazards could be associated with

1. Impact from the blast of an explosion or release of compressed liquid or gas,
2. Impact from parts of equipment that fail or any flying debris,
3. Contact with released liquid or gas, such as steam, and
4. Fire resulting from the escape of flammable liquids or gases.

5.5.6.5.1 Reduce the risk of failure

The level of risk from the failure of pressure systems and equipment depends on a number of factors including, the pressure in the system, the type of liquid or gas and its properties, the suitability of the equipment and pipework that contains it, the

138 Occupational Health and Safety in the Food and Beverage Industry

age and condition of the equipment, the complexity and control of its operation, the prevailing conditions (e.g. a process carried out at high temperature); and the skills and knowledge of the people who design, manufacture, install, maintain, test, and operate the pressure equipment and systems.

5.5.6.5.2 Provide safe and suitable equipment

1. When installing new equipment, ensure that it is suitable for its intended purpose and that it is installed correctly. This requirement can normally be met by using the appropriate design, construction with installation standards and codes.
2. The pressure system should be designed and manufactured from suitable materials. You should make sure that the vessel, pipes, and valves have been made of suitable materials for the liquids or gases they will contain.
3. Ensure the system can be operated safely without having to climb or struggle through gaps in pipework or structures, for example.
4. Be careful when repairing or modifying a pressure system. Following a major repair and/or modification, you may need to have the whole system re-examined before allowing the system to come back into use.

5.5.6.5.3 Know the operating conditions

1. Know what liquid or gas is being contained, stored, or processed, for example is it toxic/flammable?
2. Know the process conditions, such as the pressures and temperatures.
3. Know the safe operating limits of the system and any equipment directly linked to it or affected by it.
4. Ensure there is a set of operating instructions for all the equipment and for the control of the whole system including emergencies.
5. Ensure that appropriate employees have access to these instructions and are properly trained in the operation and use of the equipment or system (see the section on training below).

5.5.6.5.4 Fit suitable protective devices and ensure they function properly

1. Ensure suitable protective devices are fitted to the vessels, or pipework (e.g., safety valves and any electronic devices which cause shutdown when the pressure, temperature, or liquid or gas level exceed permissible limits).
2. Ensure the protective devices have been adjusted to the correct settings.
3. If warning devices are fitted, ensure they are noticeable, either by sight or sound.
4. Ensure protective devices are kept in good working order, at all times.
5. Ensure that, where fitted, protective devices such as safety valves and bursting disks discharge to a safe place.
6. Ensure that, protective devices cannot be altered except by an authorized person.

Occupational Hazard Origin and Preventive Measures

5.5.6.5.5 Carry out suitable maintenance

1. All pressure equipment and systems should be properly maintained. There should be a maintenance program for the system as a whole. It should take into account the system and equipment age, its uses, and the environment.
2. Look for tell-tale signs of problems with the system, such as a safety valve that repeatedly discharges. This could be an indication that either the system is over-pressurizing, or the safety valve is not working correctly.
3. Look for signs of wear and corrosion.
4. Systems should be depressurized before maintenance work is carried out.
5. Ensure there is a safe system of work, so that maintenance work is carried out properly and under suitable supervision.

5.6 ELECTRICAL SAFETY

Therefore recognizing, evaluating, and controlling hazards stands to be the most effective way.[259] A mechanical device that physically prevents the transmission or discharge of energy, including but not limited to the following[250]:

1. A manually operated electrical circuit breaker,
2. A disconnect switch,
3. A manually operated switch by which the conductors of a circuit can be disconnected from all ungrounded supply conductors and in addition no pole can be operated independently,
4. A line valve, a block, or any other device used to isolate energy.

Push buttons, selector switches, and other control circuit type devices **are not** energy isolating devices.

Increasing OSHA regulation and industry's desire to reduce accidents, injuries, and related costs has focused interest on improving industrial electrical safety performance. Efforts to improve the safety of personnel exposed to industrial electrical hazards may be considered part of an overall strategy to eliminate defects in manufacturing processes. This chapter presents a blueprint for strategy, design, and implementation of processes to link electrical safety to total quality improvement.[258]

5.6.1 ARC FLASH HAZARD

Arc flash is a dangerous condition associated with the unexpected release of tremendous amount of energy caused by an electric arc within electrical equipment. This release is in the form of intense light, heat, sound, and blast of arc products that may consist of vaporized components of enclosure material – copper, steel, or aluminum. Intense sound and pressure waves also emanate from the arc flash, which resembles a confined explosion. Arcing occurs when the insulation between the live conductors breaks down, due to aging, surface tracking, treeing phenomena, and due to human error when maintaining electrical equipment in the energized state.[553]

140   Occupational Health and Safety in the Food and Beverage Industry

5.6.2 PROTECTION AGAINST ELECTRICAL HAZARDS

As most electrical accidents result from one of the following three factors:

1. Unsafe equipment or installation,
2. Unsafe environment, or
3. Unsafe work practices.

United State Department of labor recommend asking following questions to better plan electrical safety hazards.[258]

What is the best way to protect yourself against electrical hazards?
What protection does insulation provide?
How do you identify different types of insulation?
What is guarding and what protection does it offer?
What is grounding and what protection does it offer?
What are circuit protection devices and how do they work?
What work practices help protect you against electrical hazards?
How can you protect yourself against metal parts that become energized?
How can you prevent an accidental or unexpected equipment startup?
How can you protect yourself from overhead power lines?
What protection does personal equipment offer?
What role do tools play?
What special training do employees need?
What's the value of a safety and health program in controlling electrical hazards?

Therefore, electrical accidents are largely preventable through safe work practices. Examples of these practices include the following:

1. deenergizing electric equipment before inspection or repair,
2. keeping electric tools properly maintained,
3. exercising caution when working near energized lines, and
4. using appropriate protective equipment.

Electrical shock is the sudden and accidental simulation of the body's nervous system by an electric current. Current will flow through the body when it becomes a part of an electric circuit which has a potential difference adequate to overcome the body's resistance to current flow. The causes of shock could be[257]:

1. Contact with a normally bare energized conductor.
2. Contact with an energized conductor on which the insulation has deteriorated or has been damaged so that it has lost its protective value.
3. Equipment failure, causing an open or short circuit.
4. Static electricity discharge.
5. Lightning strike.

Occupational Hazard Origin and Preventive Measures 141

Electrical hazards effects include thermal, arching, and sparking causes, accidental contact with live circuit, inadvertent activation of the product or a device, electrical system failure, explosion, lightning strike, malfunctions, and short circuiting.

People are injured when they become part of the electrical circuit. Humans are more conductive than the earth (the ground we stand on) which means if there is no other easy path, electricity will try to flow through our bodies.

There are four main types of injuries: electrocution (fatal), electric shock, burns, and falls. These injuries can happen in various ways[251]:

1. Direct contact with exposed energized conductors or circuit parts. When electrical current travels through our bodies, it can interfere with the normal electrical signals between the brain and our muscles (e.g., heart may stop beating properly, breathing may stop, or muscles may spasm).
2. When the electricity arcs (jumps, or "arcs") from an exposed energized conductor or circuit part (e.g., overhead power lines) through a gas (such as air) to a person who is grounded (that would provide an alternative route to the ground for the electrical current).
3. Thermal burns including burns from heat generated by an electric arc, and flame burns from materials that catch on fire from heating or ignition by electrical currents or an electric arc flash. Contact burns from being shocked can burn internal tissues while leaving only very small injuries on the outside of the skin.
4. Thermal burns from the heat radiated from an electric arc flash. Ultraviolet (UV) and infrared (IR) light emitted from the arc flash can also cause damage to the eyes.
5. An arc blast can include a potential pressure wave released from an arc flash. This wave can cause physical injuries, collapse your lungs, or create noise that can damage hearing.
6. Muscle contractions, or a startle reaction, can cause a person to fall from a ladder, scaffold, or aerial bucket. The fall can cause serious injuries.

5.6.3 ELECTRICAL HAZARD PREVENTION

Electric sparks and arcs occur in the normal operation of certain electrical equipment such as switches, brushes, and similar devices. They also occur during the breakdown of insulation on electrical equipment[552]. In general, following has been recommended as preventive measures when working[251]:

1. Inspect portable cord-and-plug connected equipment, extension cords, power bars, and electrical fittings for damage or wear before each use. Repair or replace damaged equipment immediately.
2. Always tape extension cords to walls or floors when necessary. Nails and staples can damage extension cords causing fire and shock hazards.

142 Occupational Health and Safety in the Food and Beverage Industry

3. Use extension cords or equipment that is rated for the level of amperage or wattage that you are using.
4. Always use the correct size fuse. Replacing a fuse with one of a larger size can cause excessive currents in the wiring and possibly start a fire.
5. Be aware that unusually warm or hot outlets may be a sign that unsafe wiring conditions exists. Unplug any cords or extension cords to these outlets and do not use until a qualified electrician has checked the wiring.
6. Always use ladders made with non-conductive side rails (e.g., fiberglass) when working with or near electricity or power lines.
7. Place halogen lights away from combustible materials such as cloths or curtains. Halogen lamps can become very hot and may be a fire hazard.
8. Risk of electric shock is greater in areas that are wet or damp. Install Ground Fault Circuit Interrupters (GFCIs) as they will interrupt the electrical circuit before a current sufficient to cause death or serious injury occurs.
9. Use a portable in-line GFCI if you are not certain that the receptacle you are plugging your extension cord into is GFCI protected.
10. Make sure that exposed receptacle boxes are made of non-conductive materials.
11. Know where the panel and circuit breakers are located in case of an emergency.
12. Label all circuit breakers and fuse boxes clearly. Each switch should be positively identified as to which outlet or appliance it is for.
13. Do not use outlets or cords that have exposed wiring.
14. Do not use portable cord-and-plug connected power tools with the guards removed.
15. Do not block access to panels and circuit breakers or fuse boxes.
16. Do not touch a person or electrical apparatus in the event of an electrical accident. Always disconnect the power source first.

5.6.4 Preventing power tool hazard

1. Switch all tools OFF before connecting them to a power supply.
2. Disconnect and lockout the power supply before completing any maintenance work tasks or making adjustments.
3. Ensure tools are properly grounded or double insulated. The grounded equipment must have an approved 3-wire cord with a 3-prong plug. This plug should be plugged in a properly grounded 3-pole outlet.
4. Test all tools for effective grounding with a continuity tester or a GFCI before use.
5. Do not bypass the on/off switch and operate the tools by connecting and disconnecting the power cord.
6. Do not use electrical equipment in wet conditions or damp locations unless the equipment is connected to a GFCI.
7. Do not clean tools with flammable or toxic solvents.

Occupational Hazard Origin and Preventive Measures

8. Do not operate tools in an area containing explosive vapors or gases, unless they are intrinsically safe and only if you follow the manufacturer's guidelines.

5.6.5 POWER CORD SAFETY

1. Keep power cords clear of tools during use.
2. Suspend extension cords temporarily during use over aisles or work areas to eliminate stumbling or tripping hazards.
3. Replace open front plugs with dead front plugs. Dead front plugs are sealed and present less danger of shock or short circuit.
4. Do not use light duty extension cords in a non-residential situation.
5. Do not carry or lift up electrical equipment by the power cord.
6. Do not tie cords in tight knots. Knots can cause short circuits and shocks. Loop the cords or use a twist lock plug.

5.6.5.1 Inerting and purging

Inerting and purging refer to replacing the atmosphere in a line, vessel, or other area with an inert atmosphere, e.g., the space above liquid fuel in a fuel tank, to reduce the likelihood of combustion.[830]

An **inert gas** is a gas which does not undergo chemical reactions under a set of given conditions. Purified nitrogen and argon gases are most commonly used as inert gases due to their high natural abundance (78% N_2, 1% AR in air) and low relative cost.[384]

Inserting or complete substitution of the air or flammable atmosphere by an inert gas such as nitrogen, CO_2, Argon, Helium, etc. is a very effective method to prevent explosion.[265] It is normally considered when the flammable or explosion hazard cannot be eliminated by other means. Typical uses are within "Storage Tanks" where:

1. A material may be above its flash point.
2. Within a system when excursions into flammable atmospheres may occur.
3. They are also used to transfer flammable liquids under pressure.

Inerting procedure is applicable to enclosed plant where there could be gas build up in a location and not for an open plant exposed to atmospheric conditions.

Risk with inserting could be:

1. Asphyxiation in confined space, and
2. Loss of inner atmosphere during maintenance or operation.

Purging is one of three main applications to improve food product quality by preventing food and drink oxidations or where food and drink are stored at or below ambient temperature 9beverages, oils, fruits, vegetables, finished products) or at elevated temperature such as deep-fat frying.[266]

5.6.5.2 Hazard preventive management

1. Review the manufacturer's operation and maintenance manual before use.
2. Before use check for loose part, damages, clean air filter, replace worn parts, etc. or lubricate if required and verify guard condition on the tools.
3. Check air pressure at the tool against the recommended one by manufacturer.
4. Provide proper PPE which could include items such as face shield, goggles, or safety glasses, aprons, footwear, helmet, and glove.
5. Inspect before use and maintain the equipment after use and store them clean.
6. Keep proper record of all required and recommended maintenance.
7. Use hoses specifically designed to resist abrasion, cutting, crushing, interior deterioration, and failure from continuous flexing.
8. Make sure connected hoses are equipped with automatic shutoff couplers or a small chain is attached to each side of the coupler to prevent whipping if the coupler fails.
9. Don't use chrome hand sockets on power wrenches which can shatter suddenly.
10. Avoid loose clothing and, also move away your hand and clothing from operating end.
11. Always use the tool handle and nowhere else for transporting or handling.
12. Shut off the air and release the pressure before disconnecting a hose.
13. Provide complete training program on all above issues for users and procedures to report problem, accidents for follow-up and improvement.

5.6.5.3 Compressed gases

Compressed gases are used in many industries and the food industry is no exception with demand for all types of produce to be supplied to every inhabited place on the globe. Regardless of season and location, everything from exotic tropical fruits to the staple diet of bread, rice, and potatoes are expected to be available all year round and in "just produced" condition at competitive, affordable prices.

Convenience and quick preparation of meals is also a high priority for the fast-paced 21st–century lifestyle. Attractively presented fresh or prepared foods and combination meals, in durable hygienic packages that offer useful shelf life under normal refrigeration, have become very popular.

Faced with these consumer preferences and growing demand for an ever-wider range of food products, retailers recognize the need for improvements in packaging technology. They need to address the spoilage problem and provide a huge diversity of new prepared foods.

Over the past three decades, a safe, tried, tested, and proven method of combating food spoilage mechanisms without the use of (or at least a substantial reduction in) undesirable preservatives, is the use of Modified Atmosphere Packaging. Sometimes referred to as "MAP" or "Gas Flushing". The shelf life of food produce is influenced by a number of factors including, Storage temperature, Quality of

Occupational Hazard Origin and Preventive Measures

raw ingredients, Product formulation, Processing method, Hygiene standards, and Packaging Material.

The main gases used are nitrogen (N_2), oxygen (O_2) and carbon dioxide (CO_2). All three occur naturally in the air we breathe but by using them individually or combining them in the food packaging process, very beneficial results can be achieved.[262]

Compressed gases are hazardous due to "high pressure" and may be hazardous because they could be: toxic, flammable (pyrophoric or explosive), reactive, and/or asphyxiant.

5.6.5.8.1 Characteristics

1. Can explode if heated or damaged.
2. Breakage of the head valve can cause the cylinder to become a missile capable of penetrating walls.
3. Sudden release of high-pressure gas streams may puncture the skin and cause fatal embolism.
4. Can displace air at a high rate and produce an oxygen deficient atmosphere or, depending on the properties of the gas, can produce a toxic, flammable, or corrosive atmosphere.

The food industry uses several types of other gases during processing at various stages such as ammonia, carbon dioxide, carbon monoxide, Freon, and chlorine[262] and as well as natural gas used for bakery production.[263]

5.6.5.8.2 Gas application in food and beverage industry

The use of gases in various processes involved are in the following processes as referred by Matheson Co. for[263]:

5.6.5.8.3 Blanketing

The use of gas to maintain an inert atmosphere above a liquid or powdered product inside a storage tank, silo, reactor, process equipment or other vessel. The inert gas will help prevent product degradation from moisture and oxygen, control volatile emissions, and safeguard against fires and explosions. Gases used are carbon dioxide and nitrogen.

5.6.5.8.4 Food chilling and freezing

Many industrial food and beverage facilities use ammonia as a refrigerant but not all have the proper equipment and protocols in place to ensure workers are kept safe when refrigeration leaks occur.

The cryogenic properties of liquid gases are used for the freezing and chilling of food products. Liquid gases can achieve a lower temperature and provide a faster cool down rate than mechanical refrigeration systems. The use of cryogenic spiral, tunnel, immersion, or batch freezer for food freezing, or snow horn or bottom injection system for chilling, will generally result in increased production, reduced dehydration losses, reduced bacteriological activity, increased shelf life, and improved efficiency versus mechanical freezers. Gases used are carbon dioxide, nitrogen.

Food and beverage workers have the right to workplace safety, and as ammonia is toxic to humans, even in low PPM levels, personal protection equipment (PPE) is needed when working with ammonia. Here is a quick overview of the impact ammonia can have on humans who breathe in anhydrous ammonia:

(1) Low levels (5 ppm) sharp odor is detectable.
(2) Moderate levels: irritation to the eyes and respiratory tract; may cause nausea as well as vomiting.
(3) High concentration levels (above 50 ppm): possible ulcerations to the eyes and severe irritation to the respiratory tract.
(4) Extremely high concentrations (300–500 ppm): potentially fatal; can cause fluid build-up in the lungs and severe shortness of breath.[806]

While the refrigerant is contained in the system, it presents no hazard to people and the environment. A hazard may be presented when there is an uncontrolled, unexpected release of refrigerant from the system. These containment failures, or refrigerant leaks, are usually the result of one of two conditions:

1) The refrigerant pressure has increased above the normal operating pressure of the container, component, or system.
2) A component, piping and/or system was compromised such that it will no longer hold the operating refrigerant pressure.[805]

5.6.5.8.5 Controlled atmosphere storage

A technique used to preserve fresh fruits and vegetables or enhance ripening by modifying the gaseous composition of the atmosphere in which the food products are stored. The atmosphere is closely monitored and precisely controlled to maintain the desired gas proportions of oxygen, nitrogen, carbon dioxide, and ethylene throughout the storage period. Gases used are carbon dioxide, nitrogen, oxygen, and ethylene

5.6.5.8.6 Grain fumigation

The use of oxygen deficient atmosphere to kill 100% of all life stages (eggs, larvae, pupae, and adults) of insects which have infested agricultural commodities such as grain, flour, oats, rice, and tobacco in silos and warehouses. The process requires that the commodity remain sealed in a carbon dioxide or nitrogen atmosphere for at least four days. Gases used are carbon dioxide and nitrogen.

5.6.5.8.7 Modified atmospheric packaging (MAP)

The use of pure gas and/or gas mixtures to prolong the shelf life of raw fresh and processed foods by retarding/inhibiting the spoilage mechanisms that affect product quality. During packaging the air is purged or evacuated from the interior of the package, replaced with a gas or gas mixture, and quickly sealed. Selection of the correct gas or gas mixture is dependent on such factors as food product, whether it is

Occupational Hazard Origin and Preventive Measures

raw or processed and packaging material. Gases used are carbon dioxide, nitrogen, and mixtures.

5.6.5.8.8 PH control

The use of carbon dioxide to control the pH levels of alkaline wastewater, cooling water, drinking water, and industrial process water systems. Carbon dioxide, a mildly acidic gas, will lower the pH of alkaline effluents without storage, handling and safety problems associated with strongly acidic chemicals, such as hydrochloric acid or sulfuric acid. Gases used are carbon dioxide, nitrogen. Gas used is carbon dioxide.

5.6.5.8.9 Sparging

The use of nitrogen reduces the concentration of dissolved gas (usually oxygen) in a liquid, or to fluff a product. Oxygen removal generally results in increased shelf life and prevents oxidation, discoloration, and flavor and aroma loss. Gas used is nitrogen.

5.6.5.8.10 Undercover gassing

Nitrogen flushes out oxygen from the headspace of lidded containers immediately before sealing. This technique can extend the shelf life of oxygen sensitive products. Undercover gassing can be used with liquids, solids, and powders packaged in cans, jars, or bottles. Examples include canned nuts, canned condensed milk, and bottled beer where type of gas used is nitrogen.

Safety precautions could include:

1. Secure gas cylinders firmly in an upright position during storage, handling, and transport using suitable racks, straps, chains, or stands.
2. Use four-wheeled hand trucks or dollies for transporting gas cylinders, do not roll or drag.
3. Leave the valve safety cap on the cylinder until the regulator is attached.
4. Store gas cylinders in an area separated from lab, only gas cylinders connected to apparatus should be present in the laboratory.
5. Locate gas cylinders in a well-ventilated room separate from the lab, if flammable gases are involved the separating wall should have a minimum fire resistance rating of one hour.
6. Verify that the regulator, its maximum pressure rating, and its connector are those specified by the manufacturer for the type of gas being used and the pressure to be delivered.
7. Never use adaptor to fit a regulator designed for one type of gas on a different type of gas cylinder.
8. Tighten regulators using the proper size of wrench.
9. Remove regulators from service if they do not thread snugly into the main cylinder outlet, do not use lubrication or sealing tape to achieve a good seal.
10. Back off the diaphragm valve completely before opening the control valve to prevent damage when the tank pressure is released.

11. Open valves slowly and stand to the side of the gauges.
12. Check the system for leak by applying a soap solution to all connectors.
13. Close the main cylinder valve and bleed residual pressure from the gas delivery system when the cylinder is not in use.
14. Remove the regulator and attach the valve safety cap for storage or transport.
15. Mark or tag empty cylinders, replace the valve safety cap, store secure in an upright position preferably separated from full cylinders for return to supplier.
16. Before installing cylinders of flammable gases or liquids (e.g. liquefied propane) check with local fire codes for installation and inspection requirement.

5.7 THERMAL HAZARDS

Most burns are minor injuries that occur at home or work. It is common to get a minor burn from hot water, a curling iron, or touching a hot stove. Home treatment is usually all that is needed for healing and to prevent other problems, such as infection.[385]

Thermal hazards are either due to:

A. Burn hazard due to heat, electrical, chemical, or radioactive burns.

Burn has been discussed previously and refers to heat, cold, electrical, radiation, and/or friction – burn. Besides breathing in hot air or gases can injure your lungs (inhalation injuries). Breathing in toxic gases, such as carbon monoxide, can cause poisoning. The deeper the burn and the larger the burned area, the more serious the burn is.[385]

B. Cold/cryogenic burn

Liquid nitrogen (LN2) is a cryogenic liquid commonly used in a variety of cooling applications such as food freezing, biological sample preservation, metal treatment and lesion removal (cryotherapy). More recently use of LN2 has become popular in the retail food and beverage industry for the preparation of novelty ice creams and cocktails. The nitrogen gas produced is a simple asphyxiant capable of displacing oxygen in air, which can lead to dizziness, nausea, vomiting, unconsciousness, or death.[749]

LN2 can also cause cold burns, frostbite, or eye damage on contact. Bare skin can stick to objects cooled with LN2. Substances such as rubber or carbon steel can become brittle and shatter. Tremendous force can be generated with rapid vaporization.

LN2 can also cause cold burns, frostbite, or eye damage on contact. Bare skin can stick to objects cooled with LN2. Substances such as rubber or carbon steel can become brittle and shatter. Tremendous force can be generated with rapid vaporization. For more prevention and management refer to injury management in food and beverage chapter.[749]

Occupational Hazard Origin and Preventive Measures

Ammonia while corrosive to the eyes may also cause blindness, skin irritation. If ingested causes burn of the upper digestive and respiratory tract. Fume inhalation can lead to coughing, choking, dizziness, weakness for several hours, or even death.[805]

5.7.1 CRYOGENIC HAZARD PREVENTION

1. Only use and store LN2 in adequately ventilated areas away from moisture and chemicals that may promote container corrosion.
2. Provide an oxygen monitoring system in areas where LN2 is stored and used, and where oxygen displacement can occur.
3. Avoid skin and eye contact with LN2. Use personal protective equipment, including loose fitting insulated (cryogenic) or loose fitting leather gloves (for quick removal), goggles and a face shield, long-sleeved shirt (or arm protection), pants without cuffs and closed-toe shoes or boots. Pants should not be tucked into boots or shoes. Canvas shoes should not be worn as they can absorb LN2.
4. Use only those containers designed for storage (e.g., cryogenic storage tank or liquid cylinder) of LN2.
5. Use proper transfer equipment (Dewar) and dispense slowly from one container to another to prevent splashing, thermal effects, and pressure build-up. Always follow the manufacturer's instructions regarding transfers. Use a dolly or handcart when moving containers.
6. Develop and implement an emergency plan and first-aid measures.

Train workers on the hazards of working with LN2, engineering controls, use of personal protective equipment, work practices, emergency response procedures, first aid, and the use and calibration of oxygen monitoring equipment.

LN2 can also cause cold burns, frostbite, or eye damage on contact. Bare skin can stick to objects cooled with LN2. Substances such as rubber or carbon steel can become brittle and shatter. Tremendous force can be generated with rapid vaporization.[749]

5.7.2 CONFINED SPACE HAZARD

A confined space is an enclosed or partially enclosed area that is big enough for a worker to enter. It is not designed for someone to work in regularly, but workers may need to enter the confined space for tasks such as inspection, cleaning, maintenance, and repair.

Confined space can carry biological, chemical, physical, and or a combination of these hazards associated with it. There could be poor lighting, high temperature, toxic chemicals, microorganisms (sewage) and danger of solid, liquid, or gas presence.

Entry into confined spaces can be very hazardous. Unless proper training, equipment, and procedures are in place, workers must not be allowed to enter such spaces.[264]

A worker is considered to have entered a confined space just by putting his or her head across the plane of the opening. If the confined space contains toxic gases, workers who are simply near the opening may be at risk. Often the toxic gases are under pressure because of heat inside the confined space or when gases are generated inside the space. As a result, the concentration of toxic gases near the entrance to the confined space can be high enough to cause death.

It is vital to identify all confined spaces in any workplace. Any area that can have a "confined" atmosphere.

In addition, an enclosed area in which workers do not normally work could have air that may be hazardous to breath or the work activity could cause a hazardous atmosphere to develop. Even though rescue would not be complicated by the design of the space (and it does not therefore fit the definition of a confined space), the space may still be deadly without air testing and a supply of clean air.[264]

Some examples of confined spaces in the food and beverage industries are process vessels, storage tanks and bins, silos, grape presses and crushers, vats, fermentation tanks, utility vaults and vessels, pits, sumps, well as wells other sources.[263,264]

5.7.3 THE RISKS

Incidents in confined spaces are not common, but when they do occur the consequences can be devastating. Confined space incidents can happen suddenly, often without any warning that something is wrong. Incidents involving atmospheric hazards (for example, toxic gases or a lack of oxygen) in confined spaces often cause serious injury or death to more than one person.[623]

Employers: As an employer, you are responsible ensuring the health and safety of your workers. This involves identifying confined space hazards in your workplace and taking the necessary steps to protect workers. Learn more about the five steps to managing confined space hazards in the workplace. Hazards in confined spaces may not be obvious, so a qualified person – someone who has proper training and experience – must look carefully at every confined space in a workplace to identify possible hazards.

Workers: As a worker, there are three steps you can take to protect yourself and others from confined space hazards. Learn more about the three steps to working safely around confined spaces.

5.7.4 HAZARDOUS ATMOSPHERES

The atmosphere in a confined space may be hazardous for several reasons. The air may have too little or too much oxygen.

Once a confined space is identified, its atmosphere must be hazard-rated as HIGH, MODERATE, or LOW. The hazard rating of a confined space must be determined by

Occupational Hazard Origin and Preventive Measures

a qualified person after considering the design, construction, and use of the confined space, the work activities to be performed, and all required engineering controls.

5.7.4.1 High-hazard atmosphere

An atmosphere that may expose a worker to risk of death, injury, or acute illness, or otherwise impair a worker's ability to escape unaided from a confined space if the ventilation system or respirator fails.

5.7.4.2 Moderate-hazard atmosphere

An atmosphere that is not clean, respirable air but is not likely to impair a worker's ability to escape unaided from a confined space if the ventilation system or respirator fails.

5.7.4.3 Low-hazard atmosphere

An atmosphere that is shown by pre-entry testing or is otherwise known to contain clean, respirable air immediately prior to entry into a confined space, and that is not likely to change during the work activity.

5.7.5 HAZARD ALERT

Example: A beverage warehouse worker was preparing to re-clean a beverage storage tank. This tank had been previously cleaned and purged with nitrogen gas. The worker crawled through a small manhole to position the spray ball in the center of the tank in order to spray the inside walls with detergent. After approximately five minutes in the tank, he was found unconscious and could not be revived.

The nitrogen had displaced oxygen in the air inside the tank, and there was not enough oxygen to sustain life. The air in the tank was not tested for oxygen content before entry.

5.7.5.1 Oxygen: too little or too much

Lack of oxygen is a leading cause of death among workers entering confined spaces. Low oxygen levels cannot be detected by sight or smell. You must test the air for this hazardous condition. A very low level of oxygen can damage the brain and cause the heart to stop after a few minutes.[264]

5.7.5.2 Toxic atmospheres

Contaminants in the air can result in an atmosphere that is toxic to workers and may result in injury or death [264].

In the past, miners would take canaries down into coal mines, since these small birds react quickly to carbon monoxide, a deadly gas. If the canaries breathed a small amount of the gas, they would sway on their perches before falling. This gave miners warning that the deadly gas was present. Today, miners have monitors to let them know when there are toxic substances in the atmosphere.

The concentration of the substance inside the confined space must be determined using a recently calibrated and properly set up air monitor with the correct sensor.

Such an air monitor may sound an alarm that will alert the worker before the allowable exposure limit is reached.

In most cases, mechanical ventilation such as fans must be used to ventilate the space, bringing in clean outside air. Additionally, the harmful substance must be eliminated wherever practicable. Air testing and ventilation are the best ways to ensure that workers are not placed at risk from hazardous atmospheres.

5.7.5.3 Explosive atmospheres

Three elements are necessary for a fire or explosion to occur: oxygen, flammable material (fuel), and an ignition source.

5.7.5.4 Oxygen

Air normally contains 20.9% oxygen, enough oxygen for a fire. However, a higher level of oxygen increases the likelihood of material burning. Air is considered oxygen-enriched at levels above 23%. Enrichment can be caused by improper isolation of oxygen lines, ventilation of the space with oxygen instead of air, or leaks from welding equipment.

5.7.5.5 Flammable material (Fuel)

Fires and explosions in confined spaces are often caused by gases or vapors igniting. Coal dust and grain dusts may explode when a certain level of dust in the air is reached.

Two or more chemicals may react with each other and become explosive.

Containers of fuels such as gasoline and propane should not be taken into a confined space as fuel can easily burn or explode.

Here are some other common substances that can cause explosions or fires in confined spaces:

1. Acetylene gas from leaking welding equipment,
2. Methane gas and hydrogen sulfide gas produced by rotting organic wastes in sewers or tanks,
3. Hydrogen gas produced by contact between aluminum or galvanized metals and corrosive liquids,
4. Grain dusts, coal dust,
5. Solvents such as acetone, ethanol, toluene, turpentine, and xylene, which may have been introduced into the space through spills or by improper use or disposal.

A trained person must test the atmosphere for gases and vapors that will burn or explode. You cannot always see or smell these dangerous gases and vapors. If any measurable explosive atmosphere is detected, the air must be further evaluated by a qualified person to ensure that it is safe to enter the confined space.

Occupational Hazard Origin and Preventive Measures

5.7.5.6 Ignition sources

Ignition sources include: open flames, sparks from metal impact, welding arcs, arcing of electrical motors, hot surfaces, discharge of static electricity, lighting, and chemical reaction.

Many processes can generate static charge, including steam cleaning, purging, and ventilation procedures. To reduce the risks from these ignition sources, use non-sparking tools and ensure all equipment is bonded or grounded properly.

Examples of some gases that may be in your workplace are indicated in Table 5.11.[264]

TABLE 5.11
The Smell of Gases and Their Main Danger in Workplaces

Contaminants	What is main danger?	What smell look like
Argon (Ar)	Displaces oxygen may accumulate at bottom	Colorless, odorless
Carbon dioxide (CO_2)	Toxic, displaces oxygen may accumulate at bottom	Colorless, odorless
Carbon monoxide (CO)	Toxic-asphyxiant (causing suffocation)	Colorless, odorless (NO WARNING)
Chlorine (Cl_2)	Toxic-lung and eye irritant may accumulate at the bottom	Greenish yellow color; sharp pungent odor
Gasoline vapors	Fire and explosion may accumulate at bottom	Colorless; sweet odor
Hydrogen sulfide (H_2S)	Extremely flammable very toxic – causes lung failure, may accumulate at the bottom	Colorless; rotten egg odor*
Methane (CH_4)	Fire and explosion may accumulate at the top	Colorless, odorless (NO WARNING)
Nitrogen (N_2)	Displaces oxygen	Colorless, odorless (NO WARNING)
Nitrogen dioxide (NO_2)	Toxic – severe lung irritant may accumulate at the bottom	Reddish brown; pungent odor
Sulfur dioxide (SO_2)	Toxic – severe lung irritant accumulates at the bottom	Colorless; rotten, suffocating odor
Oxygen (O_2)	Oxygen at low levels: asphyxiant and at high levels combustion and explosion	Colorless and odorless

5.8 PHYSICAL HAZARDS

5.8.1 LOOSE AND UNSTABLE MATERIALS

Bins and hoppers in which materials are conveyed or augured into the bin are particularly dangerous. A worker may be trapped or crushed when material is accidentally discharged into an empty bin or hopper.[264]

154 Occupational Health and Safety in the Food and Beverage Industry

The design of these confined spaces may increase the danger of being trapped or buried. For example, in an empty hopper with a floor that slopes steeply to a vertical chute, a worker can slide into the chute and become trapped there.

Wherever there are loose, unstable materials that could trap or bury you, a qualified person must inspect the space and assess the hazards. Do not enter until the hazard has been eliminated or controlled. Specific training and safety precautions must be in place before you enter.

5.8.2 SLIP, TRIP, AND FALL HAZARDS

The space you are about to enter may have a hatchway that is difficult to squeeze through, and ladders for ascending or descending. You are therefore at risk of falling while getting into the space as well as while you are inside. In addition, the flooring of tanks or other wet environments or the rungs of a ladder may be very slippery.

If the hazard cannot be eliminated and there is a danger of falling from a height, a fall protection system (such as guardrails or a harness and lifeline) may be needed.

5.8.3 FALLING OBJECTS

In a confined space there may be the danger of being struck by falling objects such as tools or equipment, particularly if access ports or workstations are located above workers.

If workers might be exposed to the hazard of falling objects, safe work procedures must be put in place to prevent this.

5.8.4 MOVING PARTS OF EQUIPMENT AND MACHINERY

Mechanical equipment such as augers, mixers, or rotating tanks can be dangerous if activated or not secured. Residual energy, such as gravity or accumulated pressure, may also pose a risk unless the equipment is locked out and de-energized. This must be done by following a written lockout procedure that is specific for each piece of equipment and that states each place where a lock must be applied. Even when the power is shut off and the equipment is locked out at control points, unsecured equipment can move, especially if it is out of balance.

Before doing any work in confined spaces. Shut the power off, ensure that the equipment is locked out at control points, test the lockout, and secure any equipment that can move, even when it has been locked out.

5.8.5 ELECTRICAL SHOCK

Electrical shock can result from defective extension cords, welding cables, or other electrical equipment. Work done in metal enclosures or in wet conditions can be particularly dangerous. Install ground fault circuit interrupters or use assured grounding

Occupational Hazard Origin and Preventive Measures   **155**

where there may be a danger of electrical shock. All electrical sources that pose a hazard to workers inside the space must be locked out following the written lockout procedure for the particular confined space.

5.8.6 SUBSTANCES ENTERING THROUGH PIPING

Piping adjacent to a confined space could contain liquids or gases or other harmful substances. If these substances enter the confined space, the hazards may include, toxic gases; (a) Burns from hot substances, (b) Drowning, and (c) Being trapped, crushed, or buried.

Substances must be prevented from entering the confined space through piping. This is done by "isolating" the piping from the confined space.

5.8.7 POOR VISIBILITY

Poor visibility increases the risk of accidents and makes it harder for a standby person to see a worker who may be in distress. If poor visibility results from inadequate lighting, the light levels should be increased (although area lighting is not always required).

If activities such as sandblasting or welding result in poor visibility, appropriate ventilation may be needed to reduce harmful substances in the air.

If portable lighting is used where there may be an explosive atmosphere, the lighting must be "explosion-proof". (The Canadian Electrical Code has a description of lighting that is approved for use in explosive atmospheres.)

Emergency lighting such as flashlights or battery-operated area units must be provided where necessary, so that workers can locate exits and escape.

5.8.8 TEMPERATURE EXTREMES

Special precautions are needed before workers enter equipment such as boilers, reaction vessels, and low-temperature systems. A qualified person must provide these procedures. Allow enough time for cooling of confined spaces that have been steam cleaned.

5.9 NOISE

Noise produced in confined spaces can be particularly harmful because of reflection off walls. Noise levels from a source inside a small confined space can be up to ten times greater than the same source placed outdoors. If the noise levels cannot be reduced, proper hearing protection must be worn where necessary.

5.9.1 RISK OF DROWNING

Confined spaces should be fully drained or dry when entered. Spaces that are not fully drained or dry may pose a risk of drowning. The risk of drowning in a vat or tank with a large amount of liquid is easily recognized. However, workers have drowned in small pools of liquid. For example, insufficient oxygen, the presence of a toxic gas, or a blow to the head can make workers unconscious. Workers who have fallen face down into a small pool of water have drowned.

5.9.2 CONFINED SPACE ENTRY PROGRAM

Before workers perform work in a confined space, the employer must prepare and implement a written confined space entry program.

The identification, evaluation, and control of confined space hazards are often quite complex. For assistance in assessing the hazards and preparing a written confined space entry program, consult a qualified occupational health and safety professional.

The health and safety professional can tell you what to do to make it safe to enter a confined space, including providing the appropriate air-testing equipment, and explaining the portable air-moving device and the personal protective equipment to be used.

The confined space entry program must include the following:

1. An assignment of responsibilities.
2. A list of each confined space or group of similar spaces, and a written hazard assessment of those spaces prepared by a qualified person.
3. Written safe work procedures for entry into and work in each of the confined spaces. Each procedure must be written specifically for each of the hazards that exist in each space during each entry.
4. The equipment necessary for each entry must also be provided, including testing devices, air-moving devices, isolation and lockout devices, and personal protective equipment.
5. A signed permit where required.
6. Training of employees.
7. A rescue plan.

5.9.3 IDENTIFYING AND ASSESSING CONFINED SPACE HAZARDS

Some factors to be considered when determining whether a space is or is likely to become confined:

5.9.4 CONSTRUCTION

1. How many sides of the space are enclosed?
2. What are the number, size, and location of openings, doors, vents, etc.?
3. How would you get out in an emergency?

Occupational Hazard Origin and Preventive Measures

5.9.5 LOCATION

1. Is the space above or below normal floor level?
2. Airborne contaminants can be lighter or heavier than air and can accumulate in the air.
3. Consider nearby operations as a source of contaminants?

5.9.6 CONTENTS

1. What is the nature of the material stored or contained in the space?
2. Do they give off flammable or toxic gases, fumes, dusts, vapors, mists, smoke, or radiation?

5.9.7 WORK ACTIVITY

1. Is the work to be done in the space likely to use up oxygen?
2. Will the activity generate heat, cold, noise, toxic, or flammable materials?
3. Will conditions change for any reason?

5.9.8 HAZARDOUS ATMOSPHERE

- Is there likely to be a hazardous atmosphere such as oxygen deficient, enriched, flammable/explosive, toxic gases, vapor dust fumes, etc.?

5.9.8.1 Preventive management and hazard control

Before a worker is permitted to enter a confined space, it must be made safe for entry and for as long as he or she remains there. If it is not possible to eliminate all actual and potential hazards, special precautions must be taken.

5.9.8.2 Pre-entry precautions
5.9.8.3 Testing the atmosphere

Before a worker enters a confined space, it must be tested, by a competent person, for oxygen level, flammability, and toxicity. Results must be recorded in a permanent record and must certify that the space is free from hazard and will remain so for as long as the worker is in the space.

5.10 PURGING

Purging with an inert gas may be necessary to reduce the risk of a fire or explosion inside or outside the space. Assess the conditions prior to entering the space (e.g. oxygen).

Ventilation may be required for any of the following reasons to:

- Provide the air necessary for normal breathing,
- Lower the concentrations of contaminants below hazardous levels,
- Heat or cool the space,
- Improve visibility.

5.11 BLANKING-OFF PROCEDURES

All line and systems that may carry hazardous materials into the confined space must be blanked off. Merely shutting off or locking or locking the valve is not enough. The blank must be able to resist both the line pressure and the material corrosiveness in the line.

5.12 CLEAN OUT

Cleaning, neutralizing, or de-scaling may be necessary to remove residual toxic or corrosive materials and prevent slippery surfaces.

5.13 LOCKOUT PROCEDURES

All mechanical equipment in the confined space, such as agitators and pumps must be disconnected from the power sources at the disconnect box and the controls tagged and locked out (de-energized). Energy sources other than electricity, such as steam, compressed air, hydraulics, etc. must be de-energized.

5.14 FIRE SAFETY

Where fire or explosion hazard may occur, all ignition sources must be eliminated. All electrical equipment used in hazardous locations must meet the requirements of the Electrical Safety Code.

5.15 ELECTRICAL SAFETY

Electrical tools and equipment that are not double insulated must be grounded. Use ground circuit interrupters when working with electrical equipment in damp or wet conditions.

5.16 EMERGENCY PROCEDURES

Before entry into a confined space, consider:

1. Size and number of openings, ladders, ramps, and platforms,
2. Fall prevention and protection (e.g. safety belts and harness, etc.),
3. Rescue (e.g. observer, alarm, first aid, rescue breathing, etc.).

5.17 ENTRY PRECAUTIONS

PERSONAL PROTECTIVE EQUIPMENT

If hazardous substances and toxic residues may exist in the space, PPE may be required to supplement other measures, such as ventilation. Depending on the

Occupational Hazard Origin and Preventive Measures 159

hazard, the worker may have to wear a respirator, supplied air breathing apparatus, chemical goggles, hard hat, gloves, safety footwear, or complete body covering.

5.18 ACCESS AND EGRESS

The size and location of all openings should be noted in case the worker needs to evacuate the space. Hinged coves and doors should be secured in the open position.

5.19 ADMINISTRATION

A company policy indicating direction, commitment, and support is required if a confined space entry program is to be effective.

The policy directive should come from the employer and state your company's objectives to:

1. Identify all confined spaces,
2. Ensure that these spaces are made safe to enter and work in, and
3. Ensure that all necessary communication and training is carried out.

5.20 TRAINING

Your employee training program should include the followings:

1. Actual and potential hazards in the confined space.
2. Proper procedures and precautions for entering and working in the space.
3. Information on the physical characteristics of the space (such as number and location of openings, ducts, vents).
4. How to carry out lockouts, blanking off supply lines and atmospheric testing.
5. The use, care, limitations, and maintenance of PPE or other safety equipment.
6. Rescue procedures (such as observe entry or help with proper PPE, first aid and cardiopulmonary resuscitation (CPR) hoist, alarm).
7. No one including the observer should enter the space without appropriate PPE.

5.21 WORK PERMIT

A work permit is a written form used to authorize jobs that expose workers to serious hazards, such as work in confined spaces. A work permit identifies:

1. The specific type of work to be done,
2. Hazards involved,
3. Necessary preparations and precautions.

160 Occupational Health and Safety in the Food and Beverage Industry

A work permit serves as a checklist to ensure that all hazards have been identified and evacuated.

The permit should be signed by the supervisor and manager in control of the system. It should be discussed with and signed by the worker (s) doing the confined space work.

5.22 SPECIFIC HAZARDS AND ELEMENTS

5.22.1 WELDING

There are many potentially severe hazards from welding activities which include[261]: burns, fire, and explosion, electrical shock and burns, highly compressed gas in cylinders.

Noise and vibration, heat, fumes, dusts, vapors and gases, radiation or welding flash (eye damage). Gas cylinders are thus pressure vessels and must be constructed and maintained to high standards. They require periodic examination and testing, which must be carried out by the owner of the cylinder; in most cases this would be the gas supplier.[261]

Hazard control measures include: 1. Engineering control for proper ventilation for proper ventilation,[268] 2. Safe work practices, and 3. PPE.[268]

5.22.2 HAZARD ASSOCIATED WITH NANOTECHNOLOGY

Hundreds of tons of metals, rare earth, carbon nanotubes, and ceramic metals are yearly produced globally. The most common nanoparticles which are used in food or packaging are metallic oxides (titanium dioxide, silicon dioxide, antimony pentoxide), metal nanoparticles (silver, magnesium, zinc), and carbon nanotubes.

Usually, nanoparticle incorporation is done in such a way that the particles are not sticking together.[543] Nanotechnology can be used to enhance food flavor and texture, to reduce fat content, or to encapsulate nutrients, such as vitamins, to ensure they do not degrade during a product's shelf life.

In intelligent food packaging, incorporating nano-sensors, could even provide consumers with information on the state of the food inside.[544]

It is widely expected that nanotechnology-derived food products will be available increasingly to consumers worldwide in the coming years.[544]

Hazard protection could be enhanced by;

- improving regulatory oversight,
- making labeling mandatory,
- providing more funds for risk research and regulation, and
- considering safety legislations.[545]

5.22.3 TEMPORARY AND SEASONAL WORKERS

The training of temporary workers in fact has become shared responsibility of staffing agency and employer both where host employer should provide temporary

Occupational Hazard Origin and Preventive Measures

workers with safety training that is identical or equivalent to that provided to the host employers own employees performing the same or similar work.[456]

5.22.4 SANITATION AND SANITARY OCCUPATIONAL HAZARDS

Sanitization could be for the purpose of facilities, floors, walls, etc. or for equipment and machines used in food processing areas. It can be done by thermal (74–165 dc) or chemical methods. The benefits of a good sanitation program include the following and all food products must be protected from contamination from receiving (and before) through distribution. Sanitation is a dynamic and ongoing function and cannot be sporadic or something that can be turned on once a day, once a week, etc. to 1. Comply with regulations, 2. Prevent a catastrophe, 3. Improving food quality and shelf life, 3. Reducing energy, maintenance, and insurance cost, and 4. Increase quality and confidence.[523, 673]

Examples of sanitizing chemicals are, bleach-type sanitizer (sodium hypochlorite), organic chlorine sanitizer (Sodium or potassium salts of isocyanuric acid), quaternary ammonium products (n-Alkyldimethylbenzyl-ammonium chloride), acid anionic sanitizer component (Sodium dodecylbenzenesulfonate), and iodophor sanitizers (Oxypolyethoxyethanol-iodine complex).[523] Occupational hazard include workers can be exposed to thermal or chemical agents, corrosivity exposure, burns and skin dryness and allergy reaction, these chemicals can cause dryness, allergic reactions and other skin or eye damage, vapor inhalation problem, high steam pressure, and machinery physical exposure.

5.22.5 PREVENTIVE MANAGEMENT AND CONTROL

- Providing all operators with adequate safety training and information on all work procedures and pressure vessel hazards.
- Providing for regular inspection and maintenance of gaskets as blow outs could cause severe steam burns.
- Having an emergency procedure in place and the operator should know the proper emergency shutdown activities such as the location of manual shut off valves for steam, air, and water.
- Establish and practice lockout procedures for repair or maintenance work.
- Steam pipes should be insulated or guarded to prevent workers or mobile equipment from contacting them.
- Small leaks should be repaired as soon as possible.
- Regular inspection and maintenance of all steam equipment.

5.22.6 PSYCHOSOCIAL HAZARDS/STRESS

Psychosocial risks and work-related stress are among the most challenging issues in occupational safety and health. They impact significantly on the health of individuals, organization, and national economies. Around half of European workers consider stress to be common in their workplace, and it contributes to around half of all

162   Occupational Health and Safety in the Food and Beverage Industry

lost working days. Like many other issues surrounding mental health, stress is often misunderstood or stigmatized. However, when viewed as an organizational issue rather than an individual fault, psychosocial risks and stress can be just as manageable as any other workplace health and safety risk.[355]

Psychosocial risks arise from poor work design, organization, and management, as well as a poor social context of work, and they may result in negative psychological, physical, and social outcomes such as work-related stress, burnout, or depression. Some examples of working conditions leading to psychosocial risks are:

- Excessive workloads,
- Conflicting demands and lack of role clarity,
- Lack of involvement in making decisions that affect the worker and lack of influence over the way the job is done,
- Poorly managed organizational change, job insecurity,
- Ineffective communication, lack of support from management or colleagues,
- Psychological and sexual harassment, third party violence.

We can basically divide these types of stress at work into two categories:

- Direct pathway/working environment/ergonomics such as fatigue, discomfort, and disease due to repetitive movement, lighting, ventilation, noise, or thermal.
- Indirect pathway/psychological work environment such as stress-related factors due to work condition (load, schedule, interpersonal relationship, career development, organizational culture, job content, machines, individual characteristics such as gender, age, education, competitiveness, etc.).

5.22.7   ENVIRONMENTAL AND STRESS HAZARD

Toxic substances have become so much an everyday fact of modern life and we must reclaim our lost history so that, going forward, we can accurately judge the steps to address the public health hazards. These hazards predictably arise as the unintended by-products of the ways in which we make and use consumer goods and produce and transport commodities and industrial products.[451]

Stressful and hazardous work environments in the food and beverage industry lead to bruised knuckles and minor cuts that reduce production rates and heighten production errors. Human error frequently results from working under hot, noisy, and extremely fast-paced conditions which are ever present in many food and beverage operations.

Health problems consisted of skin diseases and disorders, respiratory problems, poisoning, and disease from physical agents.[32]

Stress hazard roots could be A. direct pathway or B. indirect pathway[314,440]:

A. **Direct pathway/physical work environment** such as ergonomic, engineering design to reduce fatigue, noise, discomfort and disease, machine, lighting, humidity, ventilation, temperature, atmospheric pressure, etc.

Occupational Hazard Origin and Preventive Measures 163

B. **Indirect pathway/psychological work environmental**[314,440]:
1. Work-related stress such as job content, workload and schedule, organizational culture, interpersonal relationship, role in organization, career development, etc.
2. Stress reactions due to psychological, behavioral, or emotional.
3. Long-term consequences on the worker such as psychological and social (mental health) or physiological and physical (cardiovascular disease).
4. Due to individual characteristics such as gender, age, education, or competitiveness.

5.22.8 ERGONOMICS HAZARDS

Ergonomic is the science of designing work with the worker in mind. It is concerned with the way a job, a task, or workplace "fits" the worker. Some of these issues such as noise, vibration, thermal changes, etc. have been discussed earlier. Without sufficient attention to ergonomic principles, the following problems may become evident as: 1. Fatigue, 2. Repetitive motion injuries, 3. Biomedical stresses creating strains, aches, and injuries, 4. Eye strain from video display terminals (VDTs), and 5. Lowered moral.

The goal of ergonomic is to reduce the physical and mental stress associated with a job; to increase the comfort, health, and safety of a work environment, to increase productivity; to reduce associated human errors task; and to improve the work safety.[269]

Recommended ergonomic development plan include:

1. Knowing the workplace,
2. Defining the work area,
3. Initial walk-through, and
4. Developing a check list to include issues affecting ergonomic safety such as manual handling, tools, machine safety, workstation design, hazardous substances, environmental factors, welfare facilities and work organization[270] can be considered for sound OHS management and preventive measures to:
   1. Set priorities,
   2. Perform group discussion, and
   3. Manage changes.

6 Occupational Injury and Its Management in Food and Beverage Industry

6.1 INTRODUCTION

Workers in food manufacturing are more likely to be fatally injured and experience nonfatal injuries and illnesses than those in private industries as a whole. In addition, compared with other private industry workers, food manufacturing workers are much more likely to suffer an injury requiring job transfer or days away from work. Production workers constitute more than half of the employees in the entire food manufacturing industry. However, while they account for the majority of nonfatal injuries, transportation and material moving workers account for the majority of fatal injuries.[759]

Occupational harm, disease, and injuries can appear in various forms such as:[678]

- Injuries such as noise-induced hearing loss, burns, slip, trips and fall, sprain, strain, or cuts and laceration
- Disease such as solvent-induced neurotoxicity, allergic reaction, dermatitis, or infectious diseases such as Aids, tuberculosis, rabies, Lyme disease, and scabies or
- Other health conditions such as cancer, pesticide poisoning, musculoskeletal disorders, post-traumatic stress, and environmental sensitivity conditions and probable organizational conditions.[786]

In some cases not only was the lost-time injury rate more than twice as high in food industries, the occupational illness and injury was 60% higher compared to non-food industries.[665] Besides solar chemicals, radiation, night, and rotating shifts including exposure to other substances in the workplace are the top carcinogenic risks faced by workers.[447]

In general it is important to note that statistical data reported on OHS hazards in food and beverage industry sectors exist beyond those of just manufacturing and processing sectors such as other occupational hazards in postharvest of fruits and vegetables, animal and poultry breeding, egg production, slaughterhouses, bee keeping, hunting, fishing, sea food harvesting from farm to table that may be reported separately which need to be considered for more accurate data.[711,712]

Besides direct injury or fatality costs, there are always indirect costs as has been discussed previously which can vary from state or country.[674]

DOI: 10.1201/9781003303152-6

The Association of Workers Compensation Board of Canada (AWCBC) also provides some details on the analysis of injuries and fatalities considering many factors including animal product and food processing sectors.[709] One recent statistical report indicated 67 fatalities in food and beverages manufacturing sectors.[216] The cost of insurance premium may differ from one province to another. In Quebec the premium rate for insurance coverage for various food and beverage manufacturing-related sectors varies from $0.022 to $0.080 for every $100 of insurable payroll for 2021.[714]

A more detailed information in Canada reported by the Health and Safety Commission for the province of Quebec provided the percentage of various injury types covering food and beverage industry sectors from 2007 to 2016 for which we will provide such statistical data in related section in this chapter.[715]

Most food processors in the UK operate machines and prepare foods for processing using a wide range of equipment and methods and in general food processing workers may perform tasks that if not properly evaluated for OHS could lead to injuries such as rhinitis, dermatitis, and other occupational hazards.[386,387,664]

It is important to note that when analyzing statistical data to consider factors affecting changes due to local, state, province or country of origin jurisdictional, accuracy and their reliability where overall injuries and fatalities have been reported.[710]

6.1.1 INTRODUCTION: COMMON OCCUPATIONAL INJURIES

The food and drink manufacturing industry actually comprises over 30 different industries such as brewery, dairy, food additive, fruit and vegetable, margarine and edible oil, pet food, meat, poultry, flavoring, baking and confectionary, chilled and frozen products, pasta, baby food, fish, sea food, alcoholic, and non-alcoholic beverages.[675]

The injury and/or disease symptoms vary based on the hazard and or injury involved which will be discussed later in this chapter in various sectors.

6.1.2 MAIN CAUSES OF INJURY AND ACCIDENTS IN FOOD AND BEVERAGE MANUFACTURING

The food industry today has become highly diversified, with manufacturing ranging from small, traditional, family-run activities that are highly labor-intensive, to large, capital-intensive and highly mechanized industrial processes. Many food industries depend almost entirely on local agriculture or fishing. In the past, this meant seasonal production and hiring of seasonal workers. Improvements in food processing and preservation technologies have taken some of the pressure off workers to process food quickly to prevent spoilage. This has resulted in a decrease in seasonal employment fluctuations. Whereas in certain other industries such activities continue to exist, such as in fresh fruit and vegetable processing, production of baked goods, chocolate, and so forth due to holiday season necessity.

Seasonal workers are often women and foreign workers.[813] Injuries caused by knives in meat and fish preparation can be minimized by design and maintenance, adequate work areas, selection of the right knife for the job, provision of tough

Occupational Injury and Its Management in Food and Beverage Industry **167**

protective gloves and aprons, and correct training of workers on both the sharpening and the use of the knife. Mechanical cutting devices also pose a hazard, and good maintenance and adequate training of workers is critical to prevent injuries.[813]

The process involves handling and storage, extraction, production process, preservation process (radiation or antibiotic sterilization, chemical, dehydration, or refrigeration), packaging, canning, aseptic or frozen packaging where transport of frozen goods has to be in refrigerated wagons, lorries, ships, and so on with minimum exposure to heat. In beef and poultry operations, carbon dioxide is often used to cool products during shipping.[813]

The most common causes of injuries in the food industry are hand tools, especially knives; operation of machinery; collisions with moving or stationary objects falls, slips, burns, noise, and pathogenic and chemical exposures.[813] Health and Safety Executive in the UK has the following causes and main occupational health risk and recommendations for hazard prevention in various food manufacturing sectors and in each sector available statistical reports have been provided and in general include hazards associated with handling, slips, and trips, struck by moving objects or against an object, exposure to harmful substances, falls from height, machinery, noise, and transport hazards. Therefore, details of prevention methods including proper record management and evaluation for improving the system are required.[391]

In more recent accidents and injuries in food and beverage industries such statistical data may also be included with other sectors such as agriculture, fisheries, and or as part of food manufacturing.[666–668]

The beverage industry consists of two major categories and eight subgroups. The non-alcohol beverage category is made up of soft drink syrup manufacturing – soft drink and water bottling and canning; fruit juice bottling, canning, and boxing; the coffee industry, and the tea industry. Alcohol beverage categories include distilled spirits, wine, and brewing.

The Bureau of Labor Statistics (BLS) reported that workplace illnesses accounted for 4.8% of the approximately 2.9 million injury and illness cases reported by private industry employers in 2015 – roughly 140,500 cases.

Employers recorded most illness cases (62.6%) as "Other illnesses", which includes such things as musculoskeletal disorders and systemic diseases and disorders, among others. Within the beverage industry injury risks are generally well recognized and understood – manual handling and repetitive motion tasks – slips, trips, and falls; machinery-related hazards; confined space entry; and powered industrial truck-related hazards. Occupational health is generally more difficult to manage than safety. Work-related causes of ill health tend to be more difficult to spot and it often takes some time for symptoms to develop therefore the connection between cause and effect is less obvious. Recognized main occupational illness risks in beverage manufacturing include musculoskeletal disorders; noise-induced hearing loss; occupational lung disease; occupational rhinitis, and occupational skin diseases.[726]

A recent study published in the *Journal of Occupational and Environmental Medicine* in the USA as the first farm-to-table model study comparing OHS morbidity and mortality rate in "food system industries" to "non-food system industries" indicated significantly higher rate in "food system industries" than those of "non-food system industries." The overall annual economic burden of occupational morbidity and mortality in "food system industries" just in the United States is approximately $250 billion, including direct and indirect costs.[718]

Therefore considering such a higher cost within the food system indicates to further prioritize the needs for OHS hazard prevention in Canada, US, and internationally to benefit the consumer and market worldwide.

Applying the farm-to-table model is a novel construct within OHS and has the potential to reshape the understanding of how market forces in the food industry may impact producers and consumers in the food chain hazard and can assist to reshape our understanding of the burden of "foodborne" illness to include not just pathogens and toxins that are transmitted to consumers through contaminated food, but also the costs that society bears for occupational injury, illness, and death that occur in the process of producing and delivering food to consumers.[718]

Their data indicate that the differences between causes of severe injuries among food system industries and non-food system industries from the same sector vary.

Some of these differences are indicative of unique job responsibilities or conditions. For example, workers in food production industries (such as ranchers and fishermen), are more likely to be severely injured by animals compared to workers in non-food industry production sectors such as miners and forestry workers.

There are also different working environments required when food is present. One possible reason for significantly elevated rates of severe injuries due to slips, trips, and falls in food processing, storage, and retail industries compared to non-food same-sector industries may be the preponderance of refrigeration.

Many food products are processed and stored at cooler temperatures to maintain freshness, this can lead to precipitation that may make floors slippery[34] in factories, warehouses, or onsite storage areas in retail facilities.

In estimating product-specific morbidity rates, meat, fish, and dairy production and processing industries did not have significantly higher rates of occupational injury and illness than non-meat, non-dairy production, and processing industries.

However, the illnesses and injuries were more severe, suggesting that some of the differences in rates of severe injuries and illnesses between food and non-food industries may be driven by a subset of all food industries.[718]

6.1.3 MEAT, POULTRY, FISH, AND PETFOOD INDUSTRY

Meatpacking has one of the highest rates of injury in all industries. Most butchers and meat, poultry, and fish cutters and trimmers frequently work in cold, damp rooms that are refrigerated to prevent meat from spoiling. It requires physical strength to lift and carry large cuts of meat and the ability to stand for long periods. Although they may work in sanitary conditions, however, their clothing is often soiled with animal blood and the air may smell unpleasant and work with powerful cutting equipment and are susceptible to cuts on the fingers or hands.[387]

There are many serious safety and health hazards in the poultry processing industry. These hazards include exposure to high noise levels, dangerous equipment,

Occupational Injury and Its Management in Food and Beverage Industry **169**

slippery floors, musculoskeletal disorders, and hazardous chemicals (including ammonia that is used as a refrigerant). Musculoskeletal disorders are of particular concern and continue to be common among workers in the poultry processing industry. Employees can also be exposed to biological hazards associated with handling live birds or exposure to poultry feces and dust which can increase their risk for many diseases.[812]

Strong evidence indicates a risk of occupational asthma associated with poultry workers due to dust and other hazards including, manual handling (24%), slips or trips (21%), struck by objects (16%), machinery (8%), struck against objects (6%), falls from a height (5%), exposure to harmful substances (3%), or transport hazards (2%).[530,400,247]

It's no secret that the offshore fishing industry can be a dangerous place to work, especially in Alaska where 399, or one-third, of all the state's work-related deaths from 1990 to 2014 occurred in the fishing industry. Based on the study's findings, you stood the best chance of being injured while working on the production line (68 injuries, 22%); stacking blocks/bags of frozen product (50 injuries, 17%); and repairing, maintaining, and cleaning factory equipment (28 injuries, 9%).[789]

Seafood-processing workers typically labor for long hours and handle huge quantities of fish on a daily basis. Faulty equipment, lack of safety guards, fatigue, and exposure to bacteria have been reported[791] including frostbite on exposed hands and fingers occur in cold working conditions. Cuts are also common among fish processors because they work with sharp equipment: knives for manual work and blades in automatic machinery.[790]

Occupational injury could be from the fish itself, with injuries including cuts and punctures from spines and bones, or while lifting a lot of heavy cargo and carrying processed fish into freezers, which can lead to back and joint injuries.

One of the worst kinds of accidents with fish processing equipment happens when a worker gets caught in a machine. This can be fatal or cause loss of limbs and amputations.[790]

In Canada, between 1999 and 2008, an average of 14 people died in fishing accidents each year.[792] For more in-depth and case study examples refer the Appendix section.[788]

In fish and seafood processing such hazards include handling (31%), slips and trips (20%), machinery (11%), struck by moving objects (9%), strike against objects (9%), exposure to harmful substances (7%), falls (6%), and transport hazards (3%).[401]

In pet food industry causes of accidents that have been reported by Health and Safety Executive (HSE) include, slips and trips (25%), handling (19%), falls (17%), struck by falling and moving objects including tools (11%), striking against objects when moving (11%), machinery (9%), exposure to harmful substances (2%), transport (2%), and fire and explosion (1%).

Entry into silos must be avoided as a confined space hazard. Injuries also have been mainly due to manual handling, exposure to grain dust, and noise-induced hearing loss from noisy areas in a mill.[398,399]

Based on studies outside of the Gulf of Mexico in the USA, musculoskeletal disorders are likely to be one of the repeated injuries reported among fishers. Much

170 Occupational Health and Safety in the Food and Beverage Industry

less is known about other health-related problems such as poisoning from aquatic animals, dermatitis, cancer, hearing loss, and respiratory problems.[792]

A more detailed statistical data in Canada have been reported by the Health and Safety Commission (CNESST) in the province of Quebec with injury type in various food and beverage industry sectors including meat processing.[715]

Table 6.1 that follows provides general information on injury type for meat processing steps.[715,813]

The HSE has reported accident statistical data for each type of hazard associated with the poultry industry to be handling (22%), falls (17%), slips and trips (16%), struck by falling and moving objects including hand tools (13%), exposure to harmful substances (10%), machinery (9%), striking against objects when moving (7%), transport hazards (2%), fire (1%), and electrical hazards (1%).

In particular the industry poses a dust, fire, and explosion hazard and therefore the plant must be inspected regularly for signs of heating (install temperature detectors) and take measures mentioned previously for dust explosion and fire.[398] Table 6.2 provides the rate of injury in meat industries and animal food.

TABLE 6.1
Meat, Poultry, and Fish Processing Injury and Disease

Processing steps	Exposure hazard	Disease, injury, symptoms
Slaughtering, cutting, comminuting, heading, gutting, filtering, cleaning, washing and sanitizing, chiller operations, salting, smoking and cooking, refrigeration, freezing and sterilization, packaging, loose cans, and cardboard.	Dust, noise, hot and cold environment, sanitizing products, ergonomic issues, alkali, acids, Carbon monoxide (CO), ammonia, carbon dioxide (CO_2), bacteria, fungi, viruses, endotoxins, mycotoxins (fish) (Samonelle, E. coli), ergonomic issues (vibration, repetitive, forceful, and awkward movements). Cuts and lacerations when removing blades from equipment, struck by or struck against and caught in equipment, slips, trips, and falls, fire safety, electrical shock.[388] Slips, trips, and falls, poor visibility, and injured animals.[247]	Respiratory asthma problem, noise-induced hearing loss (NIHL), cumulative trauma disorders, carpal tunnel, repetitive strain injuries, fractures, cuts, lacerations, amputations, heat and cold stress, eye, nose, and skin irritation, nausea, vomiting, headaches, dizziness, Brucellosis, erysipeloid, Leptospirosis, warts, dermatophytoses, fever and tuberculosis, veterinary drugs, environmental pollutants such as dioxins, pesticides, and phthalates,[693] dust inhalation.[530,675]

Occupational Injury and Its Management in Food and Beverage Industry 171

TABLE 6.2
Percentage of Injury and Type in Meat Industry and Animal feed

Hazard	Slaughterhouse	Meat cutting operation	Animal feed
	% Injury	% Injury	% Injury
Excessive effort	23.7	19.7	28.8
Slips and trips	5.1	9.2	6.6
Body reaction	8.9	15.4	18.9
Hit by object	13	–	–
Colliding an object	5.7	5.4	–
Crush by machine	5.8	8.3	–
Transport	–	–	–
Fall from height	–	7.1	13
Repetitive movement	16.1	13.8	7.5
Violence	–	–	–
Others	21.7	21.1	18

6.1.4 GRAIN AND FLOUR MILLING INDUSTRIES

Table 6.3 provides processing steps in the milling industry with associated hazards.

Table 6.4 provides a more detailed information on injury type for the flour milling industry sector in Quebec, Canada in Canada between 2007 and 2016.[715]

TABLE 6.3
Milling (Grain) Operation Injury and Disease

Processing steps	Exposure hazard	Disease, injury, symptoms
Grinding, sifting, milling, rolling, storage and movements, silos, sacks, bags, and packaging.	Organic dust, noise, pesticides, spores from biological agents, oxygen deficiency and ergonomic hazards, manual handling and lifting, slip, trips and fall, chemical hazard vapor, machinery, confined space/silo, oxygen deficiency environment, transport, and flour dust explosion.[398,399]	Respiratory irritation, NIHL, dermatitis, dizziness, nausea, long-term kidney and nervous problems due to pesticides, fever, allergic asthma, repetitive strain injury, and death.[675]

TABLE 6.4
Percentage of Injury and Type in Flour Milling Industry

Hazard	% Injuries
Excessive effort	10.0
Slips and trips	8.7
Body reaction	12.2
Hit by object	11.0
Colliding an object	–
Crush by machine	22.5
Transport	–
Fall from height	–
Repetitive movement	–
Dust exposure	29.1
Violence	–
Others	6.2

6.1.5 BREAD, CAKE, AND BISCUIT MANUFACTURING

The main causes of injuries have been slips and trips (26%), handling (25%), struck by objects and hand tools (13%), machinery (10%), falls from height (9%), struck against objects (8%), exposure to harmful substances and hot objects (3%), and transport hazards (3 %).[394] Flour dust has been referred to as a hazardous substance for causing the second highest rate of asthma and can also cause short-term respiratory, nasal and eye symptoms. Besides bakery additives/bread improvers contain enzymes (e.g., alpha-amylase) which are potent sensitizers and their use should be minimized and can be achieved by using improvers in liquid, paste or dust-suppressed powder form.[391]

Table 6.5 provides processing steps and associated hazards in baking manufacturing.

TABLE 6.5
Baking, Biscuit, and Pasta Manufacturing Hazards and Injuries

Processing steps	Exposure hazard	Disease, injury, symptoms
Mixing, kneading, moulding, fermenting, laminating, surface treatment and seasoning, baking, drying, cutting, sacks, bags, packaging, and silos.	Flour and vegetable dust, CO_2, additives, combustion fuel, hot environment, repetitive movement, manual handling and lifting, slip, trips and fall, bagging noise, chemical hazard vapor, machinery, confined space/silo, manual handling, transport and flour dust explosion.[398,399] Manual handling, slips, trips, and fall, struck by object, machinery, transport, chemical, noise, hot objects, confined space/silos or mechanical hazards.[394]	Rhinitis, asthma, nausea, vomiting, headaches, dizziness, allergic dermatitis, heat stress, tuberculosis, or death.[675]

Occupational Injury and Its Management in Food and Beverage Industry   173

TABLE 6.6

Percentage of Injury and Type in the Baking Industry

Hazard	Frozen product	Confectionary	Catering	Wholesale	Retail
	% Injury	% Injury	% Injury	% Injury	% Injury
Excessive effort	20.9	32.9	29.1	27.9	31.2
Slips and trips	17.0	13.4	5.9	14.2	11.1
Body reaction	15.1	12.9	14.2	13.6	7.5
Hit by object	8.3	7.9	14.1	–	–
Colliding an object	–	–	17.0	–	–
Crush by machine	10.0	7.8	–	–	–
Transport	–	–	80.0	–	–
Fall from height	–	6.5	–	13.0	6.6
Repetitive movement	7.7	5.1	–	–	10.5
Violence	–	–	–	–	6.0
Others	21.0	13.5	13.1	31.4	19.3

Table 6.6 provides a more detailed information on injury type for baking industry sectors in Quebec, Canada in Canada between 2007 and 2016.[715]

6.1.6 CHOCOLATE AND SUGAR CONFECTIONARY MANUFACTURING

Cocoa is indigenous to the Amazon region of South America, and, during the first years of the 20th century, the southern region of Bahia provided the perfect conditions for its growth. The cocoa-producing region of Bahia is composed of 92 municipalities and Ileus and Itabuna are its main centers. This region accounts for 87% of the national production of cocoa in Brazil, currently, the world's second-largest producer of cocoa beans. Cocoa is also produced in about 50 other countries, with Nigeria and Ghana being the major producers.

The vast majority of this production is exported to countries like Japan, the Russian Federation, Switzerland, and the United States; half of this is sold as processed products (chocolate, vegetable fat, chocolate liquor, cocoa powder, and butter) and the rest is exported as cocoa beans.[813]

Causes of accidents vary and report from HSE refer to the main causes as handling (30%), slips and trips (22%), struck by moving objects (12%), struck against objects (9%), machinery (8%), exposure to harmful substances (6%), and transport (3%).[393]

Table 6.7 provides processing steps and associated hazards in chocolate and confectionary products.

174 Occupational Health and Safety in the Food and Beverage Industry

TABLE 6.7

Chocolate and Confectionary (Cocoa Bean, Sugar, Fat) Hazards and Injuries

Processing steps	Exposure hazard	Disease, injury, symptoms
Roasting, grinding, mixing, conching, moulding, fumigation, packaging, storage (silos, sacks), and conditioned chambers.	Organics dust, noise, aluminum phosphate (phosphine gas), ergonomiques issues, vibration, repetitive – movements.	Rhinitis, asthma, nausea, vomiting, headaches, dizziness, allergic dermatitis, heat stress, tuberculosis, death.

6.1.7 FRUIT AND VEGETABLE PROCESSING INDUSTRIES

Health and Safety Executive (HSE) reports that around 30% of manual handling injuries and the greatest proportion of musculoskeletal injuries such as back injury or work-related upper limb disorders (WRULDs) result from lifting, carrying, or moving boxes, containers, or sacks of fruit and vegetables.[675,694]

Table 6.8 shows processing steps and associated hazards in fruit and vegetable industries.

Table 6.9 provides a more detailed information on injury type for the fruits and vegetable processing sector such as canning, pickling, and drying in Quebec, Canada in Canada between 2007 and 2016.[715]

TABLE 6.8

Fruits and Vegetable Processing Hazard and Injuries

Processing steps	Exposure hazard	Disease, injury, symptoms
Grading, washing, cooling, peeling, blanching, freezing, dehydration, canning.	Manual handling, slips, trips, and fall, noise/packaging, struck by objects, machinery, chemical hazards, bleaching, sanitizer, sanitizer and ammonia leaks, transport often due to forklift trucks.[116,395]	Musculoskeletal injury from manual handling boxes, sacks, wheeled trolleys, etc. Work-related upper limb disorders, e.g. from repetitive sorting/packing work Noise-induced hearing loss from noisy machines, e.g. cleaning plant, packaging machinery. Respiratory irritation from breathing fumes such as chlorine, hypochlorite, ammonia, and sulfur dioxide Occupational dermatitis from handling fruits, vegetables, and from chemical cleaners.[675]

Occupational Injury and Its Management in Food and Beverage Industry   175

TABLE 6.9
Percentage of Injury and Type in Fruits and Vegetable Processing

Hazard	% Injuries
Excessive effort	22.1
Slips and trips	6.4
Body reaction	136.0
Hit by object	9.7
Colliding an object	–
Crush by machine	11.8
Transport	–
Fall from height	7.5
Repetitive movement	7.3
Violence	–
Others	21.7

6.1.8 DAIRY, MILK, AND CHEESE MANUFACTURING

Dairy farming is associated with higher rates of injury as compared with other industrial sectors, but a lack of work-related injury reporting continues to be an issue in several countries.

Worker fatality associated with heavy equipment use is not a new observation (e.g., tractors); however, manure-handling systems, livestock handling, and quad bike operation continue to be associated with worker injuries and fatalities on modern farms.[803] However, the follow-up manufacturing of dairy products has also been associated with a number of hazards.

The chapter presents findings from 39 detailed follow-up investigations of slips, trips, and falls (STF) incurred by individuals working in New Zealand's dairy farming industry. Key latent risk factors and their interactions identified included problems associated with time pressure and related time-saving behaviors and the presence of design errors that, for example, required workers to climb onto equipment to view aspects of the task they were working on.[804]

In the UK the statistical report released by the HSE indicates injuries have mainly been handling (31%), slips and trips (22%), struck by moving objects (10%), exposure to harmful substances (7%), falls (9%), striking against (7%), machinery (5%) and transport hazards (5%).[396,675]

Processing steps and associated hazards in dairy products is shown in Table 6.10.

Table 6.11 provides more detailed information on injury type for dairy processing such as frozen dessert and ice cream in Quebec, Canada in Canada between 2007 and 2016.[715]

TABLE 6.10
Hazards and Injuries in Dairy Product Processing Industries

Processing steps	Exposure hazard	Disease, injury, symptoms
Skimming, churning, coagulation, ripening, pasteurization, sterilization, homogenization, concentration, and desiccation.	Manual handling and lifting – especially heavy loads or sharp edges. Slips – mostly due to wet floors. Being struck by objects – mostly falling objects such as hand tools. Exposure to harmful substances – cleaning fluids, fume, splashes, CIP failures, steam, hot water Falls from height – off ladders, stairs, tanks, and from vehicles/tankers. Electrical, mold, aflatoxin, solvents, etc. Transport – including tanker movements and lift trucks Machinery – lifting machines, conveyors, packaging machines.[694,695]	Musculoskeletal injury from manual handling wheeled racks, containers etc. Work-related upper limb disorders (WRULDs), e.g. from repetitive packing work. Respiratory, skin and eye irritation, heat and cold stress. Noise-induced hearing loss from noisy plant machine such as homogenizers, bottling lines, blast chillers, pneumatics. Respiratory irritation from breathing fumes such as chlorine, hypochlorite, and ammonia from cleaning operations and refrigeration plant.[694]

TABLE 6.11
Type and Injury Percentage in Frozen Desserts and Ice Cream Industries

Hazard	Frozen product
	% Injury
Excessive effort	21.3
Slips and trips	17.3
Body reaction	11.6
Hit by object	6.9
Colliding an object	–
Crush by machine	8.2
Transport	–
Fall from height	5.8
Repetitive movement	12.5
Violence	–
Others	21.7

Occupational Injury and Its Management in Food and Beverage Industry 177

6.1.9 OILS AND FAT PROCESSING INDUSTRIES

More than 100 varieties of oil-bearing plants and animals are exploited as sources of oils and fats. The most important vegetable sources are olive, coconut, peanut, cotton seed, soya bean, rapeseed (canola oil), mustard seed, flax or linseed, palm fruit, sesame, sunflower, palm kernel, castor bean, hemp seed, tung, cocoa, and corn. Main animal sources are beef cattle, pigs and sheep, the whale, cod, and halibut.[694,695]

A 2018 report by the US Bureau of labor and Statistics indicates the total number of fatalities in oilseed seed and grain farming with 78 fatalities including 36 due to transport, 3 for fire and explosion, 7 slips, trips, and fall, 33 for exposure to harmful substances in environment, and 167 injuries due to contact with object and equipment, however in grain and oilseed milling as part of manufacturing there was only one fatality and one case of exposure to harmful substances in the environment.[696] Table 6.12 below provides processing steps and associated hazards in oil and fat processing.

Please also refer to the end of this section for further information provided as part of preventive measures and hazard management strategy.

In 2015 World Bank Group reported an inclusive report on OHS issues with recommended preventive management in vegetable oil production due to hazards such as flammable chemical (solvent and oil) which can cause fire and explosion or physical hazards such as confined space entry and electrical- or noise-induced problems.[697]

Table 6.13 demonstrates a more detailed information of injury type for snack food and specialty products such as salting, roasting, drying, cooking, dressing, and spices in Quebec, Canada in Canada between 2007 and 2016.[715]

Table 6.14 provides a more detailed information on injury type in catching, harvesting of fish and seafood products, and their preparation and packaging in Quebec, Canada in Canada between 2007 and 2016.[715]

TABLE 6.12
Oils and Fat Processing Industries Hazards and Injuries

Processing steps	Exposure hazard	Disease, injury, symptoms
Milling, solvent or steam extraction, filter press, cleaning, pasteurizing, packaging (bottles, packs, and cans, etc.).[695] In short, fats may be recovered from oil-bearing tissues by three general methods, with varying degrees of mechanical simplicity: (1) rendering, (2) pressing with mechanical presses, and (3) extracting with volatile solvents. Essential steps in the extracting and refining of edible oil from oilseed. [694]	Trichloroethylene, hexane, oils and fats, biological (molds, aflatoxin), chemical (acid and alkaline), fumes acrolein, hot environment, noise and repetitive movements.[694] Explosion and fire during extraction with hexane, mixing of incompatible materials during refining, or fire during esterification, loss of utility, and pollution have been reported.[795]	Toxic effects, respiratory irritation, dermatitis, liver disease, skin, eye and respiratory irritation, heat stress, strain injury, and NIHL.

178 Occupational Health and Safety in the Food and Beverage Industry

TABLE 6.13

Type and Percentage of Injury in Snacks and Specialty Food Products

Hazard	Snacks foods	Specialty food products
	% Injury	% Injury
Excessive effort	20.0	32.6
Slips and trips	9.3	17.5
Body reaction	15.2	18.7
Hit by object	10.9	6.3
Colliding an object	–	5.2
Crush by machine	5.5	7.5
Transport	–	–
Fall from height	19.4	6.6
Repetitive movement	10.7	–
Violence	–	–
Others	9.1	5.7

TABLE 6.14

Potential Injuries in Fish and Seafood Harvesting and Processing

Hazard	Fish and seafood harvesting	Fish and seafood processing
	% Injury	% Injury
Excessive effort	23.8	17.7
Slips and trips	14.0	17.2
Body reaction	7.2	6.5
Hit by object	15.0	9.9
Colliding an object	–	–
Crush by machine	10.5	11.6
Transport	–	–
Fall from height	8.4	–
Repetitive movement	–	7.6
Substance exposure	6.3	16.5
Violence	–	–
Others	14.8	13.1

6.1.10 CANNING

Canning is a method of preserving food from spoilage by storing it in containers that are hermetically sealed and then sterilized by heat, which was invented by Nicolas Appert for preserving food for army and navy use in France in 1809.

It was 50 years before Louis Pasteur was able to explain why the sterilized food did not spoil: as this was due to the heat that killed the microorganisms in the food,

Occupational Injury and Its Management in Food and Beverage Industry 179

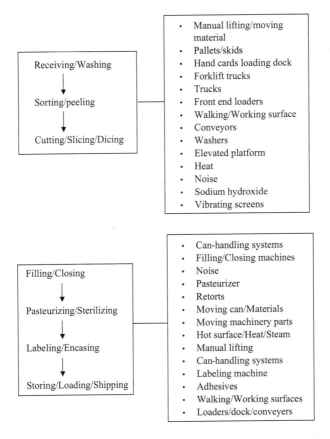

FIGURE 6.1 Flow chart of canning steps and relevant hazards. Flow chart indicates potential occupational health and safety hazards involved in a canning process operation in two steps at left side being 1. Initial receiving/washing, sorting/peeling as well as cutting/slicing/dicing and 2. Filling/closing, pasteurizing/sterilizing as well as labeling/encasing with more than a dozen different potential hazards indicated on the right side of each of these two steps.

and besides the sealing kept other microorganisms from entering the jar. The canning process itself consists of several stages: cleaning and further preparing the raw food material; blanching it; filling the containers, usually under vacuum, closing and sealing the containers; sterilizing the canned products; and labeling and warehousing the finished goods. Over time the industry has developed to include variety of food and drink with more mechanization, longer working hours, shifts, and other potential hazards.[716]

Figure 6.1 demonstrates the hazards associated with canning operation at different steps of processing.

Table 6.15 shows examples of such a process and their associated hazards that need to be addressed accordingly for their hazards as discussed earlier.[717]

180   Occupational Health and Safety in the Food and Beverage Industry

TABLE 6.15

Process Hazard Chart for Canning Operation

Processing steps	Potential hazard
Receiving	Manual lifting, pallets/skids, hand carts, forklift trucks, trucks, front-end loader, walking, and working.
Washing and Cleaning	Manual lifting, moving, conveyors, washers, elevated platforms, walking and working surfaces, heat, and noise.

6.1.11 EGG PRODUCTION

A new study finds that egg production workers are at risk for pulmonary problems due to exposure to airborne particulate matter inside hen houses.[809,811,812]

Hazards associated with egg production sectors include 1. Accidental, 2. Physical, 3. Chemical, 4. Biological, 5. Ergonomic, psychological, and organizational factors.[810]

6.1.12 FLAVOR MANUFACTURING

Royal empire has been built, unexplored land has been traversed, and great religions and philosophies have been forever changed by the spice trade. In 1492 Christopher Columbus set sail to find seasoning in three small ships toward "the east" to find lucrative lands of India and China.[802]

Way before the first artificial flavors were synthesized in the 11th century, ancient Egyptians were the first to extract flavors and scents from plants in the form of essential oils. In the 11th century, the Persian philosopher and physician Avicenna, also called Ibn Sina, figured out that oils can be distilled in much the same way alcohol is, by steaming plants, which extracts the oil and then condensing the steam back into liquid. This leads to many more essential oils, and for centuries, that industry is the flavor industry (and the scent industry).

The evolution of flavor industry followed up with fruit flavor in 1851, vanilla extraction in 1858, and in 1906 government regulation starts to appear when the U.S. flavor industry grew in 1914–1918, Flavor additives were introduced after the war (1939–1945), and flavored potato chips and snack were introduced into the market in the 1950s, followed by more regulation in 1968. Artificial vanilla starts to appear in 1982, with flavor enhancer called mono sodium glutamate (MSG) referred to as 6th flavor introduced in 1990 and then a Swiss flavor company, Evolva puts its yeast-produced biosynthetic vanillin on sale in the market in U.S.A.

There has followed some backlash from genetically modified organism (GMO) opponents – the vanillin itself isn't genetically modified, but the yeast that spits it out is – and Haagen-Dazs vows never to use it.[800]

In the mid-19th century, flavor additives – volatile organic chemicals with desirable aromatic qualities – began to be used to flavor sugary confections, carbonated beverages, and other mass-marketed delights.[801]

Many substances are used in the manufacture of flavorings. Diacetyl is a substance widely used in food and beverage flavorings. The principal types of flavorings

Occupational Injury and Its Management in Food and Beverage Industry **181**

that use diacetyl are dairy flavors (e.g., butter, cheese, sour cream, egg, and yogurt flavors) and the so-called "brown flavors" (such as caramel, butterscotch, maple, or coffee flavors). Some fruit flavors (e.g., strawberry and banana) may also contain diacetyl. The industrial sectors include candies, snack foods, prepared canned or frozen foods (sauces), some dairy products, bakeries, feed, soft drinks, and flavored cooking oil. Some foods (dairy, wine, and beer) contain naturally occurring diacetyl.

2,3 pentanedione has received attention as a flavoring substitute for diacetyl and actually both under certain conditions can cause changes in the central nervous system which also list 2,3 hexanedione and 2,3 heptanedione on the same list as well.[389]

Given the complexity of flavorings mixtures, and the lack of health data for many of the component materials, identifying the relative contributions of individual substances to causing flavoring-induced lung disease is a difficult challenge.

As noted in the NIOSH Alert: Preventing Lung Disease in Workers Who Use or Make Flavorings, the flavorings industry has estimated that over a thousand flavoring ingredients have the potential to be respiratory hazards due to possible volatility and irritant properties (alpha, beta-unsaturated aldehydes and ketones, aliphatic aldehydes, aliphatic carboxylic acids, aliphatic amines, and aliphatic-aromatic thiols and sulfides).[799]

Diacetyl is a chemical that was found to be a prominent volatile constituent in butter flavoring of microwave popcorn. It was initially investigated by the National Institute of Safety and Health (NIOSH). Diacetyl is also known as the alpha-diketone, 2,3-butanedione, document describing diacetyl and 2,3 pentanedione, where it has been deemed unsafe because of the potential risk of health effects, such as irreversible lung damage and entitled Criteria, for a Recommended Standard: Occupational Exposure to Diacetyl and 2,3-Pentanedione.[799]

Drawing a clear line between "natural" and "artificial" has never been easy because it has always been more of a cultural distinction than an actual one. As we demand more and more from our food, it's important to remember that foods can also be chemicals, that flavors have long depended on science and technology, and that the complex problem of improving our health on a healthy planet will take more than simple solutions.[801]

6.1.13 Honey production (apiculture/beekeeping)

Honey is the product of one of the most intelligent and industrious creatures whose miniature society is one of the most sophisticated in the animal kingdom It is used as a food product, in religious ceremonies, etc. and it is also known for medicinal qualities for centuries. The industrial process includes: 1. Receiving, 2. Uncapping (removing wax), 3. Extracting, 4. Purification, 5. Pasteurization, 6. Evaporation, 7. Filtering, and 8. Bottling and therefore preventive measures should reflect based on the risk involved in advance.[808]

Beekeeping is one of the high-risk activities in agriculture as the beekeeper is exposed to physical difficulties such as environmental conditions, stress, bee venom, and smoke. As a result of these factors fatigue, distractibility, psychological disorders, injuries, and/or occupational accidents may occur. It is they who need to take action in the case of an emergency health problem. Depending on the bee colony

182 Occupational Health and Safety in the Food and Beverage Industry

size, the rate of exposure to bee venom and propolis would increase. Wrong practices regarding disease control within the beehive lead to decreases in bee immunity and most importantly human health problems.[807]

6.1.14 Tobacco Manufacturing

Industries in the Beverage and Tobacco Product Manufacturing subsector manufacture beverages and tobacco products.[83]

Injury incidence rates in the tobacco industry per 100,000 workers between 2000/01 and 2002/03 were reported to be 1023.[419]

6.1.15 Catering and Food Service Industry

High levels of youth employment, workplace hazards, and characteristics unique to adolescents contribute to a relatively high incidence of injuries among teens in the restaurant industry and numerous resources available for the prevention of hazards in this sector as well as the case study section in Appendix.[89]

6.1.16 Spice Manufacturing

Spices have aromatic properties and are thus used as seasoning and food processing worldwide. In one study in India it was reported that spice grinding workers in production have been facing problems related due to spice grinding; these were burning sensation, skin diseases, coughing and sneezing, eye irritation, and breathlessness due to the presence of micro-dust in the work environment. Sweating was due to work pressure. The majority of male respondents were wearing a pant and a shirt while all female respondents were wearing salwar kameez. None of the respondents was using personnel protective devices to protect themselves.[705]

Although spices have many health benefits for human due to their enzyme-stimulating characteristics, spice particles emitted in the air during various processes and production steps are a matter of concern as these suspended particulate matters are inhaled.

These ailments may be occupational asthma due to short-term exposure and/or could lead to occupational dermatitis and rhinitis if failing to have preventive measures in place for workers. Skin contact allergy and dermatitis can occur, and their severity depends on long-term and short-term exposure to spice dust. Spice dust has been categorized based on various definitions and concluded that spice dust might also be called combustible dust due to its explosive nature.[706]

Center for Disease Control and Prevention (CDC) in another investigation has recommended following three steps for prevention:

1. Good engineering control for dust elimination,
2. Containment of dust-exposing area, and
3. Antifatigue mats and sitting stool for periodic resting of workers.[707]

Occupational Injury and Its Management in Food and Beverage Industry **183**

6.1.17 FOOD ADDITIVE

A food additive is any chemical substance that is added to food during preparation or storage and either becomes a part of the food or affects its characteristics for the purpose of achieving a particular technical effect so as to maintain its nutritive quality, enhance its keeping quality, and make it attractive or to aid in its processing, packaging, or storage are all considered to be food additives. However, some substances that aid in the processing of food, under certain conditions, are considered to be food processing aids, not always food additives. Examples of food additives include food colors (natural and synthetic), pH, adjusting agents, preservatives, bleaching agents, food enzymes glazing and polishing agents, emulsifiers, or gelling agents.[703]

Based on the food additive and process involved we must apply engineering and administrative control as discussed in previous chapters.

6.1.18 CHILLED AND FROZEN PRODUCTS

Food businesses need to ensure that the requirements of food hygiene law are achieved while maintaining a "reasonable temperature" in the workroom.[699]

6.1.19 HEALTH AND SAFETY TEMPERATURE REQUIREMENTS

Health and safety temperature requirements in open workrooms can be achieved by: maintaining a "reasonable" temperature throughout the workroom of at least 16° (or at least 13° if the work involves serious physical effort). This may mean chilling the food locally or minimizing its exposure to ambient temperature or, if this is not practical in providing warmer environment or workstations within a workroom where the overall temperature may be lower or, if this is not practical keeping the individual warm by providing suitable protective clothing, heated rest facilities, task rotation, etc.

6.1.20 WORKING IN CHILL UNITS AND FREEZERS

Health and safety temperature requirements in chill units and freezers can be met by local heating in vehicle cabs where practically keeping the individual warm by providing suitable thermal clothing, appropriate breaks to warm up, task rotation, etc.

For work in chillers around 0°C suitable personal fit clothing and normal breaks are usually sufficient practical approach. For work in blast freezers operating down to -30° C no personal protective equipment (PPE) will be sufficient and breaks at ambient temperature or in warming rooms will be needed.[699]

6.2 BOTTLED WATER ALCOHOLIC AND NON-ALCOHOLIC DRINK AND BEVERAGE INDUSTRY

6.2.1 INTRODUCTION

The beverage industry consists of two major categories and eight subgroups. The non-alcoholic category is comprised of soft drink syrup manufacture; soft drink and water bottling and canning; fruit juices bottling, canning, and boxing; the coffee industry and the tea industry. Alcoholic beverage categories include distilled spirits, wine, and brewing.[813]

The Bureau of Labor Statistics reports indicate slightly more than 1.7 million work in the food and beverage manufacturing industry. The beverage industry consists of two major categories and eight subgroups.

Within the beverage industry injury risks are generally well recognized and understood – manual handling and repetitive motion tasks; slips, trips, and falls; machinery-related hazards; confined space entry; and powered industrial truck-related hazards. Occupational health is generally more difficult to manage than safety.

Work-related causes of ill health include musculoskeletal disorders, noise-induced hearing loss, occupational lung disease, rhinitis as well as skin diseases.[700]

The Canadian bottled water industry consists of establishments primarily engaged in purifying and bottling water. The flat bottled water market is comprised of two types: spring water and processed water. Spring water is potable water obtained from an underground source and not from a municipal water supply.

It does not contain coliform bacteria, nor is its composition modified through the use of any chemicals. Processed water is water taken at a municipal source which then may have treatment such as osmosis, demineralization, or re-mineralization to obtain a desired mineral content.

Flavors, extracts, and essences derived from spice or fruit can be added to bottled water, but these additions must comprise less than 1% of the weight of the final product. Beverages containing flavors at a ratio of more than 1% by weight are considered by Statistics Canada to be soft drinks. Similarly, soda water, seltzer water, and tonic water are considered to be soft drinks.[110]

In Canada bottled water is regulated as a packaged food product by Health Canada through the Food and Drug Act. We subject our finished products and source water to microbiological analysis every day that exceeds the microbiological requirements outlined in the Safe Water Drinking Act, which governs both municipal tap water and bottled water in the province of Ontario.[114]

In the USA the Environmental Protection Agency (EPA) regulations have been successful in keeping down water borne illness. However, their standards for contaminants in tap water are too high making tap water far too toxic for regular consumption. Basic water filtration does a poor job of removing these contaminants.[111]

The global bottled water market size was valued at US$ 217.66 billion in 2020 and is expected to expand at a compound annual growth rate (CAGR) of 11.1% from 2021 to 2028. Water seems to have hit a sweet spot with health-conscious buyers. The rapid expansion of restaurants in the USA is also foreseen to create a significant demand for bottled water in the upcoming years.[624]

6.2.2 EVOLUTION OF THE INDUSTRY

Although many of these beverages, including beer, wine, and tea, have been around for thousands of years, the industry has developed only over the past few centuries.[813]

Occupational Injury and Its Management in Food and Beverage Industry   **185**

The beverage products industry, viewed as an aggregate group, is highly fragmented. This is evident by the number of manufacturers, methods of packaging, production processes, and final products. The soft drink industry is the exception to the rule, as it is quite concentrated. Although the beverage industry is fragmented, ongoing consolidation since the 1970s is changing that.

Since the early 1900s beverage companies have evolved from regional firms that mainly produced goods for local markets, to today's corporate giants that make products for international markets. This shift began when companies in this manufacturing sector adopted mass production techniques that let them expand.

Also, during this time period there were advances in product packaging and processes that greatly increased product shelf life. Air-tight containers for tea prevented the absorption of moisture, which is the principal cause of loss of flavor. In addition, the advent of refrigeration equipment enabled lager beers to be brewed during the summer months.[813]

British scientist, Joseph Priestley experimented with artificially carbonated water more than two centuries ago, leading to one of the most important nation's manufacturing sectors.[757] Today, although many of these beverages, including beer, wine, and tea, have been around for thousands of years, the industry has developed only over the past few centuries.

Since the invention of artificially carbonated water 200 years ago, soft drink manufacturing has become one of the most important sectors of the food industry.

Soft drinks are called "soft" in contrast to "hard drinks" (alcoholic beverages). Small amounts of alcohol may be present in a soft drink, but the alcohol content must be less than 0.5% of the total volume if the drink is to be considered non-alcoholic. Fruit juice, tea, and other such non-alcoholic beverages are technically soft drinks by this definition but are not generally referred to as such.[445]

Soft drinks soon outgrew their origins in the medical world and became a widely consumed beverage, available cheaply for the masses. By the 1840s, there were more than 50 soft drinks manufacturers – an increase from just ten in the previous decade.

Mixer drinks became popular in the second half of the century. Tonic water was originally quinine added to water as a prophylactic against malaria and was consumed by British officials stationed in the tropical areas of South Asia and Africa. As the quinine powder was so bitter people began mixing the powder with soda and sugar, and basic tonic water was created. The first commercial tonic water was produced in 1858.

The mixed drink gin and tonic also originated in British colonial India, when the British population would mix their medicinal quinine tonic with gin.[445]

Perhaps 8,000 to 10,000 years ago, someone discovered that fermented fruit or grain or milk or rice tasted good, made one happy, or both.

Today's consumer is likely to drink less, but be interested in higher-quality products, even if they cost more. There are establishments that specialize in wine sales, brewing and selling beer, full-bar service, and a variety of food-and-drink combinations that often include some sort of entertainment. You can buy a drink on an airplane, in a hotel room, or in your favorite neighborhood restaurant.[554]

Though the ingredients and production of beverages vary, generally the characteristics of those employed in this industry have many commonalities. The process of harvesting raw materials, whether they be coffee beans, barley, hops, or grapes, employs low-income, unskilled individuals or families. In addition to being their main source of income, the harvest determines a large part of their culture and lifestyle.

In contrast, the processing of the product involves automated and mechanized operations, usually employing a semi-skilled, blue-collar workforce. In the production facility and warehouse areas, some of the common jobs include packaging and filling machine operator, forklift operator, mechanic, and manual laborer. The training for these positions is completed onsite with extensive on-the-job instruction. As technology and automation evolve, the workforce diminishes in number and technical training becomes more important. This semi-skilled manufacturing workforce is usually supported by a highly skilled technical group consisting of industrial engineers, manufacturing managers, cost accountants, and quality assurance/food safety technicians.

The beverage industry for the most part distributes its products to wholesalers using common carriers. However, soft drink manufacturers usually employ drivers to deliver their products directly to individual retailers. These drivers-sales workers account for about one-seventh of the workers in the soft drink industry.

The more health-conscious atmosphere in Europe and North America in the 1990s has led to a flat market in the alcoholic beverage industry, with demand shifting to non-alcoholic beverages. Both alcoholic and non-alcoholic beverages, however, are expanding considerably in developing nations in Asia, South America, and to some extent Africa. Because of this expansion, numerous local jobs are being created to meet production and distribution needs.[813]

6.2.3 Main causes of injury

Due to the high demand for food and beverage production facilities, plants often run 24 hours a day and continuous operation means there is little time for maintenance and repairs to be carried out. In the food industry, it is during breakdowns that injuries occur. Workers, faced by high targets and strict deadlines, may attempt to repair the equipment themselves or even override safety guards to reach into machines and risk injury in the process. It is therefore vital that, regardless of high production targets, employees are well educated in the company's safety policies and the equipment's safety features.[753] Employers are often reluctant to spend time training staff on safety procedures, but then run the risk of having employees who are not sufficiently trained.

The UK's food processing industry employs 117,000 migrant workers from the EU, which supplies the sector with the necessary labor. However, language barriers and a high turnover of staff can indirectly create safety hazards.[753]

Typical hazards and injuries are due to manual handling, slip, trip, and fall, struck by object, machinery, noise, chemical exposure, glass breakage injuries, and

Occupational Injury and Its Management in Food and Beverage Industry **187**

transport hazard which could lead to musculoskeletal injuries work-related upper limb disorder due to repetitive packing work, noise-induced hearing loss from noisy plant environments such as casking/kegging, decorating/washing, bottling, canning and packaging machinery, and occupational lung disease from exposure to grain and malt dust, occupational lung disease and nasal cancer from exposure to hardwood dust in cooperages.[75,84,698]

Reported accidents have included, handling (35%), slips and trips (20%), struck by objects (14%), falls from height (11%), striking against objects (6%), exposure to hazardous substances (4%), transport (3%), and machinery (3%).[402]

Table 6.16 demonstrate injuries in soft drink, water bottles, and ice production.

TABLE 6.16
Injuries in Soft Drinks, Water Bottles, and Ice Production

Hazard	% Injury
Excessive effort	34.8
Slips and trips	5.2
Body reaction	7.2
Hit by object	–
Colliding an object	–
Crush by machine	10.4
Transport	–
Fall from height	8.4
Repetitive movement	–
Violence	–
Others	14.1

6.2.4 SOFT DRINKS INDUSTRY

Reported accidents have included, handling (26%), slips and trips (23%), struck by objects (13%), machinery (9%), falls from height (8%), striking against objects (8%), exposure to hazardous substances (5%), transport (3%).[403]

Table 6.16 provides a more detailed information on injury type for soft drinks, water bottle, and ice making industries in Quebec, Canada in Canada between 2007 and 2016.[715]

6.2.5 BREWING INDUSTRY

The brewing industry is the most important branch of the drinks industry. The rapid changes in this branch in recent years are partly a result of advances in biotechnology.[76]

6.2.5.1 Main cause of injury

The craft beer brewing industry has been booming lately, but OSHA has also been citing employers left and right for these common violations. Breweries are a particularly strong example of the hazards present in the food processing industry. The dust generated in the conveying, sieving, and milling of malt grains can form flammable dust clouds.

This creates a potentially explosive environment, officially classifying the environment as a hazardous area. This means that ATEX ratings must be observed on all equipment used in these facilities. Breweries are a particularly strong example of the hazards present in the food processing industry. The dust generated in the conveying, sieving, and milling of malt grains can form flammable dust clouds.

This creates a potentially explosive environment, officially classifying the environment as a hazardous area. This means that ATEX ratings must be observed on all equipment used in these facilities.[753] The term "ATEX" is derived from the French words "Atmospheres Explosible" and refers to European directives for controlling minimum safety requirements for improving health and safety of workers and the equipment and protective systems intended for use in these atmospheres.

Later down the line, carbon dioxide, a dangerous by-product of the fermentation process, can be fatal if inhaled. Workers have died while trying to perform repairs or check fermentation tanks, becoming overwhelmed by CO_2 almost immediately. This means that companies should use suitable sensors and locks to separate workers from the tanks, while also educating workers on the associated dangers.

In the beverage industry, particularly in breweries, packaging and filling are one of the most dangerous places in the processing chain. The speed of operation and high quantity of goods being moved increase the risk of things going wrong.

In the beverage industry, glass bottles are commonly filled at high speeds and at high pressure, meaning the bottles could explode if the machines are incorrectly programmed. As these beverage plants are operating under high time pressures, it is not possible to completely stop the production line for receptacles to be changed. Instead, the filler operates at a slow speed, allowing the operator to change the bottle or can. By integrating sensors that can monitor the speed of the machine, companies have the ability to implement emergency stops in the case of a breakdown or safety issue.

One important and common hazard is the presence of carbon dioxide (CO_2) which is a by-product of the fermentation process right through to packaging and bottling. In most cases, the assessment and safe working system will require atmospheric testing of CO_2

which is a common by-product of fermentation because the gas is heavier than air, collects at the bottom of containers and confined spaces such as tanks and cellars, and can even spill out of fermenting tanks and sink to the brewery floor, where it forms deadly, invisible pocket gas detection equipment.[86] Prior to entering a tank or other confined space, a "release measurement" of CO_2 must be taken using a suitable CO_2 monitor.

Some common hazards for prevention to consider as recommended by OSHA are[676]:

1. Confined space entry such as tank and presence of CO_2,

Occupational Injury and Its Management in Food and Beverage Industry   **189**

2. Ergonomic hazards such as lifting heavy kegs, sacks of grains, and other items,
3. Hazardous chemicals,
4. The control of hazardous energy (Lockout/Tagout),
5. Hazard communication for hazardous or dangerous substances and environment, and
6. Eye and face protection (PPE).

6.2.6 POTABLE SPIRITS INDUSTRY

Hazards caused by machinery especially conveyors and bottling plants, exposure to hazardous substances, transport, and striking against objects also need to be considered.

Although the order of these priorities is, slightly, different for the three main industry environments (distillation, maturation, and packaging), with appropriate attention they remain the key areas for consideration.

Many different hazards exist in this industry. For example use of chemicals such as mineral spirit, caustic, acids, cleaners, and solvents are common.[71]

Other hazards reported include manual handling (31%), slips and trips (22%), struck by objects (14%), machinery (4%), falls from height (9%), striking against objects (7%), exposure to hazardous substances (3%), and transport (3%).[404]

6.2.7 FIRE AND EXPLOSION

Few injuries have been caused by fire and explosion, yet they are ever-present risks because the industry produces, handles, and stores large amounts of highly flammable solvents (ethyl alcohol) with a flash point between 15°C and 28°C depending on the process. The avoidance of the explosion remains a high priority for the industry because of the potential damage it can create to people, process, product, property, and the environment.

Occupational health hazards include chronic health problems from heavy manual lifting such as casks, sacks, etc.; work-related upper limb disorders from repetitive handling and poor ergonomics such as in bottling lines; lung disease from exposure to grain and malt dust; nasal cancer from exposure to hardwood dust in cooperages, noise-induced hearing loss from noisy plant, bottling, and packaging machines; compressors, boilers, and steam plant in various sectors in the beverage industry.

6.2.8 FRUIT JUICE PRODUCTION AND FROZEN CONCENTRATE

Fruit juices are made from a wide variety of fruits, including oranges and other citrus fruits, apples, grapes, cranberries, pineapples, mangoes, and so forth. In many cases, various fruit juices are blended. Usually, the fruit is processed into a concentrate near where it is grown, then shipped to a fruit juice packager. Fruit juices can be sold as concentrates, frozen concentrates (especially orange juice), and as diluted juice. Often

190   Occupational Health and Safety in the Food and Beverage Industry

sugar and preservatives are added. Once received at the processing plant, the oranges are washed, graded to remove damaged fruit, accordingly by size for juice extractors.[813]

There the oils are extracted from the peel, and then the juice is extracted by crushing. The pulpy juice is screened to remove seeds and pulp, which often end up as cattle feed. If the orange juice is intended for sale as "not from concentrate", it is then pasteurized.

Otherwise, the juice is sent to evaporators, which remove most of the water by heat and vacuum, then chilled, to produce the frozen, concentrated orange juice. This process also removes many oils and essences which are blended back into the concentrate before shipping to the juice packager and finally the frozen concentrate is shipped in refrigerated trucks or tankers.[813]

6.2.9 Soft drink concentrates

Hazards in a concentrate manufacturing plant vary depending on the products manufactured and the size of the plant. Concentrate plants have a low injury rate due to a high degree of automation and mechanized handling.

Materials are handled by forklifts, and full containers are placed on pallets by automatic palletizers. Major hazards include engines and equipment in motion, objects falling from overhead containers, energy hazards in repair and maintenance, confined space hazards in cleaning mixing tanks, noise, forklift accidents, or hazardous cleaning agents. Operations in a concentrate manufacturing plant can be divided into five basic processes[813]:

1. treating water
2. receiving raw materials
3. concentrate manufacturing
4. concentrate and additives filling and
5. shipping finished products.

Each of these processes has safety hazards that must be evaluated and controlled. Water is a very important ingredient in the concentrate and it must have excellent quality. Each concentrate plant treats water until it reaches the desired quality and is free from microorganisms. Water treatment is monitored during all stages.

6.2.10 Coffee manufacturing

Coffee as a beverage was introduced in Europe during the 16th century, first in Germany and then throughout the European continent during the following century, especially in France and Holland. Afterward, it spread to the rest of the world.[813]

Coffee manufacturing is a relatively simple process, including cleaning, roasting, grinding, and packing processes, however, modern technology has led to complex processes, with an increase in the speed of production and requiring laboratories for quality-control testing of the product.[813]

Obliterative bronchiolitis, a rare, irreversible form of fixed obstructive lung disease, has been identified in workers exposed to flavoring chemicals while working

Occupational Injury and Its Management in Food and Beverage Industry **191**

in the microwave-popcorn and flavoring-manufacturing industries; the occupational risk to workers outside these industries is largely unknown.

This is an irreversible form of lung disease in which the smallest airways in the lung (the bronchioles) become scarred and constricted, blocking the movement of air, was previously identified in flavoring manufacturing workers and microwave popcorn workers who were occupationally exposed to diacetyl (2,3-butanedione) or butter flavorings containing diacetyl. Now, NIOSH research finds that workers at coffee processing facilities may also be at risk.

These cases reinforce the need for evaluating work-related exposures in all industries in which workers are exposed to diacetyl or 2,3-pentanedione. Currently, no specific federal regulations govern workers exposed to diacetyl or its substitutes.[92,302] This reinforces the need for exposure evaluation in all industries in which workers are exposed to flavoring chemicals. Additionally, a high index of suspicion is required when such workers have progressive shortness of breath.

If obliterative bronchiolitis is suspected, immediate protection from further exposure is crucial to prevent further deterioration of lung function.[95]

6.2.11 TEA MANUFACTURING

Legend tells us that tea may have been discovered almost 5,000 years ago in China by Emperor Shennong or Shen-Nung, as he discovered it when a tea leaf accidentally fell into the bowl of hot water he was drinking. "The Divine Healer" in 2737 BC. Observant of the fact that people who drank boiled water enjoyed better health, the wise emperor insisted on this precaution. When adding branches to the fire, some tea leaves accidentally fell into the boiling water. The emperor approved of the pleasing aroma and delightful flavor and tea was born.[813]

Table 6.17 provides a more detailed information on injury type for coffee, tea, and herbal teas manufacturing in Quebec, in Canada between 2007 and 2016.[715]

TABLE 6.17
Injuries Amount and Type in Coffee, Tea, and Herbal Tea Manufacturing

Hazard	% Injury
Excessive effort	14.9
Slips and trips	14.8
Body reaction	12.8
Hit by object	–
Colliding an object	7.3
Crush by machine	–
Transport	18.6
Fall from height	–
Repetitive movement	–
Violence	7.0
Others	17.0

192 Occupational Health and Safety in the Food and Beverage Industry

6.2.12 Spirit Manufacturing Hazards and Prevention

The business of manufacturing whiskey, vodka, rum, bourbon, gin, and a variety of other alcoholic drinks would not often be thought of as a dangerous enterprise. The process of producing distilled alcoholic beverages usually starts with processing and fermenting the basic ingredients. Typically, these are grains, fruits, or vegetables. Once the fermentation is complete, the percentage of alcohol in the liquid has to be increased by a distillation process. Typically, this is done with a piece of equipment called "still".[97]

6.2.13 Wine Industry

The owner of a small brew-on-premises wine-making establishment died of asphyxiation in a carbon dioxide (CO_2)-enriched, oxygen-deficient atmosphere. The CO_2 was produced by the fermentation of wine in the tightly closed, non-ventilated basement. The store did not have any natural or mechanical ventilation and was closed at the time of the incident. Fermentation of a large number of batches of wine had begun during the previous week. Two responding paramedics were also affected by the high level of CO_2 in the stairwell leading to the basement.[99]

Wine fermentation is carried out in vented glass or plastic carboys or in buckets and the CO_2 produced in the process is released into the workplace. CO_2 release is greatest early in the fermentation process when fermentation is most active.

In the absence of adequate ventilation CO_2 can accumulate, displace oxygen, and reach dangerous concentrations well in excess of the time-weighted average occupational exposure limit of 5,000 parts CO_2 per million parts of air (ppm).

The accumulation of CO_2 can occur even in shops located on the ground floor with access to the outdoors.[99]

Workplace transport accidents in the food and drink industries are the second highest cause of fatal injury, comprising over 25% of fatal accidents. During the decade between April 2000 and March 2010, 11 workers were fatally injured directly by workplace transport, for example being crushed by a vehicle.

A further ten workers were fatally injured in transport-related accidents such as falling from stationary vehicles or being struck by something falling from the vehicle.[101] Reported accidents caused by transport include also being struck by a vehicle, except forklift truck (31%), struck by a forklift truck (26%), falls from vehicles (22%), trapped between vehicles and wall or other fixed objects (6%), trapped by overturning forklift trucks (6%), trapped between two vehicles (5%), trapped by lorry tail-lift (1%), collision with wall or other fixed object (1%), technical failure (faulty brakes, etc.), and (1%) have been other safety concern in the industry overall.[405]

6.2.14 Hazard Prevention in Beverage and Drink Industries

All possible occupational health hazards and risks involved in the process must first be identified. It is important to have the senior management to commit to their implementation. Preventive measures include:

- Engineering controls – such as enclosing, containing, or isolating the section or machine and minimizing the worker exposure to hazards.

Occupational Injury and Its Management in Food and Beverage Industry 193

TABLE 6.18
Alcoholic and Non-Alcoholic Beverage and Drink Industry Hazard

Food industry source	Associated hazards
Bottled water	Manual handling, slips, trips, and fall, struck by object, machinery, forklift, chemical exposure, glass breakage injuries, musculoskeletal disorder due to repetitive work, noise-induced hearing loss due to machineries, lung disease from exposure to grain and malt dust, nasal cancer from exposure to hardwood dust in cooperages.[75,84,86]
Brewing industry	CO_2 exposure in confined space (tank), ergonomic hazards such as manual handling, hazardous chemicals, and energized machine.[676]
Soft drink industry	Manual handling, slips, trips, and fall, struck by object, transport vehicles such as forklift, noise-induced environment, and machineries.[403]
Potable spirit industry	Fire and explosion, manual handling, slips, trips, and fall, struck by object, transport vehicles such as forklift, noise-induced environment, and machineries, noise-induced hearing loss, and nasal cancer due to being exposed to hardwood dust to cooperage (barrel making).[71,404]
Soft drink, canning, and fruit juice frozen concentrate	Manual handling, engines and equipment in motion, objects falling from overhead containers, energy hazards in repair and maintenance, confined space hazards in cleaning mixing tanks, noise, fork-lift accidents, and hazardous chemical cleaning agents.[74]
Coffee industry	Flavoring chemicals, odor intensity, solvent for coffee extraction, machines, manual handling, and ergonomic issue.[96]
Tea production	Dust during blending and packaging, manual handling/ergonomic, machinery guarding, slip, trips, and fall, noise-induced hearing loss, hot water and steam, humidity, noise, hazardous chemicals such as anhydrous ammonia, chlorine, or acidic discharge as accidental release, transport forklift safety, chemical sanitizers.[99,101,405,406]
Wine industry	CO_2 and confined spaces, manual handling in the vineyard and potential animal hazard, chemical exposure.[99]

- Administrative controls involve having specific policies, procedures, and safe work practices in place to reduce exposure to occupational health hazards.
- Lastly, the identification and use of proper personal protective equipment are necessary when engineering and administrative controls are not feasible or while such controls are being implemented.[700]

Table 6.18 provides a summary of hazard origin associated with the alcoholic and non-alcoholic beverage and drink industry:

6.2.15 Injury Management and Prevention in Food and Beverage Sectors

In their last report, the International Labor Organization (ILO) reported most common injury and diseases as indicated by, Austria, Belgium, Colombia, Czechoslovakia,

194 Occupational Health and Safety in the Food and Beverage Industry

Denmark, France, Poland, Sweden, and USA were bronchitis, asthma diseases induced by inhalation of substances, diseases induced by physical agents, hearing impairment, respiratory disorders, musculoskeletal disorders, disorders associated with repeated trauma, hearing impairment by physical or by chemical agents, skin diseases, strains in various parts of the body (knees, elbows) as well as infections transmitted by animals or other toxic substance.[691]

The hazard source and risk could be

- Increased level of mechanization in industry as workers may be exposed to increased level of vibration and/or noise due to machinery such as hand-saws, grinders, or mixers, and/or
- Exposure to certain hazardous substances in solid, liquid, or gaseous form and solvents such as trichloroethylene, hexane, benzene, carbon monoxide, carbon dioxide, and polyvinyl chloride (PVC).

  Any hazard with biological, chemical, physical, or psychological origin needs to be addressed for best prevention.[691]

The overall accident, incident, injury, or disease prevention in various food and beverage sectors discussed above are handled through:

6.2.15.1 Some general OHS practices

- Make sure platforms or portable ladders are the right sizes and in good condition,
- Prohibit climbing on equipment,
- Replace drain covers as soon as the area is cleaned,
- Plan regular inspection program looking for worn or other electrical defects,
- Train employees on the hazards of the chemical cleaners used,
- Provide personal protective equipment,
- Evaluate new application and chemical delivery systems prior to use,
- Floor/platform surfaces must be non-slippery and well maintained,
- Platforms must have guard rails,
- Employees should wear non-slip safety boots,
- Knives and other cutting equipment and mechanical devices where possible,
- Machine should have appropriate guards and be kept in good operating condition,
- Train employee on OHS-related activities such as manual handling, lockout tagout, machinery, electrical safety use, and proper use of PPE,
- Prevention of repetitive strain injuries and ergonomic practices,
- Vigilant maintenance of vibrating equipment to minimize vibration,
- Avoid heavy lifting and pulling movements,
- Plan to contain unavoidable spillages so they do not contaminate the surfaces,
- Make sure there is an effective drainage,
- Set up a system for cleaning up spillages immediately,

Occupational Injury and Its Management in Food and Beverage Industry **195**

- Dry the floor right after the spill clean-up,
- Make sure cleaning is done effectively,
- Set up floor cleaning schedule which is carried out when work is not in progress or has been scheduled outside of working hours,
- Make sure the floor and surfaces are rough enough,
- Avoid obstructions in walkways and at workstations,
- Eliminate uneven floors.

6.2.15.2 Some specific OHS practices

For each specific hazard after process evaluation refers to various elements and/or route causes that have been discussed in previous chapters. As an example, injuries, caused by knives in meat and fish preparation can be minimized by design and maintenance, adequate work areas, selection of the right knife for the job, provision of tough protective gloves and aprons, and correct training of workers on both the sharpening and the use of the knife. Mechanical cutting devices also pose a hazard, and good maintenance and adequate training of workers is critical to prevent injuries.[813]

Although accidents involving transmission machinery are relatively infrequent, they are likely to be serious. Risks related to machines and handling systems must be studied individually in each industry. Handling problems can be addressed by close examination of injury history for each particular process and by the use of appropriate personal protection, such as foot and leg protection, hand and arm protection, and eye and face protection. Risks from machinery can be prevented by secure machinery guarding. Mechanical handling equipment, especially conveyors, is widely employed, and particular attention should be paid to in-running nips on such equipment. Filling and closing machines should be totally enclosed except for the intake and discharge openings. The intakes of conveyor belts and drums, as well as pulleys and gearing, should be securely protected. To prevent cuts in canning, for example, effective arrangements for clearing up a sharp tin or broken glass are required. Serious injury due to the inadvertent start-up of transmission machinery during cleaning or maintenance can be avoided by strict lockout/tagout procedures.

Falling accidents are most often caused by:

- The state of the floor. Accidents are possible when floors are uneven, wet, or made slippery by the type of surface; by-products; by fatty, oily, or dusty waste; or, in cold rooms, from humid air condensing on the floors. Anti-slip floors help to prevent slips. Finding the proper surface and cleaning regimen, along with good housekeeping and proper footwear, will help prevent many falls. Curbs around machines containing water will prevent water from flowing onto the floor. Good drainage should be provided to remove rapidly any accumulating liquids or spillage that occurs.
- Uncovered pits or drainage channels. Maintenance of covers or barricading of the hazard is necessary.

- Work at heights. Provision of safe means of access to equipment and storage areas, sound ladders, and fall protection (including body harnesses and lifelines) can prevent many hazards.
- Steam or dust. Operations that generate steam or dust may not only make the floor slippery but also prevent good visibility.
- Insufficient or inconsistent lighting. Illumination needs to be bright enough for employees to be able to observe the process. The perception of inadequate lighting occurs when warehouses appear dark compared to production areas and people's eyes do not adjust when moving from one light level to the other.

Burns and scalds from hot liquors and cooking equipment are common; similar injuries arise from steam and hot water used in equipment cleaning. Even more serious accidents can occur due to the explosion of boilers or autoclaves due to a lack of regular examination, poor employee training, poor procedures, or poor maintenance. All steam equipment needs regular and careful maintenance to prevent major explosions or minor leaks.

Electrical installations, especially in wet or damp places, require proper grounding and good maintenance to control the common hazard of electrical shock. In addition to proper grounds, outlets protected with ground fault interrupters (GFIs) are effective in protecting from electrical shock. Proper electrical classification for hazardous environments is critical. Often flavors, extracts, and dusty flammable powders such as grain dust, corn starch, or sugar (thought of as foodstuffs rather than hazardous chemicals) may require classified electrical equipment to eliminate ignition during process upsets or excursions. Fires may also occur if welding is done around explosive/combustible organic dust in grain elevators and mills. Explosions may also occur in gas or oil-fired ovens or cooking processes if they are not installed, operated, or maintained correctly; provided with the essential safety devices; or if proper safety procedures are not followed (especially in open flame operations).

Strict product sanitation control is vital at all stages of food processing, including in slaughterhouses. Personal and industrial hygiene practices are most important in guarding against infection or contamination of the products. The premises and equipment should be designed to encourage personal hygiene through good, conveniently situated, and sanitary washing facilities, shower–bath, when necessary, provision and laundering of suitable protective clothing, and provision of barrier creams and lotions, if required.

Strict equipment sanitation is also vital to all stages of food processing. During the regular operation of most facilities, safety standards are effective to control equipment hazards. During the sanitation cycle, equipment must be opened up, guards removed, and interlock systems disabled. A frustration is that the equipment is designed to run, but clean-up is often an afterthought. A disproportional share of the most serious injuries happens during this part of the process. Injuries are commonly caused by exposure to in-running nip points, hot water, chemicals, and acid or base splashes, or by cleaning moving equipment. Dangerous high-pressure hoses

Occupational Injury and Its Management in Food and Beverage Industry   **197**

which carry hot water also pose a hazard. Lack of equipment-specific procedures, lack of training, and the low experience level of the typical new employee pressed into a cleaning job can add to the problem. The hazard is increased when the equipment to be cleaned is located in areas that are not easily accessible. An effective lockout/tagout program is essential. The current best practice to help control the problem is designing of clean-in-place facilities. Some equipment is designed to be self-cleaning by use of high-pressure spray balls and self-scrubbing systems, but too often manual labor is required to address trouble spots. In the meat and poultry industries, for example, all cleaning is manual.[813]

The International Labor Organization provides detailed information on complete process analysis, hazards involved as well as their prevention for following food and beverage sectors which have mostly been discussed for information on the following topics[813]:

1. Meat packing and slaughterhouses,
2. Poultry processing,
3. Dairy products,
4. Cocoa production and chocolate manufacturing,
5. Grain, milling, and grain-based products,
6. Bakeries,
7. Sugar-beet industry,
8. Oil and fat,
9. Beverage industry (non-alcoholic and alcoholic drinks) such as fruit juices, tea, coffee, soft drinks, brewery, wine, and spirits.

6.2.16 SUGAR-BEETS INDUSTRY

The process of producing sugar from beets consists of many steps, which have been improved continuously throughout the more than century-old history of the sugar-beet industry. Sugar-beet processing facilities have become modernized and use current technology as well as current safety measures. Workers are now trained in using modern and sophisticated equipment.[813]

The sugar content of the beets ranges from 15% to 18%. They are first cleaned in a beet washer. They are then cut in beet slicers and the "cosseted" thus obtained are conveyed via a scalder into the diffuser, where most of the sugar contained in the beets is extracted in hot water. The devulgarized cosseted, called "pulps", are pressed mechanically and dried, mostly by heat. The pulps contain many nutrients and are used as animal feed.

The raw juice obtained in the diffuser, in addition to sugar, also contains non-sugar impurities which are precipitated (by adding lime and carbon dioxide) and then filtered. The raw juice thus becomes thin juice, with a sugar content of 12–14%. The thin juice is concentrated in evaporators from 65 to 70% dry matter.

This thick juice is boiled in a vacuum pan at a temperature of about 70°C until crystals form. This is then discharged into mixers, and the liquid surrounding the

198 Occupational Health and Safety in the Food and Beverage Industry

crystals is spun off. The low syrup thus separated from the sugar crystals still contains sugar which can be crystallized.

The desugaring process is continued until it is no longer economical. Molasses is the syrup left after the last crystallization. After drying and cooling, the sugar is stored in silos, where it can be kept indefinitely if adequately air-conditioned and moisture controlled.

The molasses contains approximately 60% sugar and, together with the non-sugar impurities, constitute valuable animal feed as well as an ideal culture medium for many microorganisms. For animal feed, part of the molasses is added to the sugar-exhausted pulps before they are dried. Molasses is also used for the production of yeast and alcohol.

With the help of other microorganisms, other products can be made, such as lactic acid, an important raw material for the food and pharmaceutical industries, or citric acid, which the food industry needs in great quantities. Molasses is also used in the production of antibiotics such as penicillin and streptomycin, and also of sodium glutamate.

6.2.17 WORKING CONDITIONS

In the highly mechanized sugar-beet industry, the beet is transformed into sugar during what is known as the "campaign". The campaign lasts from 3 to 4 months, during which time the processing plants operate continuously. Personnel work in rotating shifts around the clock. Additional workers may be added temporarily during peak periods.

Upon completion of the beet processing, repairs, maintenance, and updates are done in facilities.[813]

6.2.18 HAZARDS AND THEIR PREVENTION

Sugar-beet processing does not produce or involve working with toxic gases or airborne dusts. Parts of the processing facility may be extremely noisy. In areas where the noise levels cannot be brought down to the threshold limits, hearing protection needs to be provided and a hearing conservation program instituted. However, for the most part, occupationally related illnesses are rare in sugar-beet processing plants. This is partially due to the fact that the campaign is only of three to four months duration per year.

As in most food industries, contact dermatitis and skin allergies from cleaning agents used to clean vats and equipment can be a problem, requiring gloves. When entering vats for cleaning or other reasons, confined space procedures should be in effect.

Care must be taken when entering silos of stored granular sugar, due to the risk of engulfment, a hazard similar to that of grain silos. (See Section 6.1.4 in this chapter for more detailed recommendations.)

Occupational Injury and Its Management in Food and Beverage Industry **199**

Burns from steam lines and hot water are a concern. Proper maintenance, PPE, and employee training can help prevent this type of injury.

Mechanization and automation in the sugar-beet industry minimize the risk of ergonomic disorders.

Machinery must be regularly checked and routinely maintained and repaired as required. Safety guards and mechanisms must be kept in place.

Employees should have access to protective equipment and devices. Employees should be required to participate in safety training.[813]

6.3 INJURY MANAGEMENT AND OCCUPATIONAL REHABILITATION

Most deficiencies are principally due to organizational, cultural, technical, or management systems failure issues.[442]

A proper injury management task should be first prioritizing prevention and elimination of related health and safety hazards in the workplace and then plan for:

- How to eliminate these hazards or minimize their impact,
- How to develop specific procedures to do tasks safety,
- How to deal with workplace accidents and injuries,
- Effective organization for health and safety in your business.

It is often necessary and beneficial to fix the workplace as well as the worker situation when a job is clearly leading to musculoskeletal disorder through repetition and continuous risky working conditions.[442]

The key issues in managing sickness absence and return to work are[441]:

- Sickness absence is not just a matter of ill health. It is affected by a combination of health condition, personal and work-organizational factors. The last two factors get more important the longer the absence.
- Early intervention is key. The sooner action is taken, the better the chances are of an employee making a full and speedy return to work.
- A well-managed workplace is a treatment for people recovering from sickness absence and an early return to work improves both mental and physical recovery.
- Simple adjustments can enable workers to return to work safely before their symptoms completely disappear. Workers can normally return before they are 100% fit. For companies without an existing rehabilitation scheme, putting one in place is not difficult. The essential elements are to ensure:
- there is an actively managed approach with a consistently applied policy, which is regularly monitored for its effectiveness,
- employees should consider sound medical advice following an injury or illness which is keeping them from work with continued rehabilitation advice,

200 Occupational Health and Safety in the Food and Beverage Industry

- return to work plans are agreed upon between the employer and employee, and these identify the appropriate measures to be taken to assist speedy recoveries, such as provision of physiotherapy or other specialties help, and
- these plans also identify steps to be taken to assist the employee back to work as soon as possible, for example by providing modified duties or limited hours and the work environment is suitably adapted if required.

6.3.1 REHABILITATION

Rehabilitation after an injury or accident should consider the following:

- an integral part of a broad-led approach to improving performance, by having a management process that aims to ensure a healthy workforce through:
  a. reducing work-related risks to health,
  b. managing individual cases of ill health,
  c. introducing evidence-led organizational interventions to support a better physical and psychological working environment, and
  d. promoting healthy lifestyles:
- provided whether the injury/illness is work related or not,
- offered for absences arising from musculoskeletal injuries (bad backs), fractures, stress/depression, heart problems, post-operative recovery, dermatitis, asthma, etc.,
- provided for people off work and those back at work,
- considered an increasingly important issue with an aging work population.

Developing an occupational health policy includes[529]:

1. **Prevention** which is the employer's legal and moral responsibility to do whatever is reasonably practicable to prevent harm. This must also include monitoring sickness and absence, managing risk, and listening to workers. Also work design, considering duration of exposure, risk assessment, pre-employment screening, job placement, training, and monitoring.
2. **Rehabilitation** to prevent people from suffering ill health from their work, there still will be occasions where someone does become ill. Although the initial cause may not be work related, but the consequences still need to be managed.
3. **Health promotion.** The development of health and wellness program in the workplace could be a central part of its public health policy through promoting health food habits, helping staff to stop smoking, and educating about substance and drug abuse.

The Appendix section of the book provides more than twenty study cases involving various accident and injuries, their root causes and prevention recommendations for better understanding and applying safer OHS practices.

Overall injury management model should include continuously:

Occupational Injury and Its Management in Food and Beverage Industry   201

1. Gathering information about the substances, the work and working practices,
2. Evaluating the risks to health,
3. Deciding what needs to be done,
4. Recording the significant findings of your assessment, and
5. Reviewing your assessment.

7 Food Security, Safety, Biosecurity, and Defense

7.1 FOOD SECURITY

7.1.1 INTRODUCTION

An adequate presence of safe and wholesome food and water is a significant factor pointing to secure food for the public in order to prevent hunger and malnutrition which is the opposite of food security, a condition in which all household members have secure physical and economic access to adequate food.[532,313]

Food security can be improved through outdoor and indoor workers' farming, breeding, harvesting, handling, storage, preparation, and distribution of food to public.[315]

Public statistics indicate that food insecurities have been on the rise for many reasons such as recession, inflation, unemployment, lower wages, health insurance, and overall rise in the cost of living.[316]

In summary food security system is based on three fundamental elements of having adequate food and water, resources, and basic education.[280]

In fact food security is a complex topic, standing at the intersection of many disciplines.[431]

Unless humanitarian assistance is urgently provided, food security conditions could deteriorate from emergency (IPC Phase 4) to famine (IPC Phase 5) in parts of South Sudan due to continuous conflict where some 3.5 million people have been facing crisis or emergency conditions. Internally displaced people and the host communities they live in are the most affected, by intellectual property right.[430]

It is also important to note that usually where food insecurity exists, unhealthy diet, unsafe food, and unsanitary condition will be noticed.[728–733]

Agri-food industry related workers with occupational hazard factors involved, which results in accidents, injuries, and illnesses including but not limited to:

- Thermal (hot or cold) and polluted environments
- External climate extremities such as flood, tornado, drought, lightening, storm, and mental stress
- Internal environmental factors such as humidity, lighting, thermal, noise, dust, air conditioning, various repetition movements
- Chemical hazards such as pesticides
- Biological hazards such as allergens, insects, and animals

DOI: 10.1201/9781003303152-7

203

7.1.2 Effects of Food Insecurity and Risk Factors

Effects of food insecurity have been reported to be many such as malnutrition, global water crisis, land degradation, agricultural disease, and food sovereignty, while on the other hand, the risk to food security has been factors such as population growth, diminishing fossil fuel, hybridization, genetic engineering, and loss of biodiversity.[431]

The majority of food insecurities in Africa after the 1950s has been due to climate change, pests, livestock disease, lack of planning, cash crop dependence, lack of subsidies or financial assistance, increased population, terrorism, and wars, where it has been reported for its elimination requires to combat poverty, providing social assistance, improving diet, food habit quality, preventing food and water waste, as well as providing financial assistance for harvest, storage, and distribution.[530,430,677]

7.2 BIOTERRORISM AND AGRI-FOOD

With the advent of rapid transport systems, regulatory officials have seen a significant expansion in the international movement of plants and plant products from their centers of origin. Addressing the global population issue can be a factor to improve food security, which is predicted to rise to nine billion people by 2050.[533]

There are growing trends in the USA and around the world for the need for effective biosecurity measures to reduce the threat of bioterrorism and effective measures to prevent terrorists from misusing science, medicine, and biotechnology.[436]

The relatively indirect and indiscriminate nature of an agri-food terror attack requires a better food defense system.[432] In fact the poor level of security on the majority of farms today requires better protection from such attacks in future that can harm food and water supplies in the planet where agricultural bioterrorism can be as devastating as other forms of terrorism because[433]:

- It can cripple the economy of a nation.
- It can destroy the livelihood of many people.
- It puts the food supply at risk, perhaps for a long time.
- It may not be detected before it reaches difficult-to-control levels.

Agricultural targets could be animals or plants, the trucks and railroads that transport them, water supplies, farm workers, producers, grain elevators, ships or other transport means, food handlers, restaurants, grocery stores, and more as they have been previously reported to be very costly.[434]

Agri-food terrorist agents that could be used against animals will be highly contagious, virulent, and able to survive well in the environment and result in economic hardship and an import ban by other countries; however, without better biosecurity measures, expansion of military and civilian programs will increase some bioterrorism risks as more people will be trained to prevent such criminal activities worldwide also.[436]

7.2.1 Prevention Measures

Various sectors in a society and international community from academic, industry, and government require to collaborate and to develop plans for preventive measures

Food Security, Safety, Biosecurity, and Defense

which could include but not limited to agri-food safety and security and biosecurity measures, planning involving related agencies, effective communication, protection of hazardous materials involved, and physical biosecurity measures.[435]

7.3 FOOD SAFETY AND SANITATION

Today, in many countries people are concerned about the quality and safety of food and water. Many issues such as unsafe and unsanitary practices, presence of chemicals, and adulteration of such food products are of particular concern to both consumers and manufacturers. In particular during high inflation in the market and increased cost of food price adulteration and use of less-quality ingredients have tempted many from harvest farm to table to commit to such a crime worldwide. In fact adulteration has become more sophisticated by those committing to such a crime and have become costlier for government agencies for identifying such act.[775,778]

Many such incidents both fatally and nonfatally have been reported in recent years such as contaminated milk, presence of pesticides beyond permission, and presence of pathogenic organisms in meat for recall including COVID-19 food safety-related issues in various countries.[774,743,692]

One of the major causes of foodborne diseases has surely been the lack of attention to the storage and maintenance of food between 5 and 57 °C which is referred to as "Danger Zone" in order to be kept safe before consumption, particularly in public food and catering services.[625]

Other jurisdictions such as the food served aboard on aircraft although it is a heavily regulated sector of air travel have also been recommended to comply with updated food and hygiene guidelines for practice.[626,534]

Consumer role: A task force of the Council for Agricultural Sciences and Technology. A private organization of food science groups estimated in 1994 that 6.5–33 million cases of foodborne illnesses reported occurring in the USA each year.

The increased use of convenience foods, often preserved with special chemicals and processes, also complicates food safety practices. These foods provide consumers with false ideas that equivalent homemade foods are equally safe as may not necessarily be the case due to the food hygiene practices required.[65]

With the advent of meat inspection during the last century, several zoonoses, such as tuberculosis, and parasites, such as trichinosis, could be controlled; however, soon, it was recognized that food safety could not be assured by end-product testing standards such as Hazard Analysis Critical Control Point (HACCP) or Good Manufacturing Practices (GMP) only.[482]

Apart from adulteration of food and drink products which is considered an intentional crime, there is unintentional contamination of such products by biological, chemical, and physical substances from farm to table, which requires particular attention for such accidents or incidents.[311]

The very nature of our complex food industry, which provides a large retail store consumer, with the choice of over 30,000 different food products made from raw materials from all over the world, involves various steps from farm to fork with potential risks for contamination which could result in fatal or nonfatal accidents and diseases.[541]

Occupational Health and Safety in the Food and Beverage Industry

It is therefore necessary to increase food security in order to provide an efficient food safety system and an overall food defense for protecting the food and water resources.[281]

The present food situation is insecure in both rural population due to its insufficiency for production and urban area populations due to its poor food safety and quality concerns where many such accidents have been reported internationally.[282,318,319]

Therefore, it is necessary to protect the public, resources, and environment from many biological and chemical agents such as anthrax, plague or cholera; bio-toxins/ricin, botulinum, vesicants, sulfur mustard (HD), nerve agents, phosgene, hydrogen cyanide or chlorine which can be misused for intentional purposes to harm society and create major disasters.[283]

There is also an intentional/unintentional contamination that can occur through chemicals naturally present in food such as alkaloids or in environment such as heavy metals and mycotoxins that can harm the consumers where incidents have occurred because of poor harvesting or improper storage of grain, use of banned veterinary products, industrial discharges, human error and deliberate adulteration, fraud and/or food tampering where even needles and other objects have been inserted in meat packages in the supermarkets to protect consumers and manufacturers from potential risks and associated cost and damages.[286,287,288]

Tables 7.1 and 7.2 provide some historical background of food contamination in different eras.[424]

It is therefore highly significant for concerned organizations to establish and follow the recommended guidelines of Food Safety Modernization Act (FSMA) as has been endorsed and recommended by the World Health Organization (WHO).[285]

TABLE 7.1
Historical, major food contamination (1880s to 1989)

Year	Cause of food contamination
1880s	Arsenical contamination of sugar in beer
1857	Adulteration of bread with alum in London, causing rickets
1858	Sweets poisoned with arsenic in Bradford, England
1900	Beer contaminated with arsenic
1910–1945	Cadmium from mining waste contaminated rice irrigation water in Japan
1920	Bread contaminated with naturally occurring pyrrolizidine alkaloids in South Africa
1950s	Mercury poisoning in fish in Japan, contaminated by industrial discharge
1955	Milk powder contamination with sodium arsenate
1957	Chicken feed and thence chickens were contaminated with dioxins
1959	Moroccan edible oil poisoning disaster
1972	Mercury seed poisoning in Iraq
1973	Widespread feed contamination with flame retardant in the USA
1974–1976	Afghanistan: seed widespread poisoning pyrrolizidine alkaloids
1981	Spanish toxic oil syndrome
1985	Adulteration of Austrian wines with diethylene glycol
1989	Milk contamination with dioxin in Belgium

TABLE 7.2
Historical, major food contamination (1994–2020)

Year	Cause of food contamination
1994	Ground paprika in Hungary was found to be adulterated with lead oxide
1998	In India, adulteration of edible mustard oil with *Argemone mexicana* seed oil
1999	In Belgium, animal feed contaminated with dioxins and polychlorinated biphenyls
2001	Spanish olive/pomace oil contamination with polycyclic aromatic hydrocarbons
2002	In Northern Ireland, nitrofurans in chicken imported from Thailand and Brazil
2002	In the UK and Canada, banned antibiotic chloramphenicol in honey from China
2003	Banned veterinary antibiotic nitrofurans were found in chicken from Portugal
2004	New Zealand corn flour contaminated with lead
2004	Aflatoxin-contaminated maize in Kenya
2005	Farmed salmon in Canada was found to contain the banned fungicide malachite green
2007	Melamine in pet food recalls in North America, Europe, and South Africa
2008	Wheat flour contaminated with naturally occurring pyrrolizidine alkaloids
2009	Hoola Pops from Mexico contaminated with lead
2011	Meat, eggs, and egg products in Germany contaminated with dioxins
2013	October 2013 Taiwan food scandal
2020	China fermented corn noodles contaminated with Bongkrek acid

7.4 FOOD DEFENSE, BIOSECURITY, AND TERRORISM

As discussed, earlier "food defense" is the overall guidelines to be followed in order to prevent the food become the target of intentional and or unintentional contamination while "biosecurity" considers both food safety and environmental safety for a sustainable agri-food system protection in the society where there are certain guidelines and information available.[534,535]

Direct threat to food, water, and environment could have disastrous results on people, process, property, resources, and environment as a whole which is one form of bioterrorism and requires further attention internationally.[435]

Food safety can minimize both unintentional and intentional threats to food by adapting to standard procedures available such as ISO, GMP, HACCP, where such threats could have social, economic, psychological, and political implications for producers, consumers, and nations.[437]

Food defense refers to the prevention of intentional attacks on food supplies, which could be tampering or adulterating with harmful elements.

Food defense plans are steps in written documents that assess the risk areas through verification and monitoring of the system to be pursued in case of a threat by identifying vulnerable areas in advance and recommending preventive measures to minimize or eliminate the risks at the national or international level. All food manufacturers and distributors need to prepare a specific plan of action based on the type of products and facilities involved.[288]

A food defense plan assists to maintain a safe working environment through verification of potential internal and external threats from employees, contractors, and other visitors.[309]

208   Occupational Health and Safety in the Food and Beverage Industry

Food defense measures to protect premises from people who intend harm from receiving to storage, processing, and shipping. Therefore, it is essential to know the criminal background of all people that might and/or can get access to food products and facilities internally or externally.[536]

7.5 FOOD PRODUCTION SECURITY

The risk manager at H.J. Heinz (Pittsburgh), who has the global responsibility for asset conservation, facility security, environment, and business continuity planning, playing an integral role in this enterprise risk management, recommended the following steps as preventive measures for their facilities worldwide, referred by John Brown[284]:

- Identify vulnerable areas through receiving bulk ingredients, receiving and shipping areas, process areas (cooking, blanching, and mixing), and intermediate areas such as product storage and warehousing.
- Identify access points and routes which are internal or external perimeter fencing, walls, gates, building exteriors and windows, building interior and exterior doors, and people and product flow, and implement a plan to prevent unauthorized access, particularly to vulnerable areas.
- Develop a written Facility and Food Security Plan document clarifying all safety and security steps including particular areas with a commitment-based security.
- Provide security enhancers through lighting (exterior and interior), surveillance cameras, occupancy sensors, alarms, and personnel.
- Enhance security procedures through background checks on employees, use the same temporary employment agencies and contractors, have check-in and check-out procedures for visitors, vendors, and contractors, escort visitors, vendors, contractors, and delivery protocols (verification, etc.)

7.5.1 Food tampering

Food tampering is purposely contamination of food products, water, or their packages during production, storage, transportation, and/or distribution in supermarkets in order to harm people or manufacturers of such products. In Canada, the Canadian Food Inspection Agency (CFIA) is ensures the safety of food products for public consumption.[290,291]

The purpose of food tampering by individuals or groups may have different motives such as public attention, personal gains, terrorism, and revenge to inflict harm or damage a company's reputation, psychological problems, etc.

7.5.1.1 Food tampering signs and preventive measure

In general, damages to food products can be of biological, chemical, and/or physical concerns. It is therefore essential for safety reasons to avoid consuming food products that have expired, dirty, damaged, cut, punctured, torn, swollen, ripped open, missing labels, opened and resealed, leakage appearance, rusted, cracked, bent, missing safety

Food Security, Safety, Biosecurity, and Defense 209

seals, absence of vacuum in vacuum package, tampered label, bad flavor, aroma, etc. In fact such problems could have also been due to food tampering and a criminal activity in the first place for people to be aware of. It is equally important for those observing such problems or criminal activities to inform authorities for further crime prevention.

7.6 FOOD FRAUD AND ADULTERATION

7.6.1 FOOD FRAUD DEFINITION

Food fraud simply is when a food product is misrepresented for something it is not. It can have health hazards or be fatal such as having a chemical or a biological ingredient such as an allergen in it, or someone adding an adulterant or removed something for economic or other purposes.[773,292,300,427] In Europe, there is no harmonized definition for "food fraud"; however, it is considered as a consumer fraud or financial gains act based on EU Food Law.[683]

7.6.1.1 Historical perspective: the fight against food fraud and adulteration

From the beginning of civilization, people have been concerned about quality and safety as food adulteration has existed.[297]

Today's quality control of the food and drinks industry is thankful to pioneering work started by chemist Friedrich Accum and medic Arthur Hill Hassall in the 19th century.

Presently, the production of food and drink products must follow many standards and regulatory guidelines before they reach the consumer where they did not exist back in mid-1870s, when it was the turning point when chemist Friedrich Accum and medic Arthur Hill Hassall pointed to such problem.[307] In 1820 Friedrich Accum was the first scientist to bring the problem of food adulteration to public attention, where chemicals such as copperas (ferrous sulfate) or Prussian blue (ferric ferrocyanide) were used as coloring agents for food adulteration.[307]

Before 1870s, except for a few staples such as flour, most consumed food in the USA was either homemade or purchased from a neighbor; more and more food came from factories internally or overseas, so the consumers were unaware of the process, handling, and/or ingredients.[538]

About the middle of the 19th century with chemical and microscopic knowledge had reached the stage where food substances could be analyzed, and the subject of food adulteration started from the standpoint of rights and consumer welfare.[427,428]

Since the beginning of 20th century, a number of food laws have been introduced for better consumer protection in the USA while prior to that in the early 19th century it was thought that adulteration and food fraud problems were considered a public but not a government issue.[295,296]

Medicine was revolutionized in the USA between 1938 and 1951,[298] and it was in 1950 the English physician Arthur Hill Hassal used the tremendous new weapon of analytical chemistry, the microscope, to identify ground coffee adulteration and drug analysis. Of course, people consumed whatever nature provided them as food with primitive stomachs and no intentional adulteration.[294]

Thankamma reported that food adulteration is not only an intentional addition or subtraction of an ingredient but also an accidental contamination which can lead to a harmful or lower quality food product for consumers to include incidentally.[299,300,424] Table 7.3 and 7.4 representing each some examples of historical adulterants.

TABLE 7.3
Examples of adulterants used in food products as reported by Accum in 1820

Food	Adulterants
Red cheese	Colored with red lead (Pb_3O_4) and vermilion (mercury sulfide, HgS)
Cayenne pepper	Colored with red lead
Pickles	Colored green by copper salts
Vinegar	"Sharpened" with sulfuric acid; often contained tin and lead dissolved when boiled in pewter vessels
Confectionery	White comfits often included Cornish clayRed sweets were colored with vermilion and red leadGreen sweets often contained copper salts (e.g., verdigris: basic copper acetate) and Scheele's or emerald green (copper arsenite)
Olive oil	Often contained lead from the presses

TABLE 7.4
Examples of adulterants found by Hassall (1851–1854)

Food	Adulterants	Adulterants added for color, taste, and smell
Custard powders	Wheat, potato, rice flour	Lead chromate, turmeric to enhance the yellow color
Coffee	Chicory, roasted wheat, rye and potato flour, roasted beans, acorns, etc.	Burnt sugar (black jack) as a darkener
Tea	Used tea leaves, dried leaves of other plants, starch, sand, China clay, French chalk	Plumbago, gum, indigo, Prussian blue for black tea, turmeric, Chinese yellow, copper salts for green tea
Cocoa and chocolate	Arrowroot, wheat, Indian corn, sago, potato, tapioca flour, chicory	Venetian red, red ochre, iron compounds
Cayenne pepper	Ground rice, mustard seed husks, sawdust, salt	Red lead, vermilion, Venetian red, turmeric
Pickles		Copper salts for greening
Gin	Water	Cayenne, cassia, cinnamon, sugar, alum, salt of tartar (potassium tartrate)
Porter and stout	Water	Brown sugar, *Cocculus indicus*, copperas, salt, capsicum, ginger, wormwood, coriander and caraway seeds, liquor ice, honey, Nux vomica, cream of tartar, hartshorn shavings, treacle

Food Security, Safety, Biosecurity, and Defense

7.6.1.2 Food crime

Any type of tampering, adulteration in production, storage, distribution, and sales of food, beverages, as well as feed products on the supply chain is considered food crime because such activities are considered cheating and/or harming the consumer and wider food industry.[777]

7.6.1.3 Food fraud or error

The Canadian Food Inspection Agency (CFIA) considers food fraud may occur when food is misrepresented. It not only is a cheating act but also can result in serious health risks if certain biological, chemical, or physical ingredients such as an allergen or other harmful substances are present accidentally or added intentionally in food or beverages.[682]

An Interpol investigation by law enforcement agents in Europe identified fraudulent activities involving many products such as champagne, cheese, olive oil, and tea, from Bulgaria, Denmark, France, Hungary, Italy, the Netherlands, Romania, Spain, Turkey, and the UK.[302]

The spectrum of "Food Fraud" ranges from comical to sinister where people have died after eating food adulterated with engine oil and have been exposed to melamine which is known as a carcinogen.[303]

Often, food products such as olive oil, milk, honey, saffron, orange and apple juice, and coffee are targeted due to ease of adulteration and economic gains.[305] Besides financial gain, such criminal activities could be very harmful or even fatal or cause blindness, when fraudulent activities involve an alcoholic drink, such as vodka, where high levels of methanol in counterfeit versions.[301]

Motivation for financial gains has been the biggest threat noticed in food, feed, or drink fraudulent activities which has been reported at £11 billion a year just in the UK alone and a number of other countries for adulteration of olive oil, milk, pet food, beef, and peanut.[7]

7.6.1.4 Food fraud vulnerability

There are a number of factors to evaluate fraudulent activities and assess vulnerability. Each ingredient in the food manufacturer's supply chain should be assessed based on three criteria:

1. **Ease of committing adulteration based on the type of food or drink**
   Is fraud difficult to detect?
   Is there an easy way to commit fraud?
   Is the supply chain complex, making traceability difficult?
   Is the food/ingredient stored in an unsecured manner with easy access?
2. **Economic gains and incentive for committing fraud**
   Is the food/ingredient high in cost?
   Is there a large market for the food/ingredient?
   Is there a recent, significant increase in cost (such as tariffs)?
   Is there a recent, significant decrease in availability (such as low crop yield due to drought, fire, storms, flood, or pests)?

212 Occupational Health and Safety in the Food and Beverage Industry

3. Previous existence of fraud incidents

Does the food/ingredient have a history of fraud?
Does the manufacturer/supplier have a history of fraud?
Does the country/region of origin have a history of fraud?

The more "yes" answers on the list above greater the incentive or opportunity for food fraud, and thus the greater the risk of fraud.

Therefore, the above three criteria should be used to determine the vulnerability of ingredients and supply sources to economically motivated adulteration.[798]

7.6.1.5 Food adulteration in the 21st century

Historically, due to human greed adulteration has always been there and has become more sophisticated over time. One way to avoid many such adulterated food or drinks is to avoid processed food and start buying more local and fresh produce as much as possible.[542]

Other areas of concern will be but not limited to:

Organic foods are products that must be grown and manufactured in a manner that adheres to standards set by the country they are sold. However, there could be much-needed certification for all steps required, such as water, soil, chemical use, etc.

Irradiated food can be a sort of adulteration impurity due to some changes in the food product such as onion, spices, potato, rice, wheat, mango, ginger, garlic, shallot, meat and chicken, and fresh, dried, and frozen seafood and pulses.[775]

Nanotechnology is the science that takes place at small-scale dimensions called nanometer that can be utilized to design and be used for many new material application and structures normally less than 100 nm size. The main uses in the food industry could be for food fortification and modification such as seed and fertilizers, interactive smart food and packaging products, as well as food tracking.[775]

7.6.2 Development of food laws and regulations

The first Food Adulteration Act was passed in 1860, though many of Hassall's recommendations regarding the treatment of convicted adulterators of food inspection were not taken on board, but many of his recommendations appeared in the Food Acts of 1872 and following Acts between 1872 and 1899.[307]

7.6.2.1 Food legislation

One of the first food legislation dates back to the British Bread Act of 1200. The act was dealing with the quantity of bread that bakers were obliged to sell for sale which was followed by other food products such as ale, fish, meat, etc. in the 13th century. Due to the introduction of other adulterated food, the new Food and Drugs Act was introduced for considering such acts as criminal offenses for the protection of the public and food industry in 1860.

Food Security, Safety, Biosecurity, and Defense

7.6.3 PREPARING A FOOD DEFENSE PLAN

As defined in Section 7.4 "food defense" is overall guidelines to be followed in order to prevent the food become the target of intentional and/or unintentional contamination. It is, therefore, important to evaluate and identify all vulnerable and risk areas for securing food and water resources as well as emergencies such as "Recall and Trackability" when a food safety issue is involved.[537]

For example, an effective emergency plan for the food service and hospitality industry during a disaster should have been recommended, a developed plan, assigning responsibilities, conducting hazardous analysis, developing a list of tasks involved, training, practices, and updating.[450]

There is also a need to address cybercrime affecting not only food and beverage manufacturing but also other sectors of society in the age of information technology and artificial intelligence. In fact, OHS must be considered throughout the entire supply chain as monitoring, auditing, analysis, and record-keeping becomes more critical as the reliance on software and the quantity of data collected increases.

The United State Food Department of Agriculture (USDA)'s Food Safety and Inspection Services (FSIS) suggests the following guidelines and steps for food defense planning based on your organization's needs[288]:

Step 1. Conduct a food defense: Assess internal and external security risks involved.

Step 2. Develop a food defense plan: Food defense plan should include internal and external security measures for access of workers, visitors, contractors, etc.

Step 3. Implementation of the food defense plan: Implement the food defense plan such as who will manage the food defense plan to include training and evacuation.

7.6.4 FOOD BIOSECURITY AND TERRORISM

Biological security refers to the prevention of harmful microorganisms such as viruses, fungi, parasites, and bacteria in animals and plants. To prevent bioterrorism food industry needs to protect the food supply chain. The protection should include[540]:

1. **Personnel**: Food companies have increased employee screening and supervision.
2. **Product**: Food companies have established additional controls for ingredients and products during receiving, production, and distribution, to ensure a high level of food safety.
3. **Property**: Food companies have established additional controls to ensure that they have the highest barriers to guard against possible intruders.[534]
4. **Environment**: Besides the people, product, and property, the environment is also need to be protected from any threat.

Food bioterrorism has been defined as the intentional biological, chemical, and physical contamination of food and water in societies.[309]

7.6.4.1 Comparative risks of food and water as vehicles for terrorist threats

The potential of hazardous chemicals to cause harm in the wrong hands comes to mind in considering terrorism also; still chemistry also works to thwart terrorism with the use of very susceptible analytical techniques to detect harmful substances. Such sabotage could cause disruption and environmental damage to water, soil, and air.[310]

Potential effects of food terrorism impacts could be on disease and death, economic and trade, public health services, and social and political implications throughout the supply chain.[309]

7.6.5 CYBERCRIME

Cybercrime has been around for some time, but during the global COVID-19 pandemic, rates of cyber threats have risen 30,000%. The symptoms can include feeling depressed, sad, anxious, guilty, ashamed, demoralized, and anger. Common feelings can also include feeling betrayed, powerless, out of control, and vulnerable.

The cognitive symptoms can include a negative appraisal of self; perceiving the self as a failure or weak; reduced self-esteem and self-confidence; difficulty focusing or feeling scattered; increased worry and negativity; and fear for safety and feeling unsafe which could further lead to other physical and psychological health issues such as loss of appetite, obesity, headache, sleep disorder, isolation, substance abuse, etc.[734]

Today, many aspects of life have been affected by information technology, and the possibility of cybercrime has increased drastically.[735]

In fact, automation has also been rising in processing plants and food storage facilities. But while computer technology opens up new possibilities for food and beverage manufacturers, it can also open the door to significant cybercrime.[736]

Cybercrime has grown exponentially in the food and beverage industry, targeting big and small brands alike where cybercriminals are increasingly looking to exploit vulnerabilities in your multi-step manufacturing processes and vendor relationships.

Computers have helped countless food and beverage businesses achieve more and expand into new arenas. As companies begin to rely heavily on automated systems to process, store, and manage a large amount of products, the threat of cyberattacks and data breaches increases, and in some cases, there's a lot of valuable information at stake, too.

Therefore, companies need to invest in cyber defense programs to protect people, products, processes, properties, and environment. However, there are a few truths that any business owner would do well to consider[736]:

1. Any business can be targeted by cybercriminals, no matter its size or notoriety.

Food Security, Safety, Biosecurity, and Defense

2. Cyber threats can invade at many points and processes.
3. Cyber risk is always evolving.

As with many secret ingredients and formulas in the food and beverage industry and vast investment to develop a product, it will be significant to protect against cyber threats and cybercrimes by investing in better information technology security system for the prevention of such events.[737]

7.6.6 PREVENTION AND RESPONSE SYSTEMS IN THE FOOD INDUSTRY

It is therefore significant that the food industry plans to address cybercrime within the food chain with emergency measures in place.

WHO, in collaboration with Food and Agriculture Organization (FAO), can further plan, implement, and promote any safety and security measures for the prevention of cybercrime globally.[309]

7.7 FUTURE DIRECTIONS

There has been suggestions for engaging biosecurity workforces by utilizing the existing and embedded mobile technologies. The m-learning process does not impose a technology; it utilizes the accepted technologies to build bridges to a broader range of knowledge systems and ensure local people are an essential and sustainable part of knowledge transfer processes.[539]

7.8 FOOD PREPAREDNESS DURING EMERGENCIES AND DISASTERS

Research regarding human behavior in disaster and emergency scenarios contradicts the commonly held belief that there is a propensity for people to panic and perform other antisocial behaviors.[457]

Historically and in particular in recent years we have been witnessing various disasters such as floods, fires, tornadoes, blizzards, earthquakes, volcanic eruptions, sinkholes, etc. as well as war and terrorism or other unintentional accidents and disasters such as hazardous chemical spills, transportation mishaps, pollutions, and explosions worldwide.[366]

In large-scale disasters, human needs are tremendous, and social disorganization is so widespread that regular community services will be unable to respond, and only emergency social services response systems can meet urgent physical and personal needs until long-term programs are established.

7.9 EMERGENCY FOOD SERVICES

Emergency food services is an emergency response organization designed to provide food for[451]:

1. Those who cannot feed themselves, or those without food or food preparation facilities; and
2. Recovery workers and volunteers.

An effective, emergency food service program must be able to:

1. Help evacuate those affected and assist them to cope with disaster by meeting their needs physically and psychologically to maintain a feeling of well-being,
2. Meet the special food requirements for high-risk groups such as infants, children, pregnant, nursing mothers, and the elderly, and
3. Provide appropriate food service.

Some essential considerations include the following:

- Food requirements (hot or cold meals; quantities)
- Available supplies, staff, and facilities
- Season
- Religious, ethical, ethnicity or cultural needs
- Need for safe and wholesome food

Food service organization requirements are:

- Planning
- Training
- Structure
- Personal roles and responsibilities
- Back-up staff, recruiting food service staff, and mobile emergency food service team

It is essential to consider that foods and drinks have a limited shelf-life, no matter how they are stored or preserved. Food products although may not become spoiled, however, will lose much of their freshness, nutritional value, flavor, texture, and odor over time.

Many survivalists spend a lot of money on specially prepared foods they will never eat. Most food, such as dried, canned, frozen, freeze-dried, dried beans, cereals, rice, grain, etc., can be stored for emergencies at a household level.[321]

One part of emergency preparedness is to keep stocks of emergency food in strategic locations.[320]

In the USA, the Federal Emergency Management Agency (FEMA) has recommendations for family food emergency planning.[322]

The United Nations World Food Program (WFP) responds to any food emergencies as soon as the local government has requested help and crisis response mechanisms go into action. During such a crisis, saving time means saving lives, so the emergency preparedness team must ensure WFP is ready to go, anytime.

Food Security, Safety, Biosecurity, and Defense

In the early days of an emergency, while the first food supplies are being delivered, emergency assessment teams are also sent in to evaluate the quantity of food and water that may be required, for how long, and when to be delivered and quantify precisely how much food assistance is needed for how many beneficiaries and for how long. They must also determine how food can best be delivered to those in need.[429]

8 OHS from Ethics to Sustainability an Agri-food Concern

8.1 INTRODUCTION

The sustainability and prosperity of the US food system are critical to the health and prosperity of workers, employers, and consumers nationwide. To feed the nation, the US food system is a large and growing segment of the US economy and an increasingly important provider of jobs. The food production, processing, distribution, retail, and service industries collectively sell over 1.8 trillion dollars in goods and services which provides employment in grocery stores (retail), restaurants, the food service sector as well as manufacturing for workers representing significant OHS hazard.[718,720]

Figure 8.1 illustrates a modern food processing facility.

As a basic human right people are supposed to have access to adequately clean, nutritious, secure, and safe food. In fact, food and ethics intersect in everyday practices such as choice of being a vegetarian consumer and rejecting meat or adopting for indigenous or religious-based food in our daily diet.[631,632]

Today food production is more complex than ever, and numerous ethical, sustainability, and environmental issues have been raised concerning agri-food practices and the food supply chain is one reason we have been witnessing the rise of organic, holistic, functional, indigenous, religious-based and nutraceutical food even with higher cost worldwide in recent years. Therefore, familiarity with ethical issues for both OSH as well as food industry professionals is important to make appropriate ethical decisions affecting a wide variety of consumers worldwide.

With this regard, a dialogue covering the ethical implications of food production, processing, policy, supply, and consumption may help involve relevant partners and make better decisions.[4]

It is, therefore, more and more essential to promote both ethical issues on food safety and the additional regulatory laws (moral) as may be required involving risk management and OSH professionals, labor, industry, government, and academia.[633]

The 21st-century food establishments need to witness severe breaches of ethical issues in the food industry. Apart from the most essential issues such as prevention of injury or fatality, both workers and consumers face additional challenges such as lawsuits, legal and medical fees, higher insurance costs, market and product losses, time losses, and plant or environment clean up, lower morale, and many others.

DOI: 10.1201/9781003303152-8

FIGURE 8.1 Modern food service production laboratory in a university.

Negligence and lack of social responsibility are not only important for their regularity requirements but also ethical issues for a sustainable society.

Please refer to the 2008 explosion and fire at a Georgia sugar refinery among many other cases presented in the Appendix section for further details.[635 -637,138–742]

The ethical duties in business are three pillars of sustainability that include social, economic, and environmental aspects. The success of these duties depends on the management practices of an organization for their implementations, and when a society can accomplish their adaptation through laws, incentives, and enforcement.[638]

We live in a tumultuous age and change is almost a cliché. But we should recall Bertrand Russell's insight about the distinction between change and progress. The change, he said, is inevitable and scientific, and progress, however, is problematic and ethical.[639]

8.2 SUSTAINABILITY AN AGRI-FOOD CONCERN

The problems of food security, sustainability, and food safety and defense are not new, nor is the idea of convergence. One estimate reported 5,000 people die of foodborne illnesses each year in the USA and 7,000 in the European Union. In 2010, the Pew Charitable Trusts assessed the annual health-related costs from foodborne illness in the USA at $ 152 billion.[611]

Intentional destruction of crops, livestock, and water supplies goes back to ancient times; however, with the present conditions we have to be prepared for both natural and intentional disasters.

It is vital to work toward a sustainable environment and there was much enthusiasm in this regard when sustainable products started appearing in the market initially but only a handful realized their significance for the planet's survival. Sustainability has become a global issue today considering not only environmental

FIGURE 8.2 Water purification and recirculation as a sustainable practice in a food processing plant.

but also economic and societal factors – practicing and promoting depletion of natural resources in order to maintain an ecological balance while considering future generational needs.[596,598, 604]

An example of water purification and recirculation as a sustainable practice in a food processing plant is shown in Figure 8.2.

8.2.1 Sustainability Pillars

The three pillars of sustainability are environmental, economic, and social practices[596]:

1. **Environmental issues such as:**
    - Deforestation,
    - Land degradation and improper use,
    - Climate change (flood, drought, fire, etc.),
    - Global warming (greenhouse effects),
    - Pollution in land, marine, and air (heavy metals, plastic, etc.),
    - Pandemic and other potential diseases,
    - Freshwater reduction and pollution,
    - Wildlife conservation harm (land, air, marine),
    - Vanishing bees and pollination reduction/destruction,
    - Increased and novel pest infestation,
    - Lack of international regulator issues enforcement, and
    - Increased biological, chemical, physical, and psychological accidents.
2. **Economic sustainability**

Nations around the globe should be able to maintain their independence for sustainable agri-food products. However, many issues such as political, wars and natural disasters, etc. may affect their abilities to do so. The following are some examples of economic concerns in sustainability:

- Fair-trade practices,
- Lower wages, and benefits,
- Child labor practices,
- Subsidies,
- Union degradation,
- Higher competition for small businesses/farmers, and
- Trade practices for tax avoidance.

3. **Social sustainability**

Universal human rights and basic necessities are attainable by people who have access to enough resources to keep their families and communities healthy and secure. Healthy communities have just leaders who ensure personal, labor, and cultural rights are respected, and all people are protected from discrimination. Examples are:

- Increased population,
- Political issues,
- Psychological stress,
- Lack of recognition and incentive,
- Lack of ethical codes and deontology,
- New technologies such as *social media, IT; AI*, etc.,
- Occupational health and safety practices,
- Poor and unhealthy working conditions,
- Basic human needs such as shelter, water, health care, education, etc.,
- Agriculture, food safety, quality, and insecurity,
- Hunger, malnutrition, diseases such as allergy, diabetes, obesity, cancer, stress, etc.,
- Lack of early education and adequate professionals on ethical, OHS, and sustainability issues.

Many people have started to ask how they have ended up with the present situation in many aspects of their lives such as inequality, racial tension, lack or inadequate government support on many social issues, etc. witnessing the rich getting richer and the poor getting poorer shows the crumbling of sustainability in the society worldwide.[597, 603]

As an example, some areas such as laboratories in research or production have been operating inefficiently due to loss of energy through machines, thermal, and or other activities in workplaces due to bad designs, inefficient equipment, etc.[599,600]

8.2.2 Hazardous Waste Management

Hazardous waste management has become a significant issue for promoting sustainability and initially started promoting sustainability through[627,628]:

OHS from Ethics to Sustainability an Agri-food Concern

a. Reduce our dependence on fossil fuel,
b. Heavy metals,
c. Reduce synthetic chemicals harmful to nature, and
d. Reduce destruction of nature.

It is therefore vital to not only eliminate and or reduce as much as feasible the waste of all sorts, biological, chemical, physical as well as human resources but also to try to preserve and promote actions that enhance sustainability through its three main pillars, which are environmental, economic, and social responsibilities.

Promotion of sustainability are due to following actions:

1. Reduce, reuse, repair, and recycle (RRRR),
2. Make sustainable choices,
3. Grow your own chemical-free garden,
4. Eliminate or minimize waste,
5. Prevent energy waste,
6. Purchase energy-efficient products,
7. Compost any compostable product,
8. Carpool and use public transport,
9. Plant more trees,
10. Support organizations dedicated to sustainability,
11. Prevent waste of water, and
12. Avoid the use of plastic.[605,606].

8.2.3 Sustainability in the Laboratory

In 2018 McGill established the *Inaugural Sustainable Labs Award* as an integral part of McGill Vision of 2020 Climate and Sustainability Action Plan to further promote sustainability in the workplace as a Higher Education Model having close to 1000 laboratories with occupational safety and health (OSH) for prevention of many losses for best sustainable lab practices targeting the following practices[607]:

1. **Environmental practices** such as green or microchemistry, safer and sound equipment and machine, methodology, facilities, improved recycling etc.
2. **Economic practices** such as energy saving, human resources, and management practices, etc.
3. **Social and societal responsibility** such as community engagement

In fact, similar sustainable practices have also been reported previously at Harvard University and could be categorized as below[609]:

1. Green chemicals,
2. Equipment energy use information,
3. Energy saving through ultra low-temperature fridge and freezer management as well as workplace comfort and efficiency,

4. Hazardous and non-hazardous waste reduction,
5. Proper laboratory design,
6. Purchase of energy-efficient and safer equipment, and
7. Sharing of facilities and resources for cost efficiency.

8.3 ETHICAL ISSUES RELATED TO OHS AND SUSTAINABILITY

Ethics can be defined as a set of moral principles, especially ones relating to or affirming a specified group, field, or form of conduct and in the west, it has been classified into three parts virtuous, rational, and the principle of guidance for wider society's sustainable happier values. They are concerned with socially responsible behavior issues such as animal experiments, nuclear issues, tobacco products, etc.[815] Moral intensity is defined as the intensity of our feeling based on the consequences of a choice we have made based on our action.[641]

8.3.1 PURPOSES OF THE CODE OF ETHICS

The purposes of the code of ethics are:

1. Express, the core values,
2. Clarify values and rules,
3. Strengthen group identity and collegiality,
4. Foster public confidence,
5. Act as a framework for discipline, and
6. Usually is not legally enforceable.

8.3.2 ETHICS AND WASTE MANAGEMENT

Ethical objectives in waste management and in general can include but are not limited to[629]:

- Engage employees in creating a culture of compliance and ethics where their daily words and actions reflect our fundamental commitments and core values,
- Encourage employees to speak up by sharing ideas, asking questions, and reporting issues, concerns, or violations without the fear of retaliation,
- Educate employees on how to comply with waste management policies and procedures, rules, and regulations and why it's essential to do so,
- Ensure employees at all levels of the organization; comply with waste management policies and procedures and applicable external laws, rules, and regulations.

OHS from Ethics to Sustainability an Agri-food Concern

8.3.3 Is safety an ethical issue?

Significant industrial accidents have been reported due to either not following legal obligations and/or unethical behavior.

A recent dissertation explores the regulation of risk concerning the rules governing the slaughter and processing of animals for meat. However, as challenging as it is to establish OHS regulations it is even more essential to make sure these rules are being followed up ethically.[764]

Among many other significant industrial accidents, one example could be verifying the final conclusion and report on the Imperial Sugar Refinery in Georgia, USA, by the Chemical Safety Board (CSB). This will allow us to see how negligent and maybe unethical behavior led to a number of deaths, injuries, and significant lawsuits, which could have been prevented, led to such a significant disaster and, finally, creation of a standard for "Dust Explosion", noting that all employers are ethically required to provide a safe and healthy workplace for their employees.[642,738–741]

8.3.3.1 Food sovereignty

Food sovereignty is a basic human right for accessing not only the wholesome but also ethnically, indigenously, or religiously based food products for various cultures in a sustainable manner [771].

This stands in contrast to the present corporate food regime, in which corporations and market institutions control the global food system.[772]

8.3.3.2 Food ethic

As pointed earlier food ethics dispute the fact about a wide range of issues affecting the food and food industry such as inhumane labor or animal treatment, food adulteration, counterfeit or genetically modified (GMOs) products, right to indigenous, religious, or culture-based foods etc., on moral grounds, and based on what we think is right in order to redirect our thinking of appropriate food manufacturing practices from farm to table.

Today, ethical decision-making may require the example of following questions to be considered to assist us in finding the most suitable approach[641]:

1. Are our actions legal or not?
2. Are decisions made perceived ethical or not?
3. What population group will be affected?
4. What are the consequences if reported on media?
5. Which ethical framework is most appropriate?
6. What options exist? Which has the most significant ethical advantages?

Examples of practicing ethical issues in agri-food industry could be:

If we are producing vegetarian or allergen-free food products, we must make sure our production, line and storage facilities are treated properly to be free from any meat products or allergens. If we are producing food and drink products, we should consider healthier versions. If we are committed to sustainability, we should promote greener planet and promote and practice Reduce, Reuse, Repair and Recycle (RRRR).

In the past century, our way of food and drink production has been affected by science, technology, globalized trade, climate change, population, and many other factors from farm to fork which have drastically affected our approach to considering ethical issues.[15]

We do not usually see what we eat as an ethical issue. However, this was from the past when humans just ate anything, they could gather, but today many ethical issues come to mind, such as looking for the source and where it is coming from. Is it vegetarian or not, local, or from other countries? Is it wholesome, or organic, etc.?

Three ethical issues impacting food are presently as follows:

1. Using animals for food,
2. The environmental impact of our food, and
3. The effect such ethical issues may have on the world food crisis.[765]

Therefore, today many consumers consider ethical issues regarding not only the quality and safety but also the way it has been prepared for the market to make a rational ethical decision on food purchases. It could have even affected consumers in purchasing bananas as they were sprayed by a carcinogenic pesticide known as chlordecone causing prostate cancer in field workers and banned presently. These workers were neither told nor provided protection and OHS instruction during harvest or production.[766,767]

Other examples of other sustainable and ethical issues affecting the food industry from farm to fork include but are not limited to are:

- Vegetarianism versus animal protein and meat,
- Lack of animal welfare and how they are raised,
- Child or labor abuse practices,
- Lack of adequate occupational health and safety regulations,
- Unsustainable practices in agriculture and wildlife may relate to food production, hunting, fishing etc.,
- Job losses due to information technology or artificial intelligence,
- Food adulteration, fraud, tampering, and mislabeling,
- Supporting local produce,
- Organic food and fair-trade practices,
- Diseases, such as allergies, poisoning, dermatitis, rhinitis, etc.,
- Packaging and labeling issues such as mislabeling and selling using expired food,
- Ethical, indigenous, and religious issues such as novel foods, halal, kosher etc.,
- Biopiracy when research organizations take biological resources without official sanction, mainly from less affluent countries or marginalized people,[824] and

- Adapting ethical policies and laws regarding the acceptance of funding and donation.[827]

In summary, an ethical worker will deeply care for the type of their work or where their food comes from considering the welfare of the animals and people involved in making that food because that is what makes our planet a sustainable environment for future generations.[630,643]

9 The Future of OHS in Food and Beverage Industries

9.1 INTRODUCTION

In spite of strict rules, laws, and guideline developed over time in food and drink industries and production facilities within a controlled environment, it will still be risky to assume all things are safe and hygienic, and therefore many standards and OHS practices need to be maintained.

Food culture and processing have evolved and developed over centuries and with technological advances and other global challenges such as climate and environmental changes, population, diseases, etc. new OHS issues will continue to appear, which impact producers, workers, consumers, and the environment. This has led to policymakers adapting to new situation historically as follows:

1. The emergence of cities with vast trading routes for tea, spices, etc., for thousands of years.
2. The Iron Age/Roman Empire brought expanding global food systems.
3. The Middle Ages witnessed the emergence of the wealthy class with increased demand for more and different sophisticated types of food and drink products during earlier trades.
4. The Industrial Age brought further development in science and technology by the transition from manual labor to industrialization and machinery introduction into the workplace culture by the 1700s.
5. Colonization, war, natural or intentional disasters, and terrorism did bring significant political changes and issues facing food systems, and
6. Information Technology and Artificial Intelligence have been introduced in the 21st century presenting new challenges and safety issues such as cybercrime.

Traditionally, large number of workers were engaged in agriculture and farming for food production with mostly manual and lifting ergonomic problems. Industrialization lowered the number of farm workers and resulted in the expansion of food and beverage industry. These changes were also associated with the creation of other hazards such as pollution of water, air, and soil. Further advancements in science and digital technology such as information technology and artificial intelligence have brought other threats such as cybercrime to not only workers and

DOI: 10.1201/9781003303152-9

230 Occupational Health and Safety in the Food and Beverage Industry

company confidential information but also to products, property, and environment. Ultimately, they could result in food security issues by disrupting the food supply chain in a food system covering a range of issues from harvest to distribution, process, storage, and retail from farm to the table in the 21st century.[718]

As a result of these constant changes over time, food systems have been facing increased complexity. National and international food and agricultural trade will require further attention and tightening of standards. As an example, today a cheeseburger been reported to include more than 50 different ingredients coming from various continents. Recent cybercrime sabotage on the largest cream cheese manufacturer in the USA· added the chaos afflicting global food supply chains and food security challenges during the COVID-19 pandemic led to increased prices.[823]

Continuous challenges will require OHS professionals and food scientists to be adequately informed and trained to deal with not only food safety and security challenges but also protection of the food and beverage industry in future considering direct and indirect threats to the food supply.

Climate change, consumer demands, wars, terrorism, natural disasters, diseases, and inflation affect many food-commodity prices to increase affecting food safety and security through hoarding and/or food adulteration.[825,826]

Injury and accidents in the food industry require continuous improvement, report indicates that with continuous scientific and industrial advancement many workers continue to be injured in the job.[835]

The main causes of occupational ill health have been reported to be musculo-skeletal disorders, work-related stress, asthma, dermatitis, and noise in the working environment. Common hazards include slips, trips and falls, sharps instruments, lifting, repetitive work and work posture injuries, machinery, exposure to biological and chemical hazards, or thermal (heat or cold). [836]

Previously reported data showed that machinery and plant cause 30% fatal injuries, 10% major injuries requiring hospitalization, and around 7% of all injuries including significant injuries and over-three-day absence injuries.[837]

9.2 EMERGING TREND, PRIORITY AREAS, AND AN OHS MODEL

In previous chapters environmental health and safety (EHS) and health occupational health and safety (OHS) from organization and management to accident and disease prevention, food defense, biosecurity, and sustainable and ethical issues were addressed. However, for the future planning of an appropriate OHS, the following three factors must be considered:

9.2.1 EMERGING TRENDS

These are issues such as ergonomics, information technology, or artificial intelligence to be decided based on their importance, persistence, priority and public confidence, health and wellness, health and safety education, competent professionals, changing nature of the workforce and workplace, as well as response to the evolving

The Future of OHS in Food and Beverage Industries

technologies. The trend will be on the prevention of near-miss accidents, accidents prevention responsibility, engineering technics, transportation and traffic safety, financial crisis and international, and legal dispute.

9.2.2 PRIORITY AREAS

These will include but not be limited to:

- Saving lives and preventing injuries,
- Changing direction from production to service employment, from corporate to independent contractor and consultants,
- Social and cultural changes (minorities and women, higher divorce and more single-parent families, two household income),
- Stress-related factors (aging population, worker vulnerability, psychosocial stress),
- Globalization (greater demand on international standards),
- Accident Prevention (pollution, contamination, etc.),
- Traceability and recall (food safety and quality),
- Wholesome food (nutritious, organic wholistic and nutraceutical products, etc.),
- Legal and regulatory issues,
- Technology transfer from industrialized nations to developing countries,
- Ethical and sustainability issues,
- Environmental issues (climate change etc.),
- Food protection, security, and defense,
- Promote and require education on ethical issues,
- Cybercrime prevention, and
- Food ethics.

While there have been challenges to agricultural production, the pressures and constraints suggest that the current environment must consider ethical challenges accordingly. [777]

9.2.3 AN OHS MODEL

An OHS model involves planning, acting, checking, and reviewing with the following elements to be considered[657]:

9.2.3.1 Continuous Improvement Management System (CIMS)

CIMS is a systematic approach for developing Leadership, Hazard Identification/Risk Assessment, Implementation Strategy and planning, Standard Development, Procedures, Guidelines, Practice, Responsibilities, Communication, Training/Awareness, Measurement, Evaluation and Recognition, Improvement, and Correction in workplace and organization.

9.2.3.2 Program core elements to include

Accident Investigation, Contractor Program, Emergency Response, First Aid & CPR, HACCP, GMP, ISO, and other standards, Health & Safety Policy, Industrial Hygiene Monitoring Strategies, Joint Health & Safety Committees / representatives, Orientation, PPE, Preventive Maintenance, Purchasing Policy, Refusal to Work, Work Stoppage and Workplace Inspection.

9.2.3.3 Hazards to be considered

a. Intentional/deliberately (adulteration, tampering, war, terrorism, etc.), and
b. Unintentional/natural/accidental (toxicants, disasters, contamination, pollution, etc.)

The source of hazard could be biological, chemical, physical and psychological/ stress, and environmental issues related:

1. Directly due to physical work-related environment, engineering or ergonomic design, lighting, temperature, humidity, ventilation, noise, pollution, fatigue/stress-related, equipment/machinery, lack of training, skills and education, carelessness/ethical behavior, etc.), and
2. Indirectly due to work-related environment such as job content, culture, role, career, characteristic of person, schedule or workload, stress-related issues such as psychological, behavioral, emotional, cognitive reactions, mental health, medical problem, and individual characteristics impact of the worker such as age, gender, education, competitiveness, etc.

9.2.3.4 Injury Management Policy (IMP)

IMP should include, Accident Reporting, First Aid & CPR, Medical Program, Short term Disability (STD) STD, Long Term Disability (LTD), Back to Work Program, and Regulatory Health and Safety Board.

9.3 SHAPING THE FUTURE

The food and beverage manufacturing industry with numerous employees, machineries, and activities continue to be the significant sector after mining, construction, and foundries in terms of occupational health and safety hazards involved.

The introduction of the machine in Britain, and the resulting Industrial Revolution, carried many risks and hardships to workers and occurred during the second part of the 18th and the first part of the 19th centuries.[658] Industrial expansion has not been limited to only machinery but also vast expansion of transportation equipment and machinery for food, beverages, and agricultural products that require a more sustainable approach for not only in their safety features but also their energy efficiency as well as adaptation of travel efficiency for a sustainable future.[598]

The continuing burden of largely preventable occupational disease and injuries and the lack of adequate OHS services in most small and many large workplaces indicate a clear need for OHS professionals at all levels in the future.[522]*This*, in

The Future of OHS in Food and Beverage Industries

particular, is important for applying the standards and training of workers based on the market need in each sector considering cultural, indigenous, ethical, and environmental issues for such a sustainability approach in the future.[839,656]

Historically numerous outbreaks and diseases have been reported, and although the recent COVID-19 pandemic has not been a food safety issue at a global scale,[647,648,649] it did cause disruption in different food production and supply chains. As a result, a number of meat and poultry processing and production facilities were shut down due to inadequate and speedy OHS protection and prevention measures. Hackers also targeted the meat giant JBS SA and an Iowa grain cooperative, adding new challenges for OHS practices.[691] This situation led to the shortages in food supply chain for meat and poultry products in the market.[651]

It has also been reported that inadequate and poor OHS measure in workplace resulted around 2.5 million severe illnesses or deaths and 374 million of injuries apart from suffering, pain, and significant economic losses per year.[659]

It is therefore essential to plan ahead and be prepared for other similar epidemic and natural disasters in case of emergencies in the future.[650,651,652,653]

There are of course other unresolved problems and challenges to face globally such as substance abuse, AIDS, environmental pollution, psychological issues, natural disasters, and cybercrime activities to be considered for safety and security reasons. The pace and complexity of the forces for change are enormous and intimidating, yet there is a possibility for intelligent men and women to lead their societies through the complex task of preparing for the century ahead.[658]

There continues to be an increase in working hours and a lack of essential form of standards due to consumer and market demands in the workplace globally. Hybrid workplaces, digital education, online social contacts, and ordering have all become accepted norms of daily lives which could be considered essential steps in training OHS professional and skilled workers. Although some jobs may be eliminated due to a variety of factors such as inflation, information technology, and artificial intelligence, others may be created with new OHS risks, causing psychological/environmental stress development and new ethical issues.[659]

Consequently, there will be a demand for more qualified and adequate number of OHS professionals in all aspects of lives to face new hazards and challenges ahead in the workplace environment, globally.[660,663]

In addition, health and safety practitioners have more challenges to overcome due to new forms of psychological or environmental hazards such as recent pandemic of COVID-19.[828]

Agriculture, fishing, and the food industry continue to be among the highest fatal and injury-causing industry sectors after mining and construction; therefore, new training information, accreditation, and certification have been on the rise as required.

The food industry will continue to be challenged by biological, chemical, physical, and psychological/environmental hazards with new elements during such transition in the future. Moreover, market conditions, climate change, economic, social, political, ethical, cultural, labor, technological advancement in information technology and artificial intelligence will require continuous OHS criteria for not only the workers and workplace and food system from farm to fork but an overall sustainable environment.[753,754,728–737]

Accordingly, new and unknown hazards could appear in future workplaces with the advancement of science and technology, and we must, therefore, learn to adjust to new technologies, apply engineering technics, better loss management control during financial crises, adopt national and international standards and consider shifts in new workplace culture in a globalized economy.[701]

Today society's transgenerational social fabric is fraying by disconnection within family and community, making us more vulnerable to being lonely, resulting in more anxiety and depression even before the COVID-19 epidemic.

Countering dangerous workplace hazards will require not only identifying the risks but also their evaluation for the best method of prevention through social contact in today's hybrid work environment, in-person, and up-to-date training of workers for a more sustainable and ethical workplace environment and the need to ask, "am I consistent in my messaging to employees to be best prepared for the challenges ahead?"[661,829]

Today, OHS has become an essential part of any good food and beverage manufacturing activity with many benefits such as promoting a better working environment, preventing injuries and fatalities, increasing productivity, maintaining good workers, and preventing various financial losses.

At a societal level, although recognition increases human productivity, any final success would only be achievable through sharing, a healthy and equitable society, for a sustainable future.[644]

It is easy to get discouraged and overwhelmed by the many problems in our society, to be demoralized by the inequities, adversities, and other traumas that are all too pervasive in our world, but the overall trajectory of humankind has been positive in history.

Although we have advanced in science, engineering, and technology, still our way of life could be at harm and require broader OHS measures to be prepared for potential threats and hazards. These measures will allow us to make our world a safer, more just, healthier, humane, and sustainable place in the 21st century and beyond.[662, 608,645]

The role of OHS will also need to evolve with the future work based on the UN 2030 Agenda for Sustainability Development; OHS Revised Reporting Initiative; ISO 45001 standards; and International Labor Organization Management Systems.[681]

Ultimately, the new OHS updates, accreditation, and academic and industrial skill development in the field will be crucial in the future.[685]

Appendix 1
Case Studies

WHAT IS A CASE STUDY?

At face value, a case study is a deep dive into a topic. Case studies can be found in many fields including the food and beverage industry.

When you conduct a case study, you create a body of research based on an inquiry and related data for finding the root causes of an accident/incident and propose solution and what could have been the best approach and recommendation to prevent hazards and injuries to people, process, product, property, and environment.

The best approach would be to:

1. Study and investigate pattern,
2. Gather evidence,
3. Present findings, and
4. Draw a conclusion.

Workers in food manufacturing are more likely to be fatally injured and experience nonfatal injuries and illnesses than workers in private industry. Food production workers are also much more likely to suffer an injury requiring job transfer or restriction than one who requires days away from work.

Although more than 50% of workers in this industry are employed in production occupations, they operate in diverse environments that present a multitude of occupational health and safety (OHS) risks. Workers in food manufacturing are more likely to be fatally injured and experience nonfatal injuries and illnesses than those in private industry.759 In addition, food manufacturing workers are much more likely to suffer a nonfatal injury or illness requiring job transfer or restriction than one resulting in days away from work to recuperate. 759

Following, you will find more than two dozen accidental cases in the agri-food and beverage industry which have resulted in fatalities, injuries, and illnesses with biological, chemical, and/or physical root causes representing industry, academic, and government organizations nationally and internationally. The case could also be an overall overview of a sector with accidents for analyzing and proposing OHS management practices in the food and beverage industry. 782

235

236   Appendix 1

CASE # 1

They cover the main causes of injury and occupational ill health including nine real-life accidents and solutions by the Health, Safety & Environment (HSE) in the UK: 1. Manual handling and musculoskeletal injuries, 2. Slips and trips on wet or contaminated floors, 3. Fall from heights, 4. Workplace transport, 5. Struck by an object, 6. Machinery, 7. Occupational asthma, 8. Noise and 9. Occupational dermatitis. 723

CASE # 2

Provide one case study in biscuit, chocolate, gum, and toffee manufacturing in Pakistan evaluating the hazards involved and their solutions for preventive measures in process. 724

CASE # 3

Workplace Safety & Health (WSH) Council in a private firm in Singapore provides risk assessment and job analysis in six hazardous accidents with the solutions. The cases include 1. Death in a mixing tank, 2. Worker killed during machine cleaning, 3. Worker losing finger working with a bandsaw, 4. Worker being struck by a falling window sash, 5. Worker death after slip and fall in a bread factory, and 6. Worker's hand being caught in a meat mincer. 725

CASE # 4

A 35-year-old employee of Quebec's Ministry of Agriculture, Food, and Fisheries
   (MAPAQ) in Canada who died as a result of an accident in a cranberry farm, getting caught in the machinery due to loose clothing 727.

CASE # 5

Complete analysis and investigation by the USA Chemical Safety Board (CSB) being discussed with documents as well as a 10-minute video regarding the 2008 Georgia sugar refinery explosion and fire resulting in a number of fatalities, injuries, and destruction of the plant. 738 - 741

CASE # 6

Maple Leaf Canada recalling products due to a number of fatalities and illnesses due to food safety issues. 742,743 Routine application of Root Cause Analysis (RCA) principles and sharing of lessons learned can help improve food safety and prevent foodborne illnesses. 745 The case discusses 22 people who died out of 57 confirmed cases of food poisoning linked to deli meats produced at a plant owned by Maple Leaf Foods Inc. The company did recall voluntarily 220 products made at the plant

Appendix 1 237

in one of the biggest food recalls ever in Canada resulting as well in a loss of $20 million. 747, 748

CASE # 7

The case discusses a tragedy of a young female worker in Ontario, Canada who was working on a conveyor line when an article of her clothing got caught in the machinery. 746

CASE # 8

Various hazardous accident cases are discussed under "Hazard Alerts" notices regarding confined space hazards discussing accidents in a 1. Fish processing plant, 2. Beverage storage tank, 3. Chicken waste vat, 4. Cleaning operation when argon replaced oxygen, 5. Fermentation tank, 6. Explosion and hearing loss cases. 264

CASE# 9

Accidents – analysis of 257 small- and medium-size food companies in Malaysia, resulted in recommending better OHS practices compared to larger and multi-national companies based on the following typical questions for data analysis including, 1. Have you experienced any accidents in the workplace? 2. What is the level of injury that you have suffered? 3. What is the cause of the accident? 4. Were there any actions taken against you because of this accident? 5. What types of accidents that occurred in your company for the past two years? 6. What types of incentives received when complying with all safety rules and regulations? 7. What kinds of punishments an employee would receive if he/she does not comply with safety rules and regulations?[750]

CASE # 10

Fast food restaurant customer in injured in slip and fall. 751

CASE # 11

Burning hand with kitchen fryer in food service area. 752

CASE # 12

The story of temporary workers where Fiera Foods an industrial bakery co. "partner" plant in Ontario, Canada, resulted in the deaths of these two workers. 754, 755

CASE # 13

A laboratory fatal accident case in the University of California, USA. A complete analysis of the accident, failures, and lessons learned in a short – 5-minute video by the Chemical Safety Board (CSB) as well as other investigations reported. 760–763.

CASE # 14

Two cases of pesticide poisoning by organophosphate involving organophosphate in India. 789,790,791

CASE # 15

Over the course of the past ten years, there has been a significant number of food safety incidents associated with chocolate products. The presence of Salmonella enterica, Listeria monocytogenes, allergens, and foreign materials in cocoa/chocolate products have been reported on a global (European Union RASFF, Australian Competition and Consumer Commission, UK Food Standards Agency) scale. 783

CASE # 16

A global case research study analyzing the effect of the food production chain from farm practices to vegetable processing on outbreak incidence. 784

CASE # 17

Accidents and accident prevention in agriculture a review of selected case study. 784.
    The Chemical Safety Board (CSB) in the USA has resourceful material for case studies in oil refinery and chemical plant toxic releases, fires, and explosions, the CSB investigates other facilities including pharmaceutical and food processing plants for further comprehensive studies.

CASE # 18

A case study in India identifies some of the physical, chemical, biological, and psychological hazards in the seafood processing sector. Apart from the nature of the job of fish processing workers occupational hazards prevailing in the work environment contribute significantly to the occurrence of work-related diseases and prevention of such occupational hazards may help in protecting workers from occupational diseases also. 788

CASE # 19

Several accidental case studies involving fire and explosion and fatality of workers in Italy, as well as, focusing on maintenance hazards related to vegetable oil refining.[794, 795, 796, 797]

CASE # 20

Many industrial food and beverage facilities use ammonia as a refrigerant but do not always have proper equipment and protocols in place to ensure workers are kept

Appendix 1

safe when a refrigeration leak happens. 806 A ten-year (2007–2017) case study of the hazard associated with ammonia and several cryogenic gases is examined. 805 Technical Safety BC will be analyzing the ammonia leak incidents.

Appendix 2

TABLE A2.1

Examples of Exposure Agents in the Food and Beverage Industry

Irritant Substances	Sensitizers
• Ammonia	• Fermentation process, enzymes, hops
• Chlorine	• Bromelain
• Organic solvents, benzene, hexane	• Flour, wheat, grain dust
• Sulphur dioxide acids	• Tea, coffee, soya bean, sugar beat
• Detergents	• Onions, garlic, oats, barley, spices
• Hand cleaners	• Animals, mites
• Wet work pesticides	• Antibiotics, penicillin
• Plants and grains	• Disinfectants
• Meat, vegetable juices	• Detergents/enzymes
• Dry ices	• Gluteraldehyde
	• Pesticides
	• Fungicides
	• Lemon, orange, lime flavor
	• Latex
	• Formaldehyde

References

1. Occupational safety and health. Viewed October 1, 2014. http://en.wikipedia.org/wiki/Occupational_safety_and_health
2. Occupational health and safety. Raising awareness in your organization. Viewed October 1, 2014. http://ohsonline.com/articles/2014/10/01/raising-awareness-in-your-organization.aspx?admgarea=news
3. Sharon Clarke, Ronald J. Burke. 2011. *Occupational Health and Safety.* Chapter 1 Page 3, Vt.: Gower London, England.
4. Britvic. May 2012. http://en.wikipedia.org/wiki/Britvic
5. The manufacturing process of beverage production. http://www.dr-machine.com/news/item_2878.html
6. A Erickson Paul. 1996. *Practical Guide to Occupational Health and Safety.* San Diego: Academic Press. Page 3.
7. Incentives and rewards for health and safety. Viewed October 1, 2014. http://www.hse.gov.uk/construction/lwit/assets/downloads/incentives-and-rewards.pdf
8. Barbora Joseph. 2007. *Health and Safety Policy Manual.* PBG's (Pepsi Bottling Groups), Montreal, Canada.
9. Preventing injuries to restaurant and hotel workers. Viewed October 1, 2014. http://www.worksafebc.com/publications/reports/focus_reports/assets/pdf/focushotel.pdf
10. Aventis. Environmental health, safety. Global EHS standards. EHS leadership and administration. *Pocket Edition.* August 2003. Pages 3–19.
11. FAO Food and Nutrition. 2008. *Risk - Based Food Inspection Manual.* Rome, Italy: FAO. Page 89.
12. A Keith Furr. 1971. *CRC Handbook of Laboratory Safety.* 3rd Ed. Pages 1–17.
13. Wogaller Michael S. 1999. *Warning and Risk Communication.* London, Philadelphia: Taylor & Francis. Page 4.
14. American Chemical Society, Safety in Academic Chemistry Laboratories. 2002. *Accident Prevention for Faculty and Administrators.* Volume 2. 7th Ed. Pages 1–2.
15. Why does accident happen? Viewed October 1, 2014. https://www.nonprofitrisk.org/tools/workplace-safety/public-sector/concepts/why-ps.htm
16. Is it worthwhile investing in health and safety program? Institute for work and health. Viewed October 29, 2015. http://www.iwh.on.ca/sbe/is-it-worthwhile-investing-in-health-and-safety-programs
17. Health & Safety Executive (HSE). Investing in health & safety. 2010. http://www.tssa.org.uk/article-47.php3?id_article=2531
18. Food safety audit and inspections. Viewed October 1, 2014. http://www.aibonline.org/aibOnline_/GenericForm.aspx?strOpen=\www.aibonline.org\auditservices\food-safety\index.html
19. Responding to a food recall; procedures for recall of USDA food. Viewed October 1, 2014. http://www.fns.usda.gov/sites/default/files/Responding_Food_Recall_FNS_Final_May_30_2014.pdf
20. Food safety investigation and recall. Viewed October 1, 2014. http://www.inspection.gc.ca/food/information-for-consumers/food-safety-system/basic-html/eng/1374439778888/1374821384212
21. Smith Rhett. Hitting the traceability target at birds eye foods. *Food Safety Magazine,* February / March 2005. www.foodsafetymagazine.com

22. Accident prevention in food and drink industry. *Food & Drink Business Europe.* Viewed October 2, 2014. http://www.fdbusiness.com/2012/07/accident-prevention-in-the-food-and-drink-industry/

23. H Fawcett Howard. 1988. *Hazardous and Toxic Materials Safe Handling and Disposal.* Hoboken: John Wiley & Sons. Page 30.

24. OSHA. Bakery products manufacturer fined more than $270,000 for exposing workers to unsafe machinery and other hazards. Viewed November 1, 2010. https://www.osha.gov/pls/oshaweb/owadisp.show

25. Private food safety standards: Their role in food safety regulation and their impact. Viewed October 1, 2014. http://www.fao.org/docrep/016/ap236e/ap236e.pdf

26. The public health and safety organization. Viewed October 1, 2014. http://www.nsf.org/services/by-industry/food-safetyquality/processing/processing-audits

27. Introduction to HACCP. Viewed October 1, 2014. http://www.food.gov.uk/business-industry/caterers/haccp

28. Total Quality Management (OTQM). Viewed October 1, 2014. http://asq.org/learn-about-quality/total-quality-management/overview/overview.html

29. Food traceability. Viewed October 1, 2014. http://www.foodstandards.gov.au/industry/safetystandards/traceability/pages/default.aspx

30. The changing world of food traceability. Viewed October 1, 2014. http://www.foodsafetynews.com/2013/05/the-changing-world-of-food-chain-traceability/

31. Food and drug Administration (FDA). Viewed Feb 29, 2012. www.fda.gov/food/foodsafety/retailfoodprotection/fderalstatecooperativeprogram

32. Hazards in food and beverage industry. Viewed October 2, 2014. http://www.michigan.gov/documents/cis_wsh_cet0108_108504_7.htm

33. Inspection modernization. *CFIA.* Viewed October 8, 2014. http://www.inspection.gc.ca/about-the-cfia/accountability/inspection-modernization/eng/1337025084336/1337025428609

34. Safe food handling guidelines. Viewed Feb 29, 2012. www.ehs.edu

35. Naomi Rees, David Watson. 2000. *International Standards for Food Safety.* Maryland: Aspen Publisher. Chapter 1 page 1 and Chapter 3 page 29–35.

36. Ali R Taherian, Mohammad Reza Zareifard, Ebrahim Noroozi. 2010. Vegetables as ingredients in processed food products. In: *Handbook of Vegetables and Vegetable Processing*, Eds: YH Hui et al. Iowa: ISUP, John Wiley & Sons, Inc. Pages 1–23.

37. Safe maintenance; food and drink manufacture. Viewed October 2, 2014. https://osha.europa.eu/en/publications/e-facts/efact52

38. List of industrial disasters. Viewed October 2, 2014. http://en.wikipedia.org/wiki/List_of_industrial_disasters

39. Emergency response plan. Viewed October 2, 2014. http://www.ready.gov/business/implementation/emergency

40. Good manufacturing practices (GMP). Viewed October 8, 2014. http://www.hc-sc.gc.ca/dhp-mps/compli-conform/gmp-bpf/index-eng.php

41. Manitoba HACCP Advantage Program Manual; Good Manufacturing Practices (GMP) Advantage and HACCP Advantage. 2006. Manual, Federal – Provincial – Territorial Initiative. Government of Canada, Canada.

42. Personal protective equipment. Viewed October 2, 2014. http://www2.worksafebc.com/Topics/PPE/InfoSheets-Industry.asp?ReportID=35715

43. T Tweedy James. 2014. *Healthcare Hazard Control and Safety Management.* Boca Raton: CRC Press / Taylor and Francis. Chapter 2. Pages 1–3.

44. Vayrynen Seppo et al. 2015. *Integrated Occupational Safety and Health Management: Solutions and Industrial Cases.* Cham: Springer. Page 8.

45. Ron Wasik. Reviewing the recalls. *Food in Canada.* March 2010. Page 16.

References

245

46. ISO 22000, Step by step. Viewed October 8, 2014. http://www.22000-tools.com/steps-to-iso-22000.html?gclid=CJrS_va1ncECFWELMgodCBcALQ
47. Recall. Viewed October 8, 2014. http://www.consumerinformation.ca/eic/site/032.nsf/eng/00062.html
48. Personal injury. Viewed October 8, 2014. http://adviceguide.org.uk
49. Product recall. Viewed June 19, 2012. http://en.wikipedia.org/wiki/Product_recall
50. Food recall warnings, high risk. Viewed October 8, 2014. http://www.inspection.gc.ca/about-the-cfia/newsroom/food-recall-warnings/eng/1299076382077/1299076493846
51. Quest Solution. Empower your workforce. Viewed July 4, 2014. http://www.questsolution.com/food-traceability-recall/index.html
52. LJ Harris, AA Kader. 2002. *Postharvest Technology of Horticultural Crops*. Oakland: University of California, Agricultural and Natural Resources. Communication Services. Chapter 24. Pages 301–314.
53. Your online resource for recalls. Viewed June 19, 2012. http://www.recalls.gov/food.html
54. United States Department of Labor. Viewed October 8, 2014. http://www.dol.gov/owcp/dfec/regs/compliance/ca-11.htm
55. Layton Lyndsey. Traceability rule represents big adjustment for food industry. *The Washington Post*. Viewed October 8, 2014. www.washingtonpost.com
56. Food Traceability. Health & Consumer Protection Directorate – General, European Commission. Viewed October 8, 2014. http://ec.europa.eu/food/food/foodlaw/traceability/factsheet_trace_2007_en.pdf
57. Traceability. Agriculture and Agri-Food Canada. Viewed June 19, 2012. http://www.ats-sea.agr.gc.ca/trac/index-eng.htm
58. Traceability. Manitoba Agriculture, Food and Rural Initiatives. Viewed June 19, 2012. http://www.gov.mb.ca/agriculture/foodsafety/traceability
59. M Dillon, C Griffith. 2006. *Auditing in the Food Industry; from Safety and Quality to Environmental and Other Audits*. 2nd Ed. Cambridge, United Kingdom: Wood Head Publishing Limited and CRC Press. Pages 5–7.
60. Gary R Krieger, John F Montgomery. 1997. *Accident Prevention Manual for Business and Industry*. 11th Ed. National safety Council. Pages vii–viii.
61. Gene Shematek, Wayne Wood. 1993. *Laboratory Safety; CSLT Guidelines*. 3rd Ed. Page 52 –55.
62. A Recipe for Safety. Occupational health and safety in food and drink manufacture. Health and Safety Executive (HSE). 2005. Pages 1–30.
63. Food and drink injuries and ill-health survey revealed. Viewed October 8, 2014. http://www.iosh.co.uk/News/Food-and-drink-injuries-and-ill-health-survey-revealed.aspx
64. Health and safety in food and beverage industry. Viewed October 8, 2014. http://www.labour.gov.za/DOL/downloads/documents/useful-documents/occupational-health-and-safety/Useful%20Document%20-%20OHS%20-%20Occupational%20Health%20and%20Safety%20in%20the%20food%20and%20beverage%20industry.pdf
65. D Dawn Matthews. 1999. *Food Safety Sourcebook*. Michigan: Omni Graphics, Inc. Chapter 35. Page 197–199.
66. *Advance Food Safety Training in Canada. Manager Certification Course Book*. The National Restaurant Association Educational Foundation. TrainCan, Canada. 2001. Page 11–13.
67. YM Saif, AM Fadly, JR Glison, LR McDougald, LK Nolan, DE Swayne. 2008. *Disease of Poultry*. Blackwell Publishing. Pages 3–5.
68. Farber Jeff. 2014. *Retail Food Safety*. New York: Springer. Chapter 1; an introduction to retail food safety. Page 1–2.

69. DS Clark, FS Thatcher. 1968. *Microorganisms in Foods.* Toronto Canada: University of Toronto.
70. Yates Carolyn. 2007. Dinner with a dash of E. Coli; you won't find these ingredients on the label. *The McGill Tribune.* Pages 14–15.
71. *Encyclopedia of Health and Safety.* Geneva: International Labor Office. Page 65.11. 4th Ed. Volume III. Viewed October 7, 2014. http://books.google.ca/books?
72. AK Thompson. 2003. *Fruits and Vegetables; Harvesting, Handling and Storage.* Chapter 9. Oxford, UK: Blackwell Publishing Ltd. Page 80–85.
73. Hazards of confined space in food and beverage industries. Viewed October 7, 2014. http://www.worksafebc.com/publications/high_resolution_publications/assets/pdf/bk82.pdf
74. Ergonomic intervention for the soft drink beverage delivery industry. Viewed Oct 7, 2014. http://www.cdc.gov/niosh/pdfs/96-109.pdf
75. Tomoda Shizue. Occupational safety and health in the food and drink industry. United Nation, International Labor Office. Viewed Feb 19, 2016. http://www.ilo.org/wcmsp5/groups/public/---ed_protect/---protrav/---safework/documents/publication/wcms_219577.pdf
76. Institute of Food Technologists (IFT). *Food Times Gazette. The Great Food Fight.* Vol. 1 No. 1.1996.
77. Richard Holley. Smarter inspection will improve food safety in Canada. *Canadian Medical Association Journal.* 2010; 182(5). Pages 471–473.
78. Wasik, Ron. Food safety priorities for 2010. *Food in Canada.* January/ February 2010. Page 20.
79. Dermatitis. Health and Safety Executive. Viewed June 20, 2012. http://www.hse.gov.uk/food/dermatitis.htm
80. Occupational asthma. HSE. Pages 1–8. Viewed June 20, 2012. http://www.hse.gov.uk/statistics/causdis/asthma/asthma.pdf
81. Vayrynen Seppo et al. 2015. *Integrated Occupational Safety and Health Management: Solutions and Industrial Cases.* Cham: springer. Chapter 10. Environmental services and food / dietary / department safety. Page 238.
82. Vayrynen Seppo et al. 2015. *INTEGRATED OCCUPATIONAL SAFETY and health management: Solutions and industrial cases.* Cham: springer. Chapter 10. Environmental services and food / dietary / department safety.
83. International Finance Corporation. Environmental health and safety guidelines, food and beverage processing. Viewed October 7, 2014. http://www.ifc.org/wps/wcm/connect/c7bfaf0048855482b314f36a6515bb18/final+-+food+and+beverage+processing.pdf?mod=ajperes
84. Manual handling. HSE. Viewed June 20, 2012. http://www.hse.gov.uk/food/handling.htm
85. CO2 gas hazards in brewing industry. Viewed October 7, 2014. http://www.crowcon.com/article/6/co2-gas-hazards-brewing-industry
86. Nature. *The International Journal of Science* August 12, 2010; 167(7), 231.
87. AB Garcia, WB Steele, DJ Taylor. Prevalence and carcass contamination with Campylobacter in sheep sent for slaughterhouse in Scotland. *Journal of Food Safety* 2010; 30(1).
88. An injury prevention strategy for teen restaurant workers. http://www.ncbi.nlm.nih.gov/pmc/articles/PMC3061567/
89. Fall injuries prevention in workplace. Viewed October 7, 2014. http://www.cdc.gov/niosh/topics/falls/projects.html
90. Organic Gardening. 2010. Making a case for Genetic Modification (GM). Vol. 57:6. New Chapter, Inc. Page 20.

References

91. Derm Net NZ. Skin problems in food handlers and the catering industry. Viewed October 7, 2014. http://dermnetnz.org/reactions/food-handlers.html
92. Occupational dermatitis in catering and food industry. Viewed October 7, 2014. http://www.beswic.be/data/links/osh_link.2006-04-23.0322210435
93. Safe use of knife in the meat and food industry. Viewed October 7, 2014. http://www.vwa.vic.gov.au/__data/assets/pdf_file/0013/10219/Knifes_vs4.pdf
94. Guidelines on deep fryers and frying oils. Viewed October 7, 2014. http://www.bccdc.ca/NR/rdonlyres/540608BF-FBAB-4886-95FE-32BA1B465DFE/0/GuidelinesonDeepFryersandFryingOilJan13.pdf
95. Obliterative bronchiolitis in workers in a coffee-processing facility. Viewed October 7, 2014. http://www.cdc.gov/mmwr/preview/mmwrhtml/mm6216a3.htm
96. Can a distillery be a hazardous environment? Viewed October 7, 2014. http://www.arlynscales.com/can-a-distillery-be-a-hazardous-environment-s/472.htm
97. H Stadler Richard, R Lineback David. 2009. *Process-induced Food Toxicants; Occurrence, Formation, Mitigation, and Health Risks.* Danvers, MA, USA: John Wiley & Sons, Inc. Page 13–15.
98. Brew – on – premise wine making establishment; hazardous $CO2$ concentration. Viewed October 7, 2014. http://www.labour.gov.on.ca/english/hs/pubs/alerts/i35.php
99. H Schmidt Ronald, E Rodrick Gary. 2003. *Food Safety Handbook.* New York: Wiley Inter – Science. Hazard from Natural Origin. Chapter 12, Page 216.
100. Workplace transport. Viewed October 7, 2014. http://www.hse.gov.uk/food/transport.htm
101. Tennant David R. 1997. Food chemical risk analysis. *Blackie Academic & Professionals.* Page 21.
102. What makes chemical poisonous. Canadian Center for Occupational Health & Safety (CCOHS). Viewed December 1, 2015. http://www.ccohs.ca/oshanswers/chemicals/poisonou.html
103. *McGill WHMIS Handbook.* 1993. Montreal, Quebec, Canada: Environmental Health & Safety, McGill University. Pages 1–19.
104. Work safe BC. *WHMIS Logos and Symbols.* Viewed June 20, 2012. http://www2.worksafebc.com/topics/whmis/symbolsandlabels.asp?reportid=24384
105. Health Canada. Safe handling of leafy greens. *Food & Nutrition.* 03/10/2003.
106. E Nina Redman. 2000. *Food Safety.* California, USA and Oxford, England: ABC – CLIO. Chapter 4. Page 164–185.
107. Slips and trips. Viewed October 7, 2014. http://www.iosh.co.uk/Membership/Our-membership-network/Our-Groups/Food-and-Drink-Group/Recipe-for-Safety/Slips-and-Trips.aspx
108. Safety fabrications. Viewed October 7, 2014. http://www.safetyfabrications.co.uk/blog/working-height-food-and-drink-industry#a
109. Hamlet chicken processing plant fire. *Wikipedia.* Viewed June 20, 2012. http://en.wikipedia.org/wiki/Hamlet_chicken_processing_plant_fire
110. Agriculture and Agri-food Canada. Canadian bottled water industry. Viewed October 7, 2014. http://www.agr.gc.ca/eng/industry-markets-and-trade/statistics-and-market-information/by-product-sector/processed-food-and-beverages/the-canadian-bottled-water-industry/?id=1171644581795
111. Bottled water is hazardous to you and to your health. Viewed October 7, 2014. http://www.naturalnews.com/032744_bottled_water_environment.html
112. Beasley Spencer, R Colonna Guy. 2002. *Fire Protection Guide to Hazardous Materials.* 13th Ed. MA, USA: NFPA.
113. HSE, Prevention of Dust Explosion in Food Industry. Viewed June 21, 2012. http://www.hse.gov.uk/food/dustexplosion.htm

248 References

114. Nestle water. The Healthy Hydration Company. Viewed October 7, 2014. http://www.nestle-waters.ca/en/water-sources-n-quality/regulation

115. Classes of fire extinguishers. New York City fire Department. Viewed June 21, 2012. http://www.nyc.gov/html/fdny/html/safety/extinguisher/classes.shtml

116. Canadian Food Inspection Agency (CFIA). Viewed October 7, 2014. http://www.inspection.gc.ca/DAM/DAM-food-aliments/STAGING/text-texte/ffv_manual_section1_1385737617232_eng.pdf

117. D Horace Graham. *The Safety of Foods.* AVI Publishing Company Inc. 2nd Ed. 1980. Chapter 22, Page 701.

118. *Accident Prevention Manual for Industrial Operations.* 1964. Chicago: National Safety Council. 6th Ed.

119. Firenze J Robert. 1978. *The Process of Hazard Control.* Iowa: Kendall/Hunt Publishing Company.

120. Tsuguyoshi Suzuki, Nobmasa Imura, Clarkson W Thomas. 1991. *Advances in Mercury Toxicology.* New York: Plenum Press.

121. Johnson. Pat. Canadian National Summit on Food Safety. Toronto, Ontario. June 7 / 8, 2005

122. Manitoba HACCP Advantage Program Manual. Good Manufacturing Practices (GMP) and HACCP Advantage. A Federal – Provincial Territorial Initiative. 2006

123. Good manufacturing practice. Viewed Set. 3 2012. http://en.wikipedia.org/wiki/Good_manufacturing_practice

124. US Food and Drug Administration. Good Manufacturing Practices for 121st Century – Food Processing. Viewed October 1, 2012. http://www.fda.gov/food/guidancecomplianceregulatoryinformation/currentgoodmanufacturingpracticescgmps/ucm110877.htm

125. Eye and face protection. Viewed Feb 12, 2012. http://www.osha.gov/SLTC/etools/eye-andface/index.html

126. Head Protection. Viewed Feb 21s, 2012. http://www.hardhats.4ursafety.com/hard-hat-articles.html.

127. Feet protection. Viewed Feb 21, 2012. http://www.osh.net/articles/archive/osh_basics_2001_march28.htm and http://www.tdi.texas.gov/pubs/videoresource/wpfootprot.pdf

128. Respiratory protection. Viewed Feb 21, 2013. http://www.osha.gov/pls/oshaweb/owadisp.show_document?p_id=12716\

129. Anna McElhatton, J Richard Marshall. 2007. *Food Safety; A Practical and Case study Approach.* New York: Springer. Chapter 3, Pages 50 –67.

130. *Advanced Food Safety Training in Canada. Managers Certification Course Book.* 2001. The National Restaurant Association Education Foundation (NRAEF) and TRAINCAN Inc. Page 19–20.

131. *Veterinary Record (VR).* Jan 9, 2010. Vol. 166. Page 32.

132. PA Luning, F Delieghere, R Verhe. 2006. *Safety in the Agri-food Chain.* Wageningen Academic Publishers. Page 19 and 145.

133. CFIA. Viewed March 1, 2012. www.inspection.gc.ca

134. *Food Safety Code of Practice.* Canadian Restaurant and Foodservice Association. 2007. Page 18.

135. Food Safety and Hygiene. World Food Program. Viewed October 2, 2012. http://foodquality.wfp.org/FoodSafetyandHygiene/tabid/118/Default.aspx

136. Health Canada. Viewed March 20, 2012. www.hc.gc.ca

137. Histamine. *What is Food Intolerance?* Viewed October 1, 2012. http://histame.com/what-is-food-intolerance

138. P Siantar, T Darsa, W Mary, M Scott, H Peter, M Eliot. 2008. *Food Contaminants; Mycotoxins and Food Allergens.* Washington, DC: American Chemical Society. Chapter 20 Page 350.

References

139. Flanagan Simon. 2015. *Handbook of Food Allergen Detection and Control. Introduction to Food Allergy.* Cambridge; Waltham: Woodhead Publishing, an imprint of Elsevier. Chapter 1. Pages 1–15.

140. Boye Joyce, Godefroy Samuel Benrejeb. 2010. *Allergen Management in the Food Industry.* New York: Wiley. Chapter 5. Allergen management and control as part of agricultural practices. Pages 137–140.

141. Food intolerance a WebMD. Allergy Health Center. Viewed March 20, 2012. http://www.webmd.com/allergies/foods-allergy-intolerance

142. Food intolerance. *Wikipedia.* Viewed March 20, 2012. http://en.wikipedia.org/wiki/Food_intolerance

143. Food intolerance. *Food & Nutrition Australia.* Viewed March 20, 2012.

144. Manitoba food safety initiative. Fact Sheet No. 17. Manitoba HACCP Advantage. A Federal-Provincial-Territorial Initiative. Viewed March 20, 2014. www.gov.mb.ca/agriculture/foodsafety

145. *Human Resources Development Canada (HRDC). Road Map for Biotechnology Regulations.* January 1996. Page 1–1.

146. Lee Byong. 2005. *Lecture Notebook. Fundamentals of Food Biotechnology. Part II.* Chapter 11. Food safety & new biotechnology. New York: McGill University. Pages 461–463.

147. M James John, B Wesley; E Phillippe. 2012. Food allergy. *Clinical Overview of Adverse Reactions to Foods.* Edinburgh: Elsevier/Saunders. Chapter 4, pages 49–50.

148. Dermatitis. *Canadian Center for Occupational Health & Safety* Viewed March 1, 2012. www.ccoh.ca/oshansnews

149. *Food Additives.* Australia: Deakins University. Viewed October 15, 2012. www.betterhealth.vic.gov.au

150. M Burkey John. 2006. *Baby Boomers and Hearing Loss, A Guide to Prevention and Care.* New Brunswick: Rutger University Press. Page 10.

151. Effective purchasing procedures for equipment in the food and drink industries. *HSE.* Viewed November 20, 2015. Page 1. http://188.11.104.53/cciaa/data/docs/HSE%20%20Effective%20purchasing%20procedures%20x%20equipments.pdf

152. Ebrahim Noroozi. 1997. *Food Permit, Guideline and Policy Report.* Central Committee on Safety, Macdonald Campus Safety Committee McGill University, Montreal, Quebec, Canada.

153. Advanced Food Safety Training in Canada. 2001. *Managers Certification Course Book.* The National Restaurant Association Education Foundation (NRAEF) and TRAINCAN Inc. Page III–VI.

154. *Ministere Agriculture, Pecheries et Alimentation du Quebec (MAPAQ) & Institute de Technologie Agroalimentaire (ITA) Course Content.* 2008. www.mapaq.gouv.qc.ca

155. FDA, MSDS. Viewed Feb 29, 2012. www.fda.gov/food/foodsafety/retail.foodprotection

156. HSE. *Information Sheet No 23, 31 and 33.* Viewed Feb 29, 2012. www.hse.uk

157. National Research Council (U.S.), Committee for the Review of NIOSH Research Program, Institute of Medicine (U.S.). 2009. *Evaluating Occupational Health and Safety Research Programs: Framework and Next Steps.* Washington, DC: National Academies Press. Pages 1–2.

158. Occupational exposure to flavor substances. Health effects and Hazards control. US Department of Labor. Viewed April 16, 2012. www.osha.gov/dts/shib/shib10142010.html

159. Krucik George. Burns, type, treatment and more. Viewed November 27, 2015. http://www.healthline.com/health/burns#Overview1

250References

160. Philip Wexler, Jan der Kolk, Asisha Mohapatra, Ravi Agarwel. 2012. *CRC Press. Chemical, Environment, Health; A Global Management Perspective.* Pages 3, 292 and 650–651.

161. Otles Semih. 2012. *Methods of Analysis of Food Components and Additives. Determination of Pollutant in Foods.* Chapter 15. Page 405. 2nd Ed. Boca Raton: CRC Press.

162. Karl Heinz Wilm. October / November 2012. Foreign object detection: Integration in food production. *Food Safety Magazine.* Viewed Feb 19, 2016. http://www.food-safetymagazine.com/magazine-archive1/octobernovember-2012/testing-foreign-object-detection-integration-in-food-production/

163. American Chemical Society. 2003. *Safety in Academic Chemistry Laboratory. Accident Prevention for Faculty and Administrators. A Publication of the American Chemical Society.* Volume 2. 7th Ed. Page 1.

164. WHMIS. Viewed October 9, 2012. http://en.wikipedia.org/wiki/Workplace_Hazardous_Materials_Information_System

165. WHMIS. 1993. *McGill WHMIS Handbook. Prepared by the McGill University Task Force on WHMIS.* 3rd Ed. Montreal, Canada: McGill University.

166. *Laboratory Safety Manual.* 1991. 2nd Ed. Montreal, Canada: McGill University. Page 9.

167. *Material Safety Data Sheet. Prepared by the McGill University Task Force on WHMIS.* 1991. Page 1–2.

168. Material Safety Data Sheet for 1- Butanol. Catalogue No. A3834. *Fisher Scientific.* Jan 21-, 2009. 2009

169. WHMIS. *Advance Food Safety Training in Canada. Manager Certification Course Book.* 2001. The National Restaurant Association Educational Foundation. Pages 164–167.

170. *Safe Handling of Controlled Products.* 1992. Montreal, Canada: McGill University Safety Office.

171. Food manufacturing. *Preventing Fire in Food Processing Facilities.* Viewed March 6, 2012. http://www.foodmanufacturing.com/articles/2002/07/preventing-fires-food-processing-facilities

172. Scientific facts on Chernobyl Nuclear Accident. Viewed March 6, 2012. http://www.greenfacts.org/en/chernobyl/index.htm

173. Inferno: Dust explosion in imperial sugar. February 7, 2008. http://www.youtube.com/watch?v=Jg7mLSG-Yws

174. Fire extinguishers. *BFPE International.* Viewed Feb 8, 2012. http://www.bfpe.com/fireextinguishers.htm

175. Combustible dust, national emphasis program and rulemaking efforts. Viewed Oct10, 2012. http://www.nfpa.org/assets/files//PDF/Proceedings/Sanji_Kanth_Presentation.pdf

176. Hot Work. Hidden Hazards. US Chemical Safety Board. Viewed Video October 10, 2012. http://www.csb.gov/videoroom/detail.aspx?VID=65

177. Henry L Febo, PE FM Global Engineering Standards. Senior Engineering Technical Specialist, AVP Combustible dust hazard recognition –an Insurer's View. Viewed October 10, 2012. http://www.nfpa.org/assets/files//PDF/Proceedings/Febo_presentation.pdf

178. Dust explosion. US Chemical Safety Board. Viewed October 10, 2012. http://www.chemsafety.gov/index.cfm

179. HSE publication update guidance on dust explosion. Viewed October 10, 2012. http://www.hse.gov.uk/press/2003/e03221.htm

References

180. Shematek Gene, Wood Wayne. 1993. *Laboratory Safety. Fire Prevention and Control Measures.* Hamilton, On. Canada: Canadian Society of Laboratory Technologists. Pages 21 23.

181. *Pipitone David A. Safe Storage of Laboratory Chemicals.* 1991. Hoboken: John Wiley & Sons, Inc. Pages 117–125.

182. National Research Council (U.S.), Committee on Prudent Practices in the Laboratory. 2011. *Prudent Practices in the Laboratory: Handling and Management of Chemical Hazards.* Washington, DC: National Academies Press. Pages 83 –84.

183. The Chemical Institute of Canada / ORDRE DES CHIMISTES DU QUEBEC. 2010. *Laboratory Health and Safety Guidelines.* 4th Ed. Pages 17–18.

184. American Chemical Society. 2003. *Safety in Academic Chemistry Laboratory. Accident Prevention for Faculty and Administrators. A Publication of the American Chemical Society.* Volume 2. 7th Ed. Pages 2 –45.

185. American Chemical Society. 2003. *Safety in Academic Chemistry Laboratory. Accident Prevention for faculty and Administrators. A Publication of the American Chemical Society.* Volume 2. 7th Ed. Pages 17.

186. Lunn George; Sansone Eric B. *Destruction of Hazardous Chemicals in the Laboratory.* 2nd Ed. Hoboken: John Wiley & Sons. 1994. Page 343–345.

187. Clifford L Holland. 1993. *Spill Response Training Manual.* Ontario, Canada: Spill Response Management Inc.

188. *Techniques for Hazardous Chemical Control.* 1994. Raleigh, NC: L. A Weaver Co. Page 1.

189. *Chemical Spill Response Guide.* 2000. USA: Environmental Health & Safety Office, University of Indiana. Pages 2–7.

190. *Food Safety Hazards. Food Safety Code of Practice for Canada's Foodservice Industry.* Canadian Restaurant and Foodservice Association. 2007. Page 16.

191. *Chemical Hazards. Advanced Food Safety Training in Canada. Managers Certification Course Book.* The National Restaurant Association Education Foundation (NRAEF) and TRAINCAN Inc. 2001. Pages 20 –22.

192. Pest Control. 1995. *AIB Consolidated Standards for Food Safety. Department of Food Safety Hygiene.* American Institute of baking (AIB). Pages 12–13.

193. Pest Control. 2007. *Food Safety Code of Practice for Canada's Foodservice Industry.* Canadian Restaurant and Foodservices Association. Section VII, Pages 114–118.

194. Manual for Spills of Hazardous Materials. 1984. *Environmental Protection Services.* Environment Canada. Pages 1–19.

195. Gulf of Mexico Oil Spill. Viewed October 10, 2010. http://www.nola.com/news/gulf-oil -spill/

196. Bhopal Information Center. Viewed October 10, 2008. http://www.bhopal.com/

197. Bhopal. Prayer for rain. Viewed October 10, 2010. http://en.wikipedia.org/wiki/Bhopal: _Prayer_for_Rain

198. What cause slips and trips. Viewed October 11, 2010. http://www.hse.gov.uk/slips/ causes.htm

199. Slips, Trips and Fall Prevention for Healthcare workers. Center for Disease Control (CDC). National Institute for Occupational Safety and Health (NIOSH) 2010. http:// www.cdc.gov/niosh/docs/2011-123/pdfs/2011-123.pdf

200. Manahan Stanley E. 1990. *Hazardous Waste Chemistry, Toxicology and Treatment.* Lewis Publishers. Pages 1–3.

201. Reinebach, Ken, Drinkman L Diane, Wendt John, Straughn John, Luis E Fernandez, Myers Jeffrey. 1993. Chemical & Environmental Safety Program. Safety Department. University of Wisconsin-Madison.

202. Griffin Roger D. 2009. *Principles of Hazardous Materials Management*. 2nd Ed. CRC Press. Pages 159 and 173–175.
203. Hazardous Waste Management. McGill University. Viewed October 23, 2012. http://www.mcgill.ca/hwm/
204. Is that safe to eat? A guide to the funky foods in your kitchen. *Consumer Reports*. December 2006. Page 6.
205. Food Safety. Physical hazards. Viewed 8/31/ 2012. http://foodsafety.unl.edu/haccp/start/physical.html
206. Bell Jennifer. 2010. National Institute of occupational health and safety. *Slip, Trip, and Fall Prevention for Healthcare Workers*. Cincinnati, Ohio?: Dept. of Health and Human Services, Centers for Disease Control and Prevention, National Institute for Occupational Safety and Health. Pages 3–27.
207. Mark Ross. CPSC Offers Safety Tips to Prevent Ladder Injuries. U.S. *Consumer Product Safety Commission (CPSC)*. Viewed 1/09/ 2012. http://www.cpsc.gov/cpscpub/pubs/ladder.html
208. Safety in working with lift trucks. HS (G) 6, HSE (UK), 1992. Viewed 1/09/2012. http://www.ilo.org/legacy/english/protection/safework/cis/products/hdo/htm/oper_forklift.htm
209. Powered Industrial Lift Trucks. *Data Sheet I-653* Rev. 82, National Safety Council (USA). Viewed 1/09/2012. http://www.ilo.org/legacy/english/protection/safework/cis/products/hdo/htm/oper_forklift.htm
210. *Physical Hazards in Food*. Viewed Nov 22 2012. www.manitoba.ca/agriculture/foodsafety
211. Preventing slips, trips and falls from vehicles; the basics. HSE. Viewed Nov 23, 2012. http://www.hse.gov.uk/pubns/wpt01.pdf
212. Workplace Transport. *HSE. UK. Information Sheet No.30*. Viewed Nov 23, 2012. http://www.hse
213. Struck by Objects. HSE. Viewed August 14, 2014. http://www.hse.gov.uk/food/struckby.htm
214. Morra Michelle. Higher Ground, Scaffold Safety Essentials. *Canadian Occupational Safety* Viewed November 23, 2012. www.cos-mag.com
215. Kutz Mayer. 2007. *Handbook of Farm, Dairy, and Food Machinery*. Norwich: William Andrew Pub Pages 73–74.
216. Morra, Michelle. Fear Factories. The Food and Beverage Industry is trying to improve its image and safety record. *Canadian Occupational Safety*. 2007; 45(5), 18–19.
217. Packaging machinery; safeguarding, *palletizers and de-palletizers. HSE Information Sheet No. 27*.
218. Machinery Hazard. *HSE Information Sheets* No. 24, 25 and 32.
219. HSE. 2001. Safeguarding flat belt conveyors in the food and drink industries. *Food Information Sheet* No. 25.
220. Machine safety. Viewed Feb 23, 2012. http://www.machinesafety.co.uk/safeguarding-the-hazards
221. Band saw. Washington State Department of Labor. USA. Viewed Feb 23, 2012. http://www.Ini.wa.gov/Safety/topics/A toZ/MeatDepartments/default.asp
222. HSE. Viewed January 8, 2016. Food processing machinery. http://www.hse.gov.uk/food/machinery.htm
223. *Mixing Equipment and Applications in the Food Industry. A White Paper Prepared By Charles Ross & Son Company*. Pages 1–12.
224. *Guarding of Commercial Dough and Food Preparation Mixers. Work safe Alberta*. Canada: Government of Alberta. March 28, 2005. Pages 1–9.

References

253

225. Canadian Food Inspection Agency. Guide to food safety. Viewed January 14, 2016. http://www.inspection.gc.ca/food/non-federally-registered/safe-food-production/guide/eng/1352824546303/1352824822033#s1-2-3

226. Knives in the meat and food industries - Safe use and maintenance. Guidance Note. *Work Safe Victoria*. Viewed 31/ 01 / 2006. http://www.workcover.vic.gov.au/dir090/vwa/alerts.nsf/PrintVersion/4A3B3D010A0

227. HSE. Safe use of the knife in the kitchen. Viewed January 7, 2016. http://www.hse.gov.uk/catering/knives.htm

228. Laboratory refrigerator / freezer safety protocols. 2006. University of Akron. http://wwwuakron.edu/dotAsset/1973602.pdf

229. Protocols for the decontamination of refrigerators and freezers. Viewed January 14, 2016. http://wwwbiosci.ohio-state.edu/safety/refrprotocol.pdf

230. Safety: A question of life and limb; preventing guide. *Commission de la santé et de la securite du travail du Quebec, Ministere de la santé et des services sociaux*, Quebec, Canada. 2002. Pages 1–25.

231. Keller JJ *Associates*. 2008. *Employee Safety Workbook. Machine Guarding*. Neehan, Wis.: J.J. Keller & Associates. Chapter1. Page 1.

232. United States. Occupational Safety and Health Administration. 2002. *Lockout / Tagout*. Washington, D.C.]: U.S. Dept. of Labor, Occupational Safety and Health Administration. Office of Compliance. OSHA Fact Sheet.

233. Ontario Ministry of Labor. Material Handling Hazards Focus of Workplace Safety Blitz. Viewed January 15, 2016. http://www.labour.gov.on.ca/english/news/2014/bg_materialhandling20140909.php

234. Worksafe BC. Health and safety for small and medium-sized food processors. Sharp tools and broken glass. Viewed January 15, 2016. Page 8.

235. McGill Radiation Safety Policy Manual. 1991. *Environmental Health & Safety*. Montreal, Quebec, Canada: McGill University. Pages 22–23.

236. United States. Agency for Toxic Substances and Disease Registry. Division of Toxicology. 1999. *Ionizing Radiation*. Atlanta, GA: Dept. of Health and Human Services, Public Health Service, Agency for Toxic Substances and Disease Registry, Division of Toxicology. Page 2.

237. The food irradiation process. UW Food Irradiation Education Group. Viewed December 4, 2012. http://uw-food-irradiation.engr.wisc.edu/materials/FI_process_brochure.pdf

238. Food irradiation. Viewed March 29, 2012. http://sln.fi.edu/guide/wester/irradiationfac

239. Hunter Richard. The food irradiation; Questions and answers. March 30, 2012. http://www.fipa.us/q%26a.pdf

240. Expert: Japan's radioactive food contamination bans. May 20, 2011. http://content.usatoday.com/communities/sciencefair/post/2011/03/japapan-radioactive-food/1

241. Industrial intensity categories. *Queensland Government. Department of Employment, Economic Development and Innovation*. March 30, 2012. http://www.deedi.qld.gov.au/cg/industrial-intensity-categories.html

242. Noise Levels d BA / Decibels. Viewed December 4, 2012. http://www.stac-uk.com/downloads/Noise%20Levels.pdf

243. Noise hazards. March 30, 2012. http://go.hrw.com/resources/go_sc/gen/HS2LMR06.PDF

244. *Safety in Academic Chemistry Laboratories. Hearing conservation. Accident Prevention for Faculty and Administrators*. American Chemical Society (ACS). 2003. 7th Ed. Vol 2 page 6.

245. Noise Induced Hearing Loss. HSE. Viewed April 16, 2012. http://www.hse.gov.uk/food/noise.htm

246. Lee Johnny. 2005. *Addressing Domestic Violence in the Workplace*. Amherst.: HRD Press. Pages 1–3.

247. Ontario Minster of labor; United Food and Commercial Workers, Ontario Meat Council, Worker Health & Safety Center and Industrial Accident Prevention Association. Health and safety training for meat processors. *Instructor's note*. 1994. Pages 1–65.

248. Safety and maintenance of electrical stunning equipment. Viewed January 15, 2016. http://www.hsa.org.uk/downloads/technical-notes/TN5-electrical-stun-equiptment-safety-maintenance-HSA.pdf

249. Cole B C et al. 2002. Safe work in the 21st century: Education and training needs for the next decade's occupational safety and health personnel. Committee to Assess Training Needs for Occupational Safety and Health Personnel in the United States. Board on Health Sciences Staff. Abstract summary page.

250. Safety risks in the food industry. Viewed January 19, 2016. University of Puerto Rico, Río Piedras. Division of Continued Education. Susan Hardwood Grant # 16617-08-60. https://www.osha.gov/dte/grant_materials/fy07/sh-16617-07/hazardous_energy.pdf

251. Canadian Center for Occupational Health and Safety. Electrical hazards. Viewed January 19, 2016. https://www.ccohs.ca/oshanswers/safety_haz/electrical.html

252. HSE. 1999. Workplace transport safety in food and drink industry. *Food Information Sheet* No. 21.

253. The National Restaurant Association Educational Foundation. 2001. *Advance Food Safety Training in Canada*. Pages 123 –124.

254. Alli O Benjamin. 2008. *Fundamental Principles of Occupational Health and Safety*. Geneva: International Labor Office. Chapter 8. Management of occupational safety and health. Page 52.

255. What causes the whole body-vibration. Viewed January 15, 2016. http://www.hse.gov.uk/VIBRATION/wbv/exposure.htm

256. HSE. Pressure system; a guide to safety. Viewed January `15, 2016. http://www.hse.gov.uk/pubns/indg261.pdf

257. Hammer Willie. 1981. *Occupational Safety Management and Engineering. Electrical hazards*. 7th Ed. New York, USA: Prentice – Hall, Inc. Chapter 19. Pages 275–279.

258. Controlling electric hazards. Occupational health and safety administration. United State Department of Labor. Viewed January 19, 2016. https://www.osha.gov/Publications/3075.html

259. CDC / NIOSH. Electrical safety and health for electrical trades. Viewed January 19, 2016. http://www.cdc.gov/niosh/docs/2009-113/pdfs/2009-113.pdf

260. Health Canada. *International Harmonization – Workplace Hazardous Materials Information System (WHMIS); The Globally Harmonized System for the Classification and Labeling of Chemicals (GHS) - Overview 2010*. http://www.hc-sc.gc.ca/ewh-semt/occp-travail/whmis-simdut/ghs-sgh-eng.php

261. Blunt Jane; Balchin N C. 2002. *Health and Safety in Welding and Allied Processes*. Boca Raton, Fla: CRC Press; Cambridge, England: Woodhead Pub. Pages 11 –15 & 242.

262. High quality nitrogen for food industry. Viewed January 19, 2016. https://www.parker.com/literature/domnick%20hunter%20Industrial%20Division/Static%20Literature%20Files%20PDFs/IGG/174004712_EN.PDF

263. Gas applications used in food and beverages. Matheson. Viewed Dec 13, 2012. http://www.mathesongas.com/applications/FoodAndBeverage.aspx

264. Hazardous confined space for food and beverage industries. *Work Safe BC*. Viewed November 19, 2021. Pages 1–20. file:///C:/Users/enoroo/Downloads/confined_space_bk82-pdf-en.pdf

References

255

265. Inerting. Health & safety executive (HSE). Viewed April 3, 2012. http://www.hse.gov.uk/comah/sragtech/techmeasinerting.htm

266. Purging of Foodstuffs. April 3, 2012. http://www.lindegas.com/en/processes/inerting_purging_and_blanketing/purging/purging_of_foodstuffs/index.html

267. Whole body vibration. Viewed April 2, 2012. http://www.csao.org/UploadFiles/Magazine/VOL1

268. Dust Collector Service, Inc. Soup maker uses noodle to add gold series to mix. Air pollution control. Viewed April 4, 2012. http://www.dustcollectorservices.com/casestud

269. *Ergonomic. Accident Prevention Manual for Business and Industry.* National Safety Council 11th Ed. 1997. Page 314.

270. International Labor Office. 2010. *Ergonomic Checkpoints: Practical and Easy-to-Implement Solutions for Improving Safety, Health and Working Conditions.* Geneva: International Labor Office.

271. *First Aid in the Workplace.* 2002. 5th Ed. Canada: Publication Quebec. Page 4–9.

272. Rickey Smith, Mobley Keith R. 2003. The Risk, performing risk assessment (the preventive maintenance tool). Industrial machinery repair; maintenance and best practices pocket guide. *Preventive maintenance. Elsevier Inc. Page 52.*

273. Effective purchasing procedures for equipment in the food and drink industries. *HSE.* Viewed November 20, 2015. Page 2–5. http://188.11.104.53/cciaa/data/docs/HSE%20%20Effective%20purchasing%20procedures%20x%20equipments.pdf

274. The right to refuse work. Ontario Ministry of labor. Viewed November 20, 2015. http://www.labour.gov.on.ca/english/hs/pubs/farming/ohsa/ohsa_8.php

275. CSA Group, Bureau de normalisation du Québec, Standards Council of Canada, Mental Health Commission of Canada. 2013–2014. *Psychological Health and Safety in the Workplace: Prevention, Promotion, and Guidance to Staged Implementation.* Mississauga, Ontario: CSA Group; Québec, Québec: Bureau de normalisation du Québec, Standard Council of Canada. Ottawa, Ontario: Canadian Electronic Library. Page 1.

276. Effective workplace inspection. Canadian Center for Occupational Health & Safety. Viewed November 23. 2015. http://www.ccohs.ca/oshanswers/prevention/effectiv.html

277. Hostile Intruder Protocol. Viewed December 19, 2012. http://www.mcgill.ca/safety/alerts/hostile

278. E-facts 52. Safe maintenance; food and drink manufacture. Viewed January 20, 2016. http://www.osha.mddsz.gov.si/resources/files/pdf/52_hazards-risks-manual-handling.pdf

279. Occupational Health & Safety Information System (OHSIS). Viewed December 19, 2012. http://products.ihs.com/Ohsis-SEO/OHSIS-Food-Drink-Safety.html

280. Food security; trade, foreign policy, diplomacy and health. Viewed October 7, 2014. http://www.who.int/trade/glossary/story028/en/

281. FAO / WHO guidance to governments on the application of HACCP in small and or less developed food businesses. 2006. Food and Nutrition Paper No. 86 (ftp://ftp.fao.org/docrep/fao/009/a0799e00.pdf).

282. Khanna, Sri Ram; Saxena Madhu. 2003. *Food Standards and Safety in Globalized World. The Impact of WTO and Codex.* Delhi, India: New Century Publication. Chapter 2.0 Page 49.

283. Riviere, Jim. 2002. Chemical Food Safety. A Scientist Perspective. *Iowa State Press.* Chapter 10 pages 135–144.

284. Brown, John. Food Security – Best practices, benchmarks and implementing your action plan. Canadian National Summit on Food Safety. Toronto, Ontario, Canada. June 7 & 8, 2005.

285. Terrorist Threats to Food - Guidelines for Establishing and Strengthening Prevention and Response Systems. May 2008. http://www.who.int/foodsafety/publications/fs_management/terrorism/en/

286. Regina grocery stores reviewing safety practices following food-tampering cases in Calgary. Viewed April 5, 2010. http://www.leaderpost.com/news/Regina+grocery+stores+reviewing+safety+practices+following+food+tampering+cases+Calgary/2766397/story.html

287. Calgary food tampering work of 'idiots': Chief. Viewed April 2, 2010. http://www.cbc.ca/canada/calgary/story/2010/04/02/calgary-food-tampering-grocery-pins-metal-police-12.html

288. Food defense; your responsibility; a guide to food defense for food processors and distributors. California Department of Public Health, Food and Drug Branch. Viewed April 19, 2012. http://www.cdph.ca.gov/pubsforms/Guidelines/Documents/fdb%20Def%20REV%20EN2.pdf

289. Sherwood Erica. 2006. *Aviation Food Safety*. Blackwell Publishing. Page 1–4.

290. Food Tampering; An extra Ounce of Caution. *Food Safety News Flash from FDA*. Viewed April 24, 2012. http://www.vidyya.com/6images/v6i211.gif

291. *Food Tampering*. Canadian Food Inspection Agency. Viewed April 23, 2012. http://www.inspection.gc.ca

292. Food Adulteration. Viewed April 23, 2012. http://doctor.ndtv.com/storypage/ndtv/id/37331/type/feature/Food_adulteration.html

293. Okun, Michell. 1986. *Fairy Play in the Marketplace, The First Battle for Pure Food and Drugs*. Deklab: *Northern Illinois University Press*. Page 1 and 15.

294. Ferrieres, Madeleine. 2005. *Sacred Cow, Mad Cow, A History of Food Fears*. New York: Columbia University Press. Chapter 3. Page 47.

295. Early History of Food regulation in United State. Viewed Jan 3, 2013. http://en.wikipedia.org/wiki/Early_history_of_food_regulation_in_the_United_States

296. Milestone in Food and Drug Law history. Viewed Jan 3, 2013. http://www.fda.gov/AboutFDA/WhatWeDo/History/Milestones/default.htm

297. Fortin, JD Neal, D Food *Regulation, Law, Science, Policy, and Practice*. 2009. Wiley & Sons Inc. Publication.

298. Hilts, J Philip. *Protecting America's Health, The FDA, Business, and One Hundred Years of Regulation*. Alfred A. Knopf. New York. 2003. Chapter six Page 95.

299. Thankama, Jacob. *Food Adulteration*. Published by S G Wasani for the Macmillan Company of India Limited, New Delhi, India. 1976. Chapter 1. Page 3.

300. *Food Adulteration. The Columbia Encyclopedia*, 6th Ed. Viewed March 11, 2012. http://www.encyclopedia.com/topic/food_adulteration.aspx

301. Poulter Sean. Exposed; the great "luxury" food fraud. *Daily Mail*. UK. Viewed April 14, 2008. http://www.thisismoney.co.uk/bargains-anoffs/article.html?in_article_id=440733&in_page_id=5

302. Holbrook Emily. Food fraud: What is for dinner? Risk Management Magazine. Risk and Insurance Management Society (RIMS). http://rmmag.com/MGTemplate.cfm?Section=RMMagazine&NavMenuID=128&template=/Magazine/DisplayMagazines.cfm&IssueID=364&AID=4517&Volume=59&ShowArticle=1

303. Food Cheats leave a bad taste. *Chemistry and Industry*. Viewed August 22, 2011. http://www.highbeam.com/doc/1G1-267808596.html

304. Wilson Bee. Swindled; The Dark History of Food Fraud from Poisoned Candy to Counterfeit Coffee. Viewed April 24, 2012. http://press.princeton.edu/titles/8723.html

305. Food product design; 7 food ingredients most prone to food fraud. April 6, 2012. Viewed April 24, 2012. http://www.foodproductdesign.com/news/2012/04/7-food-ingredients-most-prone-for-food-fraud.aspx

References

306. Adulterant. Viewed Jan 3, 2013. http://en.wikipedia.org/wiki/Adulterant
307. Coley Noel. The fight against food adulteration. *RSC Advancing the Chemical Sciences.* Viewed Jan 3, 2013. http://www.rsc.org/education/eic/issues/2005mar/thefightagainst foodadulteration.asp
308. Smith, BL. Food Legislation. The Canadian Encyclopedia. Viewed April 26, 2012. http://www.thecanadianencyclopedia.com/index.cfm?PgNm=TCE&Params =A1ARTA0002885
309. *Terrorist Threats to Food. Guidance for Establishing and Strengthening Prevention and Response Systems.* 2008. Department of Food Safety, Zoonoses and Foodborne Diseases Cluster on Health Security and Environment. World Health Organization (WHO). Pages 43–62.
310. Manahan, SE. *Environmental Chemistry.* 2005. CRC Press. 9th Ed. Pages 1–21.
311. Food safety. Viewed October 7, 2014. http://en.wikipedia.org/wiki/Food_safety
312. Lorenzen Carol L. Food defense: Protecting the food supply from intentional harm. Viewed 4/19/2012. http://extension.missouri.edu/publications/displayPub.aspx ?MP914
313. Townsend – Drake Alex. Food insecurity. Viewed October 8, 2014. http://www.post-conflictdev.org/sites/default/files/food-insecurity-alex-research.pdf
314. The causes of food insecurity in rural areas. Viewed October 8, 2014. http://www.fao .org/docrep/003/x8406e/X8406e02.htm
315. Food security and political stability in the Asia-pacific region. Viewed October 8, 2014. http://www.apcss.org/Publications/Report_Food_Security_98.html
316. Hunger and food insecurity. National coalition for the homeless. Viewed October 8, 2014. http://www.nationalhomeless.org/factsheets/hunger.html
317. Food security. Planted a community food network. Viewed October 8, 2014. http://www .plantednetwork.ca/page.aspx?pageId=4&gclid=CO316fOdncECFe0-Mgod7VcAWw
318. Recipe for disasters; keep safe from food poisoning. Viewed October 8, 2014. http:// www.cdc.gov/features/RecipesForDisaster/
319. Tragic school meal food poisoning disaster reveals widespread government food safety issues Safety matters. Viewed October 8, 2014. http://www.lrbconsulting.co.uk/blog /tragic-school-meal-food-poisoning-disaster-reveals-widespread-government-food -safety-issues/
320. Emergency food – The importance of preparedness. Viewed October 8, 2014. http:// www.gcrieber-compact.com/page/2613/Emergency_food?gclid=CLez3rOjncECFe0 -Mgod7VcAWw
321. Disaster preparedness and self -reliance. *The New Survivalist.* Viewed October 8, 2014. http://www.thenewsurvivalist.com/golden_rule_of_food_storage.html
322. Food and water in an emergency. Viewed October 8, 2014. http://www.fema.gov/pdf/ library/f&web.pdf
323. Food and water needs; preparing for a disaster or emergency. Viewed October 8, 2014. http://emergency.cdc.gov/disasters/foodwater/prepare.asp
324. Berger, Michael. *The Promise of Food Nanotechnology.* Viewed October 9, 2014. http://www.nanowork.com
325. Schmidt Ronald H, Rodrick Gary E. 2003. *Food Safety Handbook.* Page 89. Wiley – Inter-Science, A John Wiley & Sons Publication.
326. Hammer. Willie. 1981. *Occupational Safety Management and Engineering.* 2nd Ed. Pages 5–6.
327. CSST, Main risk of injury in food and beverage processing in Quebec. Viewed October21, 2013. http://www.csst.qc.ca/prevention/risques/pages/selectionsecteur.aspx
328. Safety. Food and Drug Administration. Viewed July 4, 2014. http://www.fda.gov/Safety /Recalls/ucm165546.htm

329. Olishifski, Julian B, McElroy Frank E. 1971. *Fundamental of Industrial Hygiene.* National Safety Council. Pages 1–4.

330. Mannuel Fred A. *Innovative in Safety Management.* Wiley – Inter Science. A John Wiley & Sons Inc. Publication. 2001. Page 72–73.

331. Fleming, Diane O. 1995. *Laboratory Safety Principles and Practices.* 2nd Ed. Washington, DC: ASM Press.

332. Wargniez, Aylen Badilla et al. 2012. Improving laboratory safety through mini-scale experiments: A case study of New Jersey City University. *Journal of Chemical Health & Safety,* November / December 2012.

333. Perkins, Jimmy L. 1997. *Modern Industrial Hygiene.* Vol 1. Van Nostrand Reinhold.

334. E - Medicine Health. *Terrorism and Other Public Health Threats.* Viewed July 16, 2014. http://www.emedicinehealth.com/terrorism_and_other_public_health_threats-health/page3_em.htm

335. World health Organization (WHO). 1993. *Laboratory Biosafety Manual.* Geneva: WHO. Page 1–2.

336. ABSA. Risk group classifications for infectious agents. Viewed July 16, 2014. http://www.absa.org/riskgroups/

337. Medicine Net. $_{Com}$. Definition of food borne illnesses. Viewed July 16, 2014. http://www.medterms.com/script/main/art.asp?articlekey=25399

338. CDC, Center for Disease Control and Prevention. http://www.cdc.gov/foodborneburden/estimates-overview.html

339. WHO Initiative to Estimate the Global Burden of Foodborne Diseases A summary document. Viewed July 16, 2014. http://www.who.int/foodsafety/foodborne_disease/Summary_Doc.pdf?ua=1

340. Foodborne disease. Health Canada. Viewed July 16, 2014. www.hc-Sc.gc.ca

341. CFIA. List of ingredients and allergens. Viewed Feb 24, 2016. http://www.inspection.gc.ca/food/labelling/food-labelling-for-industry/list-of-ingredients-and-allergens/eng/1383612857522/1383612932341

342. WHMIS, A retrospective look. *Consumer and Corporate Affairs Canada, Product Safety Branch, WHMIS division. Marc* 14, 1989

343. HSE. Contact dermatitis. Viewed July 16, 2014. http://www.hse.gov.uk/skin/professional/causes/dermatitis.htm

344. HSE. *Preventing Contact Dermatitis at Work.* Published by Health and Safety Executive INDG233 (rev1). 09/11.

345. McGill WHMIS Handbook. *Prepared by the McGill University Task Force on WHMIS* 2nd Ed. Page1. 1991.

346. WHMIS. Viewed January 25, 2011. http://en.wikipedia.org/wiki/Workplace_Hazardous_Materials_Information_System

347. Globally Harmonized System (GHS). Viewed January 27, 2013. http://www.ccohs.ca/oshanswers/chemicals/ghs.html

348. *WHMIS,* University of Ontario Institute of Technology. November 2004.

349. GHS Saskatchewan 2013. Viewed December 12, 2015. http://www.lrws.gov.sk.ca/globally-harmonized-system-classification-labeling-chemicals

350. GHS Classification and Labeling of Chemicals. New York & Geneva: United Nations. 2011. http://www.unece.org/fileadmin/DAM/trans/danger/publi/ghs/ghs_rev04/English/ST-SG-AC10-30-Rev4e.pdf

351. Lim CJ. *Food Cities.* Routledge, Taylor & Francis. 2014.

352. Bird Frank, E, Germain. George. 1990. *L. Practical Loss Control Leadership.* Loganville: International Loss Control Institute. 30249. Pages 1–3 & 20–21.

353. The Pure Food and Drug Act. *Wikipedia.* Viewed August 13, 2014. http://en.wikipedia.org/wiki/Pure_Food_and_Drug_Act

References

354. Bird Frank, E, Germain. George. 1990. *L. Practical Loss Control Leadership.* Loganville: International Loss Control Institute. 30249. Page 396.

355. Psychosocial risks and stress at work. European Agency for Safety and Health at Work. Viewed August 13, 2014. https://osha.europa.eu/en/topics/stress/index_html

356. Lieberman Phil, Arnold John A. 2014. *Allergic Disease, Diagnostic and Treatment.* Page 255. Berlin: Springer.

357. Naidu NVR et al. 2006. *Total Quality Management.* New Delhi: New Age International. Page 1.

358. Food allergy or something else. Viewed Nov 21, 2013. http://wwwwebmd.com/allergies /foods-allergy-intolerance

359. Bird Frank, E, Germain. George. L. Occupational health. 1990. *Practical Loss Control Leadership.* Loganville, Georgia: International Loss Control Institute. 30249. Pages 347–348.

360. Bird Frank, E, Germain. George. 1990. *L. Practical Loss Control Leadership.* Loganville: International Loss Control Institute. 30249. 1990. Pages 417 –432.

361. Furr, A Keith. *CRC Handbook of Laboratory Safety.* 1989. Pages 19–21. 3rd Ed. CRC Press.

362. Furr, A Keith. *CRC Handbook of Laboratory Safety.* 1989. Pages 61 –103. 3rd Ed. CRC Press.

363. Food futures. *Food Australia.* Volume 66. Issue 4. Page 9. August / September 2014.

364. Shridhara Bhat K. 2010. *Total Quality Management: Text and Cases.* Mumbai: Himalaya Pub. House. Page 2.

365. Safety in field work. *McGill University.* Viewed August 21, 2014. www.mcgill.ca/ehs/ safetycommittees/fieldworksafety

366. Emergency food service: Planning for disasters. *Public Heath Canada.* 2007. Pages 11 –32.

367. Fire. *Wikipedia.* Viewed August 29, 2014. http://simple.wikipedia.org/wiki/Fire

368. Bird Franc E, Germain George L. *Fire Loss Control. Practical Loss Control Leadership; The Conservation of People, Property, Process, and Profit.*1986. Pages 381–383.

369. Classification of fire and hazard type as per NFPA. Viewed December 17, 2015. http:// www.enggcyclopedia.com/2011/11/classification-fires-hazards-nfpa/

370. *Fire Safety Information Sheet.* Fire Prevention Office. McGill University. 19/11/1999.

371. National Fire Protection Association (NFPA). NFPA 654: Standard for the prevention of fire and dust explosions from the manufacturing, processing, and handling of combustible particulate solids. Viewed September 2, 2014. http://www.nfpa.org/codes-and -standards/document-information-pages?mode=code&code=654

372. Weaver LA. 1996.*Techniques for Hazardous Chemical and Waste Spill Control.* L.A Weaver Company. 7th Ed. Page 2.

373. *Laboratory Safety Manual. Storage of Chemicals.* McGill University. 3rd Ed. 1994. Pages 9 –18.

374. Chemical Storage. Chemical hygiene safety plan. Viewed September 3, 2014. http:// www2.lbl.gov/ehs/chsp/html/storage.shtml

375. American Chemical Society. 2003. Inventory management, storage and disposal. Safety in academic chemistry laboratories. Accident prevention for faculty and administrators. *A Publication of the American Chemical Society Joint Board – Council Committee on Chemical Safety.* Volume 2. 7th Ed. Pages 37 –40.

376. *Cleaning and Sanitizing. Food Safety Code Of Practice; for Canada's Foodservice Industry.* 2007 Ed. Pages 104–105.

377. Irving N Sax, Lewis J Richard. 1987. *Condensed Chemical Dictionary.* 11th Ed. New York: Van Nostrand Reinhold Company Inc.

378. Control physical hazards. Viewed September 4, 2014. http://www.ukessays.co.uk/essays/education/control-physical-hazards.php
379. Foreign objects direction: Integration in food production. Viewed September 4, 2014. http://www.foodsafetymagazine.com/magazine-archive1/octobernovember-2012/testing-foreign-object-detection-integration-in-food-production/
380. Stier F Richard. April / May 2014. Foreign Material Control: Food Quality, Safety or Both? Viewed Feb 19, 2016. http://www.foodsafetymagazine.com/magazine-archive1/aprilmay-2014/foreign-material-control-food-quality-safety-or-both/
381. Machine safety. Safe use of machinery among top ten most common inspection orders by MOL. *Canadian Occupational Safety.* Viewed September 4, 2014. http://www.cos-mag.com/legal/legal-stories/4090-safe-use-of-machinery-among-top-10-most-common-inspection-orders-by-ontario-mol.html?
382. Inventory management, storage and disposal. Safety in academic chemistry laboratories. Accident prevention for faculty and administrators. *A Publication of the American Chemical Society Joint Board – Council Committee on Chemical Safety.* 2003. Volume 2. 7th Ed. Page 20.
383. Prevention of sharp object induced injuries. Viewed September 4, 2014. http://www.ab.ust.hk/hseo/sftywise/200508/page3.htm
384. Inert Gas. *Wikipedia.* Viewed September 9, 2014. http://en.wikipedia.org/wiki/Inert_gas
385. Burns, first aid and emergency. *WebMD.* Viewed September 9, 2014. http://www.webmd.com/first-aid/tc/burns-topic-overview
386. Food Process Worker. Viewed September 16, 2014. http://www.jobguide.thegoodguides.com.au/occupation/Food-Process-Worker
387. Spellman R Frank, Bieber M Revonna. 2008. *Occupational Safety and Health Simplified for the Food Manufacturing Industry.* The Scarecrow Press, Inc.UK. Pages 10–11.
388. Poultry processing industry e Tool. Viewed 7/ Oct/ 2004. whttp://ww.osha.gov/SLTC/etools/poultry/followup.html
389. Occupational exposure to flavor substances; health effects and hazards control. US Department of Labor. Viewed April 16, 2012. http://www.osha.gov/dts/shib/shib10142010.html
390. *Health and Safety Training for Meat Processors. Ontario Minister of Labor.* United Food and Commercial Workers. Ontario Meat Council. Workers Health and Safety Center. Industrial Accident Prevention Association. Canada. 1994. Pages 14–15.
391. Health and safety in food and drink manufacture. Viewed March 7, 2005. http://www.hse.gov.uk/food/slaughter.htm
392. Priorities for health and safety in the biscuit manufacturing industry. Food Sheet No. 10. *Printed and Published by Health and Safety Executive (HSE)* 6/02.
393. Priorities for health and safety in cocoa, chocolate, and sugar confectionary industries. Food Sheet No. 9. *Printed and Published by Health and Safety Executive (HSE).* 1/97
394. Priorities for health and safety in bakery and flour confectionary industries. Food Sheet No. 4. *Printed and Published by Health and Safety Executive (HSE).* 4/97
395. Priorities for health and safety in the fruit and vegetable industry. Food sheet No. 5. *Printed and Published by Health and Safety Executive (HSE).* 9/98
396. Priorities for health and safety in the dairy industry. HSE. Food Sheet No. *Printed and published by health and safety Executive (HSE).* 8.4/02.
397. Priorities for health and safety in the poultry processing industry. Food Sheet No. 11. *Printed and published by health and safety Executive (HSE).* 1/97
398. Priorities for health and safety in the flour and grain milling industries. Food Sheet No. 13. *Printed and published by health and safety Executive (HSE).* 1/97

References 261

399. Priorities for health and safety in the compound animal feed industry. Food Sheet No. 12. *Printed and published by health and safety Executive (HSE)*. 1/97

400. Health and safety priorities in the meat processing industry. Food Sheet No. 15. *Printed and published by health and safety Executive (HSE)*. 5/97

401. Priorities for health and safety in the fish processing industry. Food Sheet No. 16. *Printed and Published by Health and Safety Executive (HSE)*. 5/97

402. Priorities for health and safety in the brewing industry. Food Sheet No. 18. *Printed and Published by Health and Safety Executive (HSE)*. 5/99

403. Priorities for health and safety in the soft drinks industry. Food Sheet No. 19. *Printed and Published by Health and Safety Executive (HSE)*. 5/99

404. Priorities for health and safety in the potable spirit industry. Food Sheet No. 20. *Printed and Published by Health and Safety Executive (HSE)*. 5/99

405. Workplace transport safety in food and drink premises. Food Information Sheet No. 21. *Printed and Published by Health and Safety Executive (HSE)*. 5/99

406. Reducing noise exposure in the food and drink industries. Food Information Sheet No. 32. *Printed and Published by Health and Safety Executive (HSE)*. 03 /02

407. Slips and trips: Summary guidance for the food industry. Food Sheet No. 6. *Printed and Published by Health and Safety Executive (HSE)*. 9/98

408. Preventing falls from height in the food and drink industries. Food Information Sheet No. 30. *Printed and Published by Health and Safety Executive (HSE)*. 07/01

409. Roll cages and wheeled racks in the food and drink industries: Reducing manual handling injuries. Food Information Sheet No.33. *Printed and Published by Health and Safety Executive (HSE)*. 04/03

410. Injuries and ill health caused by handling in the food and drink industries. Food Information Sheet No. 23. *Printed and Published by Health and Safety Executive (HSE)*. 3/00

411. Reducing injuries caused by sack handling in the food and drink industries. Food Information Sheet No. 31. *Printed and Published by Health and Safety Executive (HSE)*. 11/01

412. How do I carry a risk assessment Viewed August 31, 2005. http://www.hse.gov.uk/contact/faqs/riskassess.htm

413. Catering and hospitality. Frequently asked questions. Viewed September 22, 2014. http://www.hse.gov.uk/catering/faqs.htm

414. Maintenance priority in catering. Catering Sheet No. 12. *Printed and Published by Health and Safety Executive (HSE)*. 9/00

415. Safe use of cleaning substances in the hospitality industry. Viewed September 22, 2014. http://www.hse.gov.uk/pubns/cais22.pdf

416. Work related contact dermatitis. Viewed September 22, 2014. http://www.hse.gov.uk/catering/dermatitis.htm

417. Occupational dermatitis in the catering and food industries. Food Sheet No. 17. *Printed and Published by Health and Safety Executive (HSE)*. 1/00

418. Safe use of knives in the kitchen. Viewed September 22nd, 2014. http://www.hse.gov.uk/catering/knives.htm

419. Tobacco manufacturing. Viewed July 3, 2005. http://www.hse.gov.uk/food/tobacco.htm

420. Managing the health and safety of catering equipment and workplace. Catering Information No. 8. *Printed and Published by Health and Safety Executive (HSE)*. 1/00

421. Planning for health and safety when selecting and using catering equipment and workplace. Catering Information Sheet No. 9. *Printed and Published by Health and Safety Executive (HSE)*. 1/97

422. Precautions at manually ignited gas-fired equipment. Catering Information No. 3. *Printed and Published by Health and Safety Executive (HSE).* 2/95
423. Safety during emptying and cleaning of fryers. Catering Information Sheet No. 17. *Printed and Published by Health and Safety Executive (HSE).* 11/03
424. List of food contamination incidents. July 15, 2021. http://en.wikipedia.org/wiki/List _of_food_contamination_incidents
425. Food safety incidents in China. Viewed September 23, 2014. http://en.wikipedia.org/ wiki/Food_safety_incidents_in_China
426. Radioactive contamination in Japanese product. Viewed July 15, 2021. https://sbfsg .rocks/archive/index.php/t-440595.html
427. Adulteration of food. Viewed September 23, 2014. http://www.encyclopedia.com/topic /food_adulteration.aspx
428. Food adulteration. Viewed September 23, 2014. http://www.infoplease.com/encyclope-dia/society/food-adulteration.html
429. Responding to emergencies. World Food Program. Viewed October 8, 2014. http:// www.wfp.org/emergencies
430. What causes hunger? World Food Program. Viewed September 24, 2014. http://www .wfp.org/hunger/causes
431. Food security. Viewed September 24, 2014. http://en.wikipedia.org/wiki/Food_security
432. Current Trends in Agro-terrorism (Anti-livestock, Anti-crop, and Anti-soil Bio-agricultural Terrorism) and Their Potential Impact on Food Security. Studies in conflict and terrorism. *McGill University*, Volume 24, issue 2, September 2001. Viewed September 24, 2014. http://www.tandfonline.com/doi/abs/10.1080/10576100151101623
433. Radford G Davis. Agriculture bioterrorism. Professor of Public Health at Iowa State University, College of Veterinary Medicine, Department of Veterinary Microbiology and Preventive Medicine. *Action Bioscience.* Viewed September 25, 2014. http://www .actionbioscience.org/biotechnology/davis.html
434. Bioterrorism; agricultural shock. *Nature* 421, 106 –108. Viewed September 25, 2014. http://www.nature.com/nature/journal/v421/n6919/full/421106a.html
435. Meyerson Laura A, Reaser Jamie K Bioinvasions, bioterrorism, and biosecurity. Viewed September 25, 2014. http://www.esajournals.org/doi/abs/10.1890 /15409295(2003)001%5B0307:BBAB%5D2.0.CO%3B2
436. Biosecurity measures for preventing bioterrorism. Viewed September 25, 2014. http:// cns.miis.edu/archive/cbw/biosec/pdfs/biosec.pdf
437. Food and Drug Administration. Guidance for industry: retail food stores and food service establishments: food security preventive measures guidance. Viewed September 25, 2014. http://www.fda.gov/Food/GuidanceRegulation/GuidanceDocumentsRegula toryInformation/FoodDefense/ucm082751.htm
438. Food defense: Your responsibility. A guide to food defense for Food Processors and distributors. Food and Drug Branch, California Department of Public Health. Viewed April, 12, 2012. http://www.cdph.ca.gov/pubforms/guidelines/documents/pdf.pdf
439. Welch RW, Mitchell PC. Food processing a century of change. Viewed November 25, 2010. http://w2ww.bmb.oxfordjournal.org
440. Mental health at work. Canadian Center for Occupational Health and Safety (CCOHS). Viewed September 30, 2014. http://www.ccohs.ca/oshanswers/psychosocial/mental-health_work.html
441. Occupational rehabilitation in food manufacturing. Viewed January 22, 2016. http:// www.hse.gov.uk/food/rehabilitation/
442. Injury prevention resources for tourism & hospitality – Food and beverage industry. Viewed September 30 2014. http://www2.worksafebc.com/Portals/Tourism/Prevention -FoodBeverage.asp

References

263

443. Rees Andy. 2006. *Genetically Modified Food: A Short Guide for the Confused*. Page 3.

444. Intelex. Food and beverage. Food safety management system. Viewed October 7, 2014. http://www.intelex.com/Food_and_Beverage-20industry.aspx

445. Soft Drink. Viewed Feb 19, 2016. https://en.wikipedia.org/wiki/Soft_drink

446. Tutelyan Victor. 2013. *Genetically Modified Food Sources: Safety Assessment and Control*. Amsterdam: Elsevier/Academic Press. Chapter 8 – Information Service for the Use of Novel Biotechnologies in the Food Industry. Pages 329–331.

447. *Report identifies top 10 cancer risks for workers*. Page 7. Canadian Occupational Safety. October 2014.

448. Protective footwear. *Canadian Occupational Safety*. Pages 13–14. October 2014.

449. Ototoxic chemicals. Some solvents can cause permanent hearing damage. Pages 22–23. *Canadian Occupational Safety*. October 2014.

450. Khan A Mahmood et al. Dealings with disasters. Page 382, chapter 18. *Food Safety, Researching the Hazard in Hazardous Food*. 2014

451. Blanc Paul D. *How Everyday Product Make People Sick; Toxins at Home and Workplace*. Pages 5–7. University of California Press. 2007

452. Good Manufacturing Practices (GMP) Guidelines. *Health Products & Food Branch Inspectorate (HPFB)*. Health Canada. 2009. Edition Version 2. Pages 4 –7.

453. Workplace violence. Canada Bread frozen bakeries ltd. Occupational health and safety program manual. Element 34. Section 44. April 23, 2005.

454. Workplace violence. *OSHA Fact Sheet*. Viewed November 10, 2014. https://www.osha .gov/OshDoc/data_General_Facts/factsheet-workplace-violence.pdf

455. Inspection policy. Canada Bread Frozen Bakeries td. Occupational health and safety program manual. Element 39. Section 49. April 23, 2006

456. OSHA introduces best practices to protect temporary workers. *Canadian Occupational Safety*. November 2014. Page 6.

457. Gantt Paul and Gantt Ron. Disaster psychology; dispelling the myths of panic. *Professional Safety. Journal of American Society of Safety Engineers*. August 2012. Page 19.

458. Safety matters. U-TECK offers field safety tips. *Professional Safety*. August 2012. Page 14.

459. Holiday bulletin. McGill University. Viewed November 26, 2014. http://www.mcgill.ca /university-services/holiday-bulletin

460. Cyber security during the holidays. Viewed November 26, 2014. http://resources .infosecinstitute.com/cyber-security-holidays/

461. Amendments to the Canada Labor Code impact the right to refuse dangerous work. *Canadian Occupational Safety*. Viewed November 28, 2014. http://www.cos-mag.com

462. Doering L Ronald. Get out of jail free; the due diligence defense. Food Law. Food in Canada. *Canada's Food & Beverage Processing Magazine*. Page 28. October 2015.

463. Weekes Joanna. Six elements of effective safety management system. Viewed October 29, 2015. http://www.healthandsafetyhandbook.com.au/6-elements-of-an-effective -safety-management-system/

464. Core elements of managing for health and safety. HSE. Viewed October 29, 2015. http:// www.hse.gov.uk/managing/core-elements.htm

465. Bird E Frank, Germain George L. Loss control leadership. *The Conservation of People, Property, Process and Profits*. 1990. Pages 1–14.

466. Bird E Frank, Germain George L. Loss control leadership. *The Conservation of People, Property, Process and Profits*. 1990. Page 25.

467. General Principles of food hygiene. *Codex Alimentaire, WHO & FAO*. Viewed October 29, 2015. http://www.codexalimentarius.org/search-results/?cx=018170620143701

104933%3Ai-zresgmxec&cof=FORID%3A11&q=HACCP&siteurl=http%3A%2F%2Fwww.codexalimentarius.org%2F&sa.x=0&sa.y=0

468. Bird E Frank, Germain George L. *Loss Control Leadership. The Conservation of People, Property, Process and Profits.* Pages 17–38.

469. Contractor safety program. *Office of Environmental Health & Safety, University of Toronto.* Viewed October 30, 2015. Pages 1–16. http://www.ehs.utoronto.ca/Assets/ehs+Digital+Assets/ehs3/Biosafety/Contractor+Safety+Program/Contractor+Safety+Program.pdf

470. Krieger Gary R, Montgomery John F. *Industrial Hygienist. Accident Prevention Manual for Business and Industry.* 11th Ed. National Safety Council. 1996. Pages 170–174.

471. Krieger Gary R, Montgomery John F. *Panning for Inspection. Accident Prevention Manual for Business and Industry.* 11th Ed. National Safety Council. Page 162–164. 1996.

472. Krieger Gary R, Montgomery John F. *Personal Protective Equipment. Accident Prevention Manual for Business and Industry.* 11th Ed. National Safety Council. Page 461. 1996.

473. Bird E Frank, Germain George L. *Loss Control Leadership. The Conservation of People, Property, Process and Profits.* Pages 358 –359. 1990.

474. Personal protective equipment. *OSHA Fact Sheet.* Viewed November 6, 2015. http://permanent.access.gpo.gov/gpo14984/ppe-factsheet.pdf

475. *Personal Protective Equipment.* OSHA Occupational Safety and Health Administration. Pages 1–43. 2003.

476. Lewis Bernard T, Payant Richard P. 2003.The facility manager's emergency preparedness handbook. *New York AMOCOM.* Page Xi.

477. *Respiratory Protection.* 2002. OSHA 3079. Page 1.

478. Mobley, Keith, Higgins, Lindley R, Wikoff, Darrin J. 2008. *Maintenance Engineering Handbook.* 7th Ed. McGraw Hill. Page 1.5.

479. Working alone, a handbook for small business. *Work safe BC.* Viewed November 23, 2015. http://www.worksafebc.com/publications/health_and_safety/by_topic/assets/pdf/bk131.pdf

480. Doering Ronald L. October 2015. Get out of jail free; the due diligence defense. Food law. *Food in Canada.* Page 28.

481. Stranks Jerrem. 2007. W. A-Z of food safety. *Thorogood Publication.* Page 102.

482. Sava Buncic. 2006. *Integrated Food Safety and Veterinary Public Health. School of Veterinary, University of Bristol.* Page 54 –55.

483. Corby Joe, Klein Ron, Elliott Gary, Ryan John. 2015. Integrated Food Safety System (IFSS) Orientation. Page 3 7–38. *New York Springer.*

484. Weasel Lisa H. 2009. *Food Fray, Inside the Controversy Over Genetically Modified Food.* New York: Amacom – American Management Association. Abstract.

485. Heinrich Dickel et al. Importance of Irritant Contact Dermatitis in Occupational Skin Disease. American Journal of Clinical Dermatology. 2002; 3(4), 283–289. Springer Link.

486. Delga Celine. June 2015. Ergonomie. *TMS dans l'industrie alimentaire. Travail et Sante.* Vol. 31 N.2.

487. Bertheau Yves; Davison J. 2012. *Genetically Modified and Non-genetically Modified Food Supply Chains: Co-existence and Traceability.* Chichester, West Sussex, UK; Ames, Iowa, USA: Wiley-Blackwell. Page 3.

488. REACH. The new European chemical legislation. Viewed December 1, 2015. http://www.bfr.bund.de/en/reach___the_new_european_chemicals_legislation-9749.html

References

489. Cheshire Godfrey. The human experiment. Viewed December 1, 2015. http://www.rogerebert.com/reviews/the-human-experiment-2015

490. GHS pictogram and hazards. Viewed December 2, 2015. http://www.ccohs.ca/products/posters/ghs_pictograms/

491. Sogaard Ingmar and Krogh Hans. 2010. *Safety risk in society; fire safety.* Page 9–10. Nova Science Publication Inc.

492. Stec Anna, Hull Richard. 2010. Fire Toxicity. Fire scenarios and combustion conditions. Page 26–27. *Oxford CRC Press.*

493. United States. Chemical Safety and Hazard Investigation Board. 2008. *Barton Solvents: Static Spark Ignites Explosion Inside Flammable Liquid Storage Tank.* Washington, DC: U.S. Chemical Safety and Hazard Investigation Board. Pages 4–5.

494. Mixing and heating a flammable liquid in an open top tank. 2006. *Case Study.* Bellwood: Universal Form Clamp, Inc. Washington, DC: U.S. Chemical Safety and Hazard Investigation Board.

495. Drysdale Dougal. 2011. *An Introduction to Fire Dynamics. Fire Science and Combustion.* Chapter 1 Pages 1–2. Chichester, West Sussex: Wiley.

496. Flammable and combustible liquids – Hazards. Viewed December 17, 2015. Canadian Center for Occupational Health and Safety. http://www.ccohs.ca/oshanswers/chemicals/flammable/flam.html

497. *Fast Facts. Portable Fire Extinguishers Are Often the First Line of Defense for Fire Suppression.* 2009. Washington, DC: Office of Compliance.

498. United States. Occupational Safety and Health Administration. Combustible dust. 2011. Washington, DC: U.S. Dept. of Labor, Occupational Safety and Health Administration.

499. Nolan Dennis P. 2011. *Handbook of Fire and Explosion Protection Engineering Principles: For Oil, Gas, Chemical and Related Facilities.* Chapter 8, Segregation, separation and storage. Pages 91 –92. Burlington, MA. Elsevier Inc.

500. United State General Accounting Office. 2000. DOD competitive sourcing: Potential impact on emergency response operations at chemical storage facilities is minimal: Report to congressional committees. Washington, D.C. (P.O. Box 37050, Washington 20013): U.S. General Accounting Office. Page 5.

501. American Chemical Society. 2003. *Safety in Academic Chemistry Laboratories.* Volume 2. 7th Ed. Washington, DC. Pages 37–38.

502. United States. Environmental Protection Agency, United States. Occupational Safety and Health Administration, United States. Bureau of Alcohol, Tobacco, Firearms, and Explosives, 2013. *Chemical Advisory: Safe Storage, Handling, and Management of Ammonium Nitrate.* Washington, D.C.: United States Environmental Protection Agency. Pages 2–3.

503. Films for the Humanities & Sciences (Firm); Films Media Group. 2012. *Preventing Slips, Trips and Falls. E- Video: Clipart/Images / Graphics: English.* Montreal, Canada: McGill University Library.

504. Bell Jennifer; National Institute for Occupational Safety and Health. 2010. *Slip, Trip, and Fall Prevention for Healthcare Workers.* Cincinnati: Department of Health and Human Services, Centers for Disease Control and Prevention, National Institute for Occupational Safety and Health, Pages 1–41. https://www.cdc.gov/niosh/docs/2011-123/pdfs/2011-123.pdf

505. Jacqueline Coutts, Richard Fielder, Wiley Inter - Science (Online Service). 2009. *Management of Food Allergens.* Chichester, West Sussex; Ames, Iowa: Blackwell. Page 114.

506. Bobrowsky Peter T. 2013. *Encyclopedia of Natural Hazards. AA-LAVA.* Dordrecht; New York: Springer. Pages 3–4.

507. Schmidt Ronald H, Rodrick Gary E. 2003. Food safety handbook. Characterization of food hazards. *Wiley Inter-Science*. Page 12.

508. Films for the Humanities & Sciences (Firm); Films Media Group; National Association of Home Builders of the United States. 2008. *Fall Protection*. New York: Films Media Group.

509. Lee T, Ostrom Lee T, Wilhelmsen Cheryl A.2012. *Risk assessment: Tools, Techniques, and Their Applications*. Hoboken: Wiley. Page 7.

510. Shafiur Rahman Shafiur. 2012. *Wiley Inter-Science (Online service). Handbook of Food Process Design*. Hoboken: Wiley-Blackwell. Pages 1407 –1408.

511. Burnham John, C. 2009. *Accident Prone: A History of Technology, Psychology, and Misfits of the Machine Age*. Chicago, London: University of Chicago Press. Page 1.

512. US Department of Labor. Occupational Safety and Health Administration. Safeguarding Equipment and Protecting Workers from Amputations. *Small Business Safety and Health Management Series*. Viewed January 14, 2016. https://www.osha.gov/Publications/OSHA3170/osha3170.html

513. HSE. Are you a purchaser of work- equipment. Viewed January 14, 2016. http://www.hse.gov.uk/work-equipment-machinery/purchaser.htm

514. Bendix Corporation.; National Institute for Occupational Safety and Health. Division of Laboratories and Criteria Development. *Machine Guarding: Assessment of Need*. Page 1.

515. United States. Office of Compliance. 2008. Lockout / tagout. FAST FACTS (*United States. Office of Compliance*). Washington, DC: Office of Compliance.

516. Keller JJ & Associates. 2008. Lockout / tagout. *Employee Safety Workbook*. Neenah: J.J. Keller & Associates. Page 1.

517. Sutton Ian. 2015. *Plant Design and Operation*. Waltham, Massachusetts; Oxford, England: GPP. Chapter 3. Page 1.

518. Keller KJ. 2010. *Electrical Safety Code Manual: A Plain Language Guide to National Electrical Code, OSHA, and NFPA 70E*. Amsterdam; Boston: Butterworth-Heinemann. Pages 65 –66.

519. Worksafe BC. Health and safety for small and medium-sized food processors. Viewed January 15, 2016. Pages 3 –38. http://www.worksafebc.com/publications/high_resolution_publications/assets/pdf/BK128.pdf

520. Exposure in glass manufacturing industry. International Agency for Research on Cancer (IARC) - Summaries & Evaluations. Viewed January 15, 2016. http://www.inchem.org/documents/iarc/vol58/mono58-4.html

521. Sam Mannan; Frank P Lees. *Lees' Loss Prevention in the Process Industries: Hazard Identification, Assessment, and Control*. Amsterdam; Boston: Elsevier Butterworth-Heinemann. Pages 1–12.

522. Institute of Medicine (U.S.). Committee to Assess Training Needs for Occupational Safety and Health Personnel in the United States. 2000. *Safe Work in the 21st Century: Education and Training Needs for the Next Decade's Occupational Safety and Health*. Washington, DC. National Academy Press. Pages 1–2.

523. Marriott Norman G, Robertson Gill. 1997. *Essentials of Food Sanitation*. Boston: Springer US: Imprint: Springer. Pages 5–6.

524. Delga Celine. Ergonomie. TMS dans l'industrie alimentaire. *Travail et Sante*. June 2015. Vol. 31. No.2. Page 16–19.

525. Koningsveld EAP et al. 2007. Meeting Diversity in Ergonomic. Amsterdam: Elsevier. Page 3–12.

526. Krieger Gary R, Montgomery John F. 1997. *Accident Prevention Manual. Employee Assistance Program*. Itasca: National safety Council. Pages 343 –346.

527. Krieger Gary R, Montgomery John F. 1997. *Accident Prevention Manual. Employee Assistance Program*. Itasca: National safety Council. Pages 470–475.

References

528. Krieger Gary R; Montgomery John F. 1997. *Accident Prevention Manual. Employee Assistance Program.* Itasca: National safety Council. Pages 83–84 and 481.
529. Health and safety Executive (HSE). 2005. *A Recipe for Safety. Occupational Health and Safety in Food and Drink Manufacture.* HSE Books. Pages 1–26.
530. HSE. Respiratory hazards of poultry dust. Viewed January 21, 2016. http://www.hse.gov.uk/pubns/web40.pdf
531. Health and safety guidance note for the meat industry. February 1, 2014. *British Meat Processing Council (BMPC).* Viewed January 21, 2016. Pages 101 –121.
532. National Research Council (US). 2012. Committee on food security for all as a sustainability challenge. *A Sustainability Challenge: Food Security for All: Report of Two Workshops.* Washington, DC: National Academies Press. Page 15–16 and Page 133.
533. Gordh Gordon, Simon McKirdy. 2014. *The Handbook of Plant Biosecurity: Principles and Practices for the Identification, Containment and Control of Organisms that Threaten Agriculture and the Environment Globally.* Dordrecht: Springer. Page 1.
534. G Marriott Norman G, B Gravani Robert. 2006. *Principal of Food Sanitation. The Relationship of Biosecurity to Sanitation.* New York: Springer. Pages 16–25.
535. United States. Food Safety and Inspection Service. 2013. *Food Defense Guidelines for the Transportation and Distribution of Meat, Poultry, and Processed Egg Products.* Washington, DC: United States Department of Agriculture, Food Safety and Inspection Service. Pages 1–6.
536. United States. Food Safety and Inspection Services. 2008. *Guide to Food Defense in Slaughter and Processing Facilities: Protect Your Customers, Your Employees and Your Business.* Washington, DC: U.S. Department of Agriculture, Food Safety and Inspection Service.
537. Doeg Colin. 2005. *Crisis Management in the Food and Drinks Industry: A Practical Approach.* New York: Springer. Pages 23 –32.
538. Committee on the Review of the Use of Scientific Criteria and Performance Standards for Safe Food; Institute of Medicine (U.S.). 2003. *Scientific Criteria to Ensure Safe Food.* Washington, DC: National Academies Press. Pages 11–13.
539. Falk Ian et al. 2011. *Managing Biosecurity Across Borders.* Dordrecht: Springer. Pages 199–211.
540. Ryan Brunette. 2013. *Biosecurity: Understanding, Assessing, and Preventing the Threat.* Hoboken: The Institute of Electrical and Electronics Engineers, Inc. Published 2013 by John Wiley & Sons, Inc. Pages 1–3.
541. M Edwards. 2008. *Detecting Foreign Bodies in Food.* Cambridge England: Woodhead Publication Ltd. Page 5–6.
542. Wilson Bee. 2008. *Swindlwed; The Dark History of Food Fraud, from Poisoned Candy to Counterfeit Coffee.* Princeton & Oxford: Princeton University Press. Pages 322–323.
543. MH Van de Voorde, N Christian. 2013. Nanotechnology in a nutshell: from simple to complex systems. Chapter 26: *Risks and Toxicity of Nanoparticles.* Paris: Atlantis. Pages 439–448.
544. Sekhon bhpinder S. Food nonotechnology – An overview. *Nanotechnol Sci Appl.* 2010; 3: 1–15. Institute of Pharmacy and Department of Biotechnology, Punjab College of technical Education, Jhande, Ludhiana, India. Published online 2010 May 4. PMCID: PMC3781769
545. Consumer reports. *Nanotechnology: Untold Promise, Unknown Risk.* 2007 Jul; 72(7): 40–5.
546. Ronald J. Burke, Sharon Clarke and Cary L. Cooper. 2011. *Occupational Health and Safety.* Farnham, England; Burlington, Vt: Gower. Page 325.
547. BO Alli. 2001. Fundamental principles of occupational health and safety. *Management of Occupational Health and Safety.* Geneva: International Labor Office. Page 49. Viewed December 6, 2021. https://www.ilo.org/wcmsp5/groups/public/@dgreports/@dcomm/@publ/documents/publication/wcms_093550.pdf

548. Lees Michele. 2003. *Food Authenticity and Traceability. Humber Institute of Food and Fisheries.* Boca Raton: CRC Press; Cambridge: Woodhead. Chapter 23. Page 496.
549. Loui McCurley. 2013. *Falls from Height: A Guide to Rescue Planning.* Hoboken: Wiley. Pages 7–8.
550. International Labor Office. 2011. Safety And Health In Agriculture: ILO Code Of Practice. Geneva: International Labor Office. Pages 227 –228.
551. C Burnham John. 2009. *Accident Prone: A History of Technology, Psychology, and Misfits of the Machine Age.* Chicago, London: University of Chicago Press. Chapter 1& 2.
552. BP International, Institution of Chemical Engineers (Great Britain). 2006. Hazards of electricity and static electricity. Chapter 2. *Sparks, Arcs and Ignition Energy.* Rugby: Institution of Chemical Engineers. Page 1.
553. JC Das. 2012. *Arc Flash Hazard Analysis and Mitigation.* Hoboken: John Wiley & Sons, Inc. Pages 1–2.
554. The beverage industry past and present. Viewed Feb 19, 2016. Pages 1–22. http://catalogimages.wiley.com/images/db/pdf/0471362468.01.pdf
555. History of tea background. Tea Association of Canada. Viewed Feb 19, 2016. http://www.tea.ca/about-us/media-kit/history-of-tea-backgrounder/
556. M Cramer Michael. 2013. *Food Plant Sanitation: Design, Maintenance, and Good Manufacturing Practices.* Boca Raton: CRC Press, Taylor & Francis Group. Page 2.
557. Joseph D Nally. 2007. *Good Manufacturing Practices for Pharmaceuticals.* New York: Informa Healthcare. Page 7.
558. Wareing Peter. 2010. *HACCP: A Toolkit for Implementation.* Leatherhead, Surrey: Leatherhead Pub.; Cambridge: Royal Society of Chemistry. Page 1–2.
559. Mortimore Sara, Wallace Carol. 2001. *HACCP.* Oxford; Malden, MA: Blackwell Science. Page 2.
560. Christian Coff. 2008. Ethical traceability and communicating food. Chapter 12. *Interpreting Traceability: Improving the Democratic Quality of Traceability.* Berlin: Springer. Page 267.
561. National Mine Health and Safety Academy. 2003. *Accident Prevention.* Washington, DC: U.S. Dept. of Labor, Mine Safety and Health Administration, National Mine Health and Safety Academy. Page 3.
562. K Eckhnoffr Rolf. 2003. *Dust Explosions in the Process Industries.* Amsterdam, Boston: Gulf Professional Pub. Pages 199–200.
563. Canada, Transport Canada. 2004. *Aviation: Occupational Health and Safety: Right to Refuse Dangerous Work on Board Aircraft While in Operation.* Ottawa: Transport Canada.
564. Ekblom Paul. 2010. *Crime Prevention, Security and Community Safety Using the 5Is Framework.* Houndmills, Basingstoke; New York: Palgrave Macmillan. Page 138.
565. H Schmidt Ronald, E Rodrick Gary. 2003. *Food Safety Handbook. Chemical and Physical Hazards Produced During Food Processing.* New York: Wiley Inter-Science. Pages 224, 255 and 768.
566. United Nations, Economic Commission for Europe, Secretariat. 2011. *Globally Harmonized System of Classification and Labelling of Chemicals (GHS).* New York: United Nation. Chapter 1. Page 3.
567. International Program on Chemical Safety, Inter-Organization Program for the Sound Management of Chemicals, World Health Organization. 2010. *WHO Recommended Classification of Pesticides by Hazard and Guidelines to Classification 2009.* Geneva: World Health Organization. Page 1.
568. Bryson Bryan. 2004. A short history of nearly everything. Anchor Canada. Page 111.
569. Jeffrey M Farber, C Todd Ewen. 2000. Safe handling of foods. Boca Raton, FL, USA: CRC Press. Chapter 14. Page 415.

References

570. L Goetsch David, Ozon Gene, CRSP. 2018. Occupational health and safety for technologists, engineers and managers. North York, ON, Canada: Pearson. Pages 2–19 and 436–448.

571. CD Ewen Todd 2000. Food safety information for those in recreational activities or hazardous occupations or situations. Chapter 14 in *Safe Handling of Foods*. Pages 415–454.

572. R Spellman Frank, M Bieber Revonna. 2008. Occupational safety & health simplified for the food manufacturing industry. Introduction. Pages 1–15; 77–91; 93–108.

573. R Spellman Frank, M Bieber Revonna. 2008. Occupational safety & health simplified for the food manufacturing industry. Radiation usage in food industry. Pages 109–119.

574. R Spellman Frank, M Bieber Revonna. 2008. Occupational safety & health simplified for the food manufacturing industry. Ergonomic and food manufacturing. Pages 121–129.

575. R Spellman Frank, M Bieber Revonna. 2008. Occupational safety & health simplified for the food manufacturing industry. Hazards in the poultry processing industry. Pages 41–75.

574. R Spellman Frank, M Bieber Revonna. 2008. Occupational safety & health simplified for the food manufacturing industry. Hazards in the poultry processing industry. Hazards in the meat packing industry. Pages 27–40.

575. Chu Will. *Forced Labor in Food Industry, Progress Has Been Slow Says Report.* https://www.foodnavigator.com/Article/2016/11/02/Forced-labour-in-food-industry-Progress-has-been-slow-says-report

576. Stones Mike. Safety is no accident in food factories. https://www.confectionerynews.com/Article/2009/07/13/Safety-is-no-accident-in-food-factories

577. Maslow's hierarchy of needs. https://en.wikipedia.org/wiki/Maslow's_hierarchy_of_needs

578. Lynch Donna, CSP, AOEE. 9 Health and safety statistics food and beverage companies need to know. December 12, 2016. https://us.anteagroup.com/en-us/blog/9-health-safety-statistics-food-and-beverage-companies-need-know

579. Hazards in food & beverage manufacturing are found in many workplaces. *Industrial Safety & Hygiene News.* Viewed November 28, 2019. https://www.ishn.com/articles/106363-hazards-in-food-beverage-manufacturing-are-found-in-many-workplaces

580. Quebec government worker dies on cranberry farm. Woman fell from platform while filming at site, scarf got caught on moving machinery. By the Canadian Press. October 7, 2019. https://www.cos-mag.com/personal-process-safety/41248-quebec-government-worker-dies-on-cranberry-farm/?utm_source=GA&utm_medium=20191010&utm_campaign=COS-Newsletter&utm_content=6051ED26-27AA

581. Center for Disease Control (CDC). National Institute for Occupational safety and Health (NIOSH). Washington, USA. Two teens farm workers asphyxiate in an agricultural Silo. March 31, 2008. https://www.cdc.gov/niosh/face/stateface/wa/03wa038.html

582. Cloud Meghan, Neal James. Recent foster farms salmonella outbreak illustrates USDA food safety regime. November 19, 2013. https://www.foodsafetymagazine.com/enewsletter/recent-foster-farms-salmonella-outbreak-illustrates-usda-food-safety-regime/

583. R Spellman Frank, M Bieber Revonna. 2008. Occupational safety & health simplified for the food manufacturing industry. Hazards in the poultry processing industry. Pages 41–75.

584. R Spellman Frank, M Bieber Revonna. 2008. Occupational safety & health simplified for the food manufacturing industry. The food flavoring industry and popcorn workers' lung. Government Institutes, NW, USA. Pages 93–108.

585. Imperial sugar company dust explosion and fire investigation report by U.S Chemical Safety Board. USA. Septembre 2009. http://www.csb.gov/userfiles/file/imperial%20sugar%20report%20final%20-%20to%20post.pdf

References

586. Spolar Matthew. July 8, 2009. Worker dies after fall into Hershey’ s-bound chocolate. https://www.inquirer.com/philly/news/breaking/20090708_Worker_falls_into_chocolate_vat__dies.html http://www.cnn.com/2009/US/07/09/new.jersey.chocolate.death/, http://news.bbc.co.uk/2/hi/8141612.stm

587. Death by chocolate mum of two, 24, dies after falling into a tank of boiling liquid chocolate at sweet factory where she worked. December 15, 2016. https://www.thesun.co.uk/news/2397935/mum-of-two-dies-tank-liquid-chocolate-factory-moscow /

588. Workers baked alive in bread factory horror. *Daily Mail Reporter.* Viewed November 27, 2019. https://www.dailymail.co.uk/news/article-60734/Workers-baked-alive-bread-factory-horror.html

589. Spinner Jenny. July 7, 2014. Kellog's worker dies due to fall from a ladder. https://www.bakeryandsnacks.com/Article/2014/07/08/Food-processing-plant-accident-leads-to-Kellogg-s-worker-death

590. Shelf collapse leads to a spill and a spill/explosion at separate institutions. Viewed Nov 28, 2019. http://www.umdnj.edu/eohssweb/aiha/accidents/fire.htm#Butyl

591. CBC News. Saskatoon, Saskatchewan. Canada. July 9, 2018. Fatal crash for Elmira dairy farmers on cross-Canada tractor trip. https://www.cbc.ca/news/canada/kitchener-waterloo/elmira-henk-bettina-schuurmans-accident-saskatoon-1.4739990

592. Tegna. Pennsylvania. USA. April 23, 2019. Worker killed in meat grinder accident at Pennsylvania plant. https://www.13newsnow.com/article/news/nation-world/worker-killed-in-meat-grinder-accident-at-pennsylvania-plant/291-7f1c5250-dc35-4913-83a2-3c90cf5be1f1 https://www.foodprocessing.com/industrynews/2019/woman-killed-in-meat-grinder/

593. Food service accidents: Causes & prevention. Viewed November 28, 2019. https://study.com/academy/lesson/food-service-accidents-causes-prevention.html

594. Zukowski Dan. Nov 16, 2016. Oil from BP spill has officially entered the food chain. Viewed November 27, 2019. https://readersupportednews.org/news-section2/318-66/40325-oil-from-bp-spill-has-officially-entered-the-food-chain.

612. Sean Tucker, Anya Keefe. 2020 *Report on Work Fatality and Injury Rates in Canada.* Faculty of Business Administration. University of Regina. Canada.https://www.uregina.ca/business/faculty-staff/faculty/file_download/2020-Report-on-Workplace-Fatalities-and-Injuries.pdf.pdf

613. Canadian center for Occupational Health & Safety. https://www.ccohs.ca/search/?q=food+nd+beverage+industry+accident+rate&section=oshanswers

614. Statistic Canada. https://www150.statcan.gc.ca/n1/pub/82-003-x/2006007/article/injuries-blessures/4149017-eng.htm

615. Young retail workers, slips, trips and falls. CDC. National Institute for Occupational Safety & Health. 2018. Viewed May 18, 2021. https://www.cdc.gov/niosh/topics/retail/slips.html

616. News Release. Bureau of Labor Statistics, US Department of Labor. Viewed May 18, 2021. https://www.bls.gov/news.release/pdf/cfoi.pdf

617. Foodborne pathogen. AIMS Microbiol. 2017; 3(3): 529–563.Viewed May 18, 2021. https://www.ncbi.nlm.nih.gov/pmc/articles/PMC6604998/

618. Foodborne pathogen. *Partnership for Food Safety Education.* Viewed May 18, 2021. https://www.fightbac.org/food-poisoning/foodborne-pathogens/

619. Food safety. World Health Organization (WHO). Viewed May 18, 2021. https://www.who.int/news-room/fact-sheets/detail/food-safety

620. Machine guarding. Ministry of Labor. Ontario, Canada. Viewed May 18, 2021. https://www.ontario.ca/page/compliance-initiative-results-machine-guarding

\622. Saurabh Ranjan. RLS Human Care Knowledge of safety. July 11, 2019. Viewed May 18, 2021. https://rlsdhamal.com/types-of-machine-guarding/

References

271

623. Confined space. Work Safe BC. Viewed May 19, 2021. https://www.worksafebc.com/en/health-safety/hazards-exposures/confined-spaces
624. Bottled water market size, share & trends analysis report by product (purified, mineral, spring, sparkling, distilled), by region (North America, Asia Pacific, Europe, CSA, MEA), and segment forecasts, 2021–2028. Grand View Research. Viewed May 19, 2021. https://www.grandviewresearch.com/industry-analysis/bottled-water-market
625. In-flight food safety. Airline Routes and Ground Services (ARGS). March 1, 2020. Viewed May 20, 2021. https://airlinergs.com/issue-article/in-flight-food-safety/
626. Smart Sense. Food safety in the air. December 24, 2019. Viewed May 20, 2021. https://blog.smartsense.co/food-safety-in-the-air
627. John Gertsakis and Helen Lewis. Sustainability and the waste management hierarchy. Viewed May 20,2021. file:///C:/Users/Ferry/Downloads/Publications%20Towards%20Zero%20Waste%20Sustainability%20and%20the%20Waste%20Hierarchy%202003%20(2).pdf
628. Sustainability explained through animation. Viewed May 20, 2021. https://www.youtube.com/watch?v=-g21O7a9280
629. Waste management, compliance & ethics. Viewed December 16, 2020. www.Wm.com/ca/en/inside-wm/who-we-are/compliance-ethics
630. Ethical advocate. 7 Biggest ethical issues facing the agricultural industry. December 3, 2020. https://www.ethicaladvocate.com/7-biggest-ethical-issues-facing-agricultural-industry/
631. Peter Obervasch. Ethics in food safety practices. Global Harmonization Initiative. December 3, 2020. https://www.globalharmonization.net/wg-ethics
632. What is food ethic? Penn State College of Liberal Arts. Rock Ethic Institute. Viewed December 3, 2020. https://rockethics.psu.edu/initiatives/bioethics/programs/food-ethics/what-is-food-ethics
633. FAO. Executive summary. Viewed Dec 3, 2020. http://www.fao.org/3/j0776e/j0776e01.htm
634. Ethics in food safety management. Chapter 46. Viewed Dec 3, 2020. https://www.sciencedirect.com/science/article/pii/B9780123815040000469
635. What led to the imperial sugar explosion video. CSB. Viewed Dec 3, 2020. https://www.youtube.com/watch?v=fI-jlNqpCQ8
636. 2008. Georgia sugar refinery explosion. Wikipedia. Viewed Dec 3, 2020. https://en.wikipedia.org/wiki/2008_Georgia_sugar_refinery_explosion
637. Gavin Van Horn. Center for human and nature. Ethics and sustainability. November 26, 2020. https://iseethics.files.wordpress.com/2013/09/ethics_and_sustainability_primer.pdf
638. Bruce Jennings. Ethical aspects of sustainability. November 26, 2020. https://www.humansandnature.org/ethical-aspects-of-sustainability
639. David Morris. Ethics and sustainability. Institute for Local Self- reliance. November 26, 2020. https://ilsr.org/presentation-ethics-sustainability/
640. E Noroozi. Fall 2019. *Sustainability Form Lab to Labor.* Sustainability Committee, McGill University, Macdonald Campus.
641. Stewart Sampson. New wave. Ethics for occupational health & safety services. Ethics for occupational health & safety professional course Note. *6 Mountain Ash Court.* Dartmouth, Nova Scotia B2Y4J8. December 9, 2020. Newwaveohs2016@gmail.com
642. US Chemical Safety & Hazard Investigation Board (CSB). Imperial sugar refinery explosion. Georgia, USA. February 7, 2008. https://www.csb.gov/assets/1/20/imperial_sugar_report_final_updated.pdf?13902, https://en.wikipedia.org/wiki/2008_Georgia_sugar_refinery_explosion, https://www.csb.gov/videos/inferno-dust-explosion-at-imperial-sugar/

643. Allie Sibole. The ethics of sustainability. Markkula Center for Applied Ethics. Santa Clara University. November 26,2020. https://www.scu.edu/environmental-ethics/resources/the-ethics-of-sustainability/

644. A quote from Rumi. Viewed June 1, 2021. https://www.pinterest.ca/pin/596586281858494299/#:~:text=More%20information-,Rumi's%20answer%20to%20questions%20asked%20by%20a%20disciple%20%2D%20What%20is,fear.....%3F

645. Douglas Malloch (1877–1938). Be the best of what ever you are. Viewed June 1, 2021. https://www.great-inspirational-quotes.com/be-the-best-of-whatever-you-are.html

646. Food and beverages global market report 2021: COVID-19 impact and recovery to 2030. The Business Research Group. Viewed June 1, 2021. https://www.thebusinessresearchcompany.com/report/food-and-beverages-global-market-report#:~:text=The%20global%20food%20and%20beverages,(CAGR)%20of%206.1%25.

647. Ebrahim Noroozi. Food safety and occupational health & safety during COVID-19. COS. June 5, 2020. https://www.thesafetymag.com/ca/news/opinion/food-safety-and-ohs-matters-during-covid-19/224343

648. Maia Foulis. COS Magazine editor interview article. Viewed June 3, 2021. https://www.thesafetymag.com/ca/topics/occupational-hygiene/what-you-need-to-know-about-food-safety/223007

649. Is it safe to gift homemade holiday baked goods during the pandemic? CTV News Interview & Contribution from E. Noroozi. November 28, 2020. https://www.ctvnews.ca/health/coronavirus/is-it-safe-to-gift-homemade-holiday-baked-goods-during-the-pandemic-1.5208749?cache=yes

650. Meat plants—A new front line in the covid-19 pandemic. Viewed June 3, 2021. https://www.bmj.com/content/370/bmj.m2716

651. Impact of the COVID-19 pandemic on the meat industry in Canada. Viewed June 3, 2021. https://en.wikipedia.org/wiki/Impact_of_the_COVID19_pandemic_on_the_meat_industry_in_Canada

652. These are the meat plants in Canada affected by the coronavirus outbreak. Viewed June 3. https://www.ctvnews.ca/health/coronavirus/these-are-the-meat-plants-in-canada-affected-by-the-coronavirus-outbreak-1.4916957

653. How COVID-19 spreads in meat processing plants. Viewed June 3, 2021. https://www.chemistryviews.org/details/news/11276227/How_COVID-19_Spreads_in_Meat_Processing_Plants.html

654. Globally Harmonized System (GHS). Canadian center for occupational health & safety. July 1, 2016. http://www.ccohs.ca/oshanswers/chemicals/ghs.html

655. WHMIS. 2015. WorkSafe BC. March 15, 2021. https://www.worksafebc.com/en/health-safety/hazards-exposures/whmis/whmis-2015

656. R Spellman Frank and M Bieber Revonna. 2008. Occupational safety and health simplified for the food manufacturing. Chapter 10 page 194. https://hsseworld.com/wpcontent/uploads/2019/08/Occupational_Safety_and_Health_Simplified_for_the_Food_Manufacturing_Industry.pdf

657. *Food and Beverage Sector Health & Safety Guide.* 2002. Industrial Accident Prevention Association (IAPA), ON, Canada. Pages i–iii.

658. JS Felton. Occupational safety in US in 21st century. Viewed June 8, 2021. https://watermark.silverchair.com/50-7-523.pdf?token=AQECAHi208BE49Ooan9kkhW_Ercy7Dm3ZL_9Cf3qfKAc485ysgAAArswggK3BgkqhkiG9w0BBwagggKoMIICpA IBADCCAp0GCSqGSIb3DQEHATAeBglghkgBZQMEAS4wEQQM

659. International Labour Organization (ILO). 1919–2019. Viewed June 8, 2021. Safety and health at the heart of the future work. https://www.ilo.org/wcmsp5/groups/public/---ed_protect/---protrav/---safework/documents/publication/wcms_678357.pdf

References

273

660. Dee - anne Durbin. The Associated Press. Meat company JBS Foods confirms it paid US$11M ransom in cyberattack. Viewed June 10, 2021. https://globalnews.ca/news/7936930/jbs-foods-ransomware-attack-paid/

661. D Perry Bruce and O Winfrey. 2021. What Happened to You: Conversations on Trauma, Resilience, and Healing. New York, USA: Flatiron Books: An Oprah Book. Chapter 9, Pages 247–270.

662. D Perry Bruce and O Winfrey. 2021. What Happened to You: Conversations on Trauma, Resilience, and Healing. New York, USA: Flatiron Books: An Oprah Book. Chapter 10, Pages 275–287.

663. Seven challenges and trends the food industry can expect in 2021. Viewed July 1, 2021. https://www.newfoodmagazine.com/article/129788/trends-and-challenges-2021/

664. Food and drink manufacture. Viewed July 1, 2021. https://www.hse.gov.uk/food/

665. Study confirms higher injury rates in food industry. Viewed July 8, 2021. https://ohsonline.com/articles/2015/07/16/study-confirms-high-injury-rates-in-food-industry.aspx

666. National census of fatal occupational injuries in 2015–2019 (USA). https://www.bls.gov/news.release/pdf/cfoi.pdf

667. Injury and illness rates higher in special food services than in broader food services industry. October 30, 2019. https://www.bls.gov/opub/ted/2019/injury-and-illness-rates-higher-in-special-food-services-than-in-broader-food-services-industry.htm?view_full

668. Health and safety at work, summary of accident, injury and OHS illnesses with some data compared to EU. 2020. Viewed October 31, 2019. https://www.hse.gov.uk/statistics/overall/hssh1920.pdf, https://www.hse.gov.uk/statistics/industry/manufacturing.pdf

669. Nine (9) health & safety statistics food and beverage companies need to know. Antea Group. October 25, 2020. https://us.anteagroup.com/news-events/blog/9-health-safety-statistics-food-and-beverage-companies-need-know

670. Food recall incidents and food recalls. Viewed October 26, 2021. https://inspection.canada.ca/food-safety-for-consumers/canada-s-food-safety-system/food-recall-incidents-and-food-recalls/eng/1348756225655/1348756345745

671. Joint health and safety committee - What is a joint health and safety committee? Viewed October 28, 2021. https://www.ccohs.ca/oshanswers/hsprograms/hscommittees/whatisa.html

672. Health and safety committee representative. Viewed October 28, 2021. https://www.canada.ca/en/employment-social-development/services/health-safety/committess.html 673. *Basic Elements of a Sanitation Program for Food Processing and Food Handling.* Viewed October 28, 2021. https://ucfoodsafety.ucdavis.edu/sites/g/files/dgvnsk7366/files/inline-files/26500.pdf

674. Injury rates for 17 food & drink manufacturing sectors HSE, UK (2010/11). Viewed October 28, 2021. https://www.hse.gov.uk/food/injury-rate-comparison.htm

675. HSE. Viewed October 28, 2021. https://www.hse.gov.uk/food/fruitveg.htm

676. Six common OSHA Violations in the beer brewing industry. Viewed October 29, 2021. https://www.hsewatch.com/six-common-osha-violations-in-the-beer-brewing-industry

677. Dominic Patton. China to strengthen push to reduce food waste. Viewed November 1, 2021. www.pmtoday.co.uk

678. Food and Beverage Sector; Health & Safety Guide. 2002. *Industrial Accident Prevention Association (IAPA). Part III.* ON, Canada. Page 25.

679. Occupational health and safety in food, drink and tobacco sector. International Labor Organization. Viewed November 1, 2021. https://www.ilo.org/global/topics/safety-and-health-at-work/industries-sectors/WCMS_219018/lang--en/index.htm

680. ISO 45001:2018. Occupational health and safety management systems — Requirements with guidance for use. https://www.iso.org/standard/63787.html

681. The bottom line: The future of work: OHS & Well-being Hazards - VelocityEHS. Viewed November 2, 2021. http://www.ehs.com/2021/01/the-bottom-line-the-future-of-work-ohs-well-being-hazards/

682. Food fraud. Viewed November 2, 2021. https://inspection.canada.ca/food-label-requirements/labelling/consumers/food-fraud/eng/1548444446366/1548444516192

683. Food safety. European Commission. Viewed November 2, 2021. https://ec.europa.eu/food/safety/agri-food-fraud/food-fraud-what-does-it-mean_en#:~:text=Food%20fraud%20is%20about%20%E2%80%9Cany,%2Dfood%20chain%20legislation)%E2%80%9D

684. Maslow hierarchy of needs. 1943. Viewed November 2, 2021. https://www.simplypsychology.org/maslow.html

685. Overview of the global food system: changes over time/space and lessons for future food safety. Viewed November 2, 2021. https://www.ncbi.nlm.nih.gov/books/NBK114491/

686. FSMA's produce safety rule now available in Chinese, Portuguese and Spanish. Viewed November 2, 2021. https://www.fda.gov/about-fda/office-global-operations/china-office

687. National Medical Product Administration. Viewed November 3, 2021. https://en.wikipedia.org/wiki/National_Medical?Products_Administration

688. Standard Council of Canada (SCC). Viewed November 3, 2021. https://www.scc.ca/

689. Best practices for Food Recall Responses & Prevention. Viewed November 3, 2021. https://blog.smartsense.co/food-recall-response-and-prevention

690. Occupational health and safety management system. Viewed November 3, 2021. https://hsseworld.com/occupational-safety-health-management-systems/

691. Occupational health and safety in food and drink industries. ILO. Viewed November 4, 2021. https://www.ilo.org/wcmsp5/groups/public/---ed_protect/---protrav/---safework/documents/publication/wcms_219577.pdf

692. *Shenmiao (Ivy) Li, Ph.D. in Food Safety Research Presentation.* Department of Food Science. McGill University. November 4, 2021.

693. Chemical contaminants in meat and poultry products. Viewed November 4, 2021. https://www.intechopen.com/chapters/52046

694. Fats and oil processing chemistry. Viewed November 5, 2021. https://www.britannica.com/science/fat-processing

695. Food and Beverage Sector, Health & Safety Guide. 2002. *Industrial Accident Prevention Association (IAPA).* Appendix 3. Page 7.

696. Injuries, illness and fatalities. Un Bureau of Labour Statistic. Viewed November 5, 2021. https://www.bls.gov/iif/oshwc/cfoi/cftb0322.htm

697. Environmental, Health, and Safety Guidelines Vegetable Oil Production And Processing. World Bank Group. Viewed November 5, 2021. https://www.ifc.org/wps/wcm/connect/6a6f9bc9-09dc-4d89-a003-82ee3749ad4c/FINAL_Feb+2015_Vegetable+Oil+Processing+EHS+Guideline.pdf?MOD=AJPERES&CVID=kU6XfbQ

698. Alcoholic and soft drink. Viewed November 5, 2021. https://www.hse.gov.uk/food/drink.htm

699. HSE. Chilled and frozen products. Viewed November 5, 2021. https://www.hse.gov.uk/food/chilled.htm

700. Understanding occupational health risks within beverage manufacturing. Viewed November 5, 2021. https://www.bevindustry.com/articles/90282-understanding-occupational-health-risks-within-beverage-manufacturing

701. E Noroozi. Feb 2020. *OHS Lecture Presentation in Professional Practice Course.* Food Science Department. McGill University. Canada.

702. Coffee workers at risk for lung disease. Viewed November 8, 2021. https://blogs.cdc.gov/niosh-science-blog/2016/01/25/coffee-workers/

References

703. Allergenic Hazard, Food Safety Hazard, Government of Canada. Viewed November 9, 2021. https://inspection.canada.ca/food-safety-for-industry/archived-food-guidance/non-federally-registered/product-inspection/inspection-manual/eng/1393949957029/1393950086417?chap=5#s26c5

704. Natural toxins in food. Viewed November 9, 2021. https://www.who.int/news-room/fact-sheets/detail/natural-toxins-in-food#:~:text=All%20solanacea%20plants%2C%20which%20include,well%20as%20in%20green%20tomatoes.

705. Occupational health hazards encountered by the workers of Spice manufacturing units in Ludhiana. Viewed November 9, 2021. https://www.proquest.com/docview/1986556421?pq-origsite=gscholar&fromopenview=true

706. Assessment of occupational health hazards due to particulate matter originated from spices. Viewed November 9, 2021. https://www.ncbi.nlm.nih.gov/pmc/articles/PMC6538991/

707. Evaluation of dust exposure in spice shop. *CDC*. Viewed November 9, 2021. https://www.cdc.gov/niosh/hhe/reports/pdfs/2016-0098-3288.pdf

708. Food, drink and tobacco sector. Viewed November 9, 2021. https://www.ilo.org/global/industries-and-sectors/food-drink-tobacco/lang--en/index.htm

709. National work injury, disease and fatality statistics. Sections 51 and 52. Pages 370–372. Viewed November 9, 2021. https://awcbc.org/wp-content/uploads/2021/04/National-Work-Injury-Disease-and-Fatality-Statistics-2017-2019.pdf

710. 2020. *Report on Work Fatality and Injury Rates in Canada*. Viewed November 9, 2021. https://www.uregina.ca/business/faculty-staff/faculty/file_download/2020-Report-on-Workplace-Fatalities-and-Injuries.pdf.pdf

711. Agriculture related fatalities in Canada (CASA). Viewed November 9, 2021. https://www.casa-acsa.ca/en/cair/reports/

712. Commercial Fishing Safety. Transportation Safety Board of Canada. Viewed November 9, 2021. https://www.bst-tsb.gc.ca/eng/surveillance-watchlist/marine/2020/marine-01.html

713. 2020. Annual report. New-found land and labrador. Fish Harvest safety Association. Viewed November 9, 2021. https://ccd719b3-08ef-42f1-aaff-4dff3ec34751.filesusr.com/ugd/6533c9_7c91118e7fa94737b0354e52a9e572cb.pdf

714. Occupational Health and Safety. 2021. Schedule for rate classification unit. Viewed November 10, 2021. https://www.cnesst.gouv.qc.ca/sites/default/files/documents/dc200-414aweb.pdf

715. Principaux risques de lésion par secteur d'activité. Commission des normes, de l'équité, de la santé et de la sécurité du travail (CNESST). Viewed November 11, 2021. https://risquesdelesions.cnesst.gouv.qc.ca/Pages/vueensemble.aspx?SCIAN=312110&vue=PME

716. Canning. Viewed November 12, 2021. https://www.britannica.com/topic/canning-food-processing

717. Food and Beverage Sector, Health & Safety Guide. 2002. *Industrial Accident Prevention Association (IAPA)*. Appendix 3. Page 5–7.

718. Kira L Newman, Juan S Leon, Lee S Newman. Estimating occupational illness, injury, and mortality in food production in the United States. A farm-to-table analysis. Viewed November 15, 2021. https://journals.lww.com/joem/Fulltext/2015/07000/Estimating_Occupational_Illness,_Injury,_and.3.aspx

719. J Paul Leigh. Economic burden of occupational injury and illness in the United States. Viewed November 15, 2021. https://www.ncbi.nlm.nih.gov/pmc/articles/PMC3250639/

720. The hands that feeds us. The food chain workers alliance. Viewed November 15, 2021. http://foodchainworkers.org/wp-content/uploads/2012/06/Hands-That-Feed-Us-Report.pdf

References

721. J Paul Leigh, James P Marcisn, Ted R Miller. An estimate of the U.S. Government's undercount of nonfatal occupational injuries. Viewed November 15, 2021. https://pubmed.ncbi.nlm.nih.gov/14724473/

722. Workplace injuries significantly higher in food and drink. Viewed November 16, 2021. https://www.foodmanufacture.co.uk/Article/2020/11/06/Workplace-injuries-significantly-higher-in-food-and-drink

723. Food industry case studies. Viewed November 16, 2021. https://www.hse.gov.uk/food/experience.htm#machinery

724. OHS practices in food industry in Pakistan. A case study in biscuit manufacturing. Viewed November 16, 2021. http://iepkarachi.org.pk/OHS%20Practice%20in%20food%20industry%20-%20Sajeela%20Ghaffar.pdf

725. Case studies in food manufacturing. Viewed November 16, 2021. https://www.tal.sg/wshc/-/media/TAL/Wshc/Resources/Training-Materials/PDF/Case-Studies_Food-Manufacturing.pdf

726. Understanding occupational health risks within beverage manufacturing. Viewed November 16, 2021. https://www.bevindustry.com/articles/90282-understanding-occupational-health-risks-within-beverage-manufacturing

727. Quebec government worker dies in workplace accident on cranberry farm. Viewed November 16, 2020. https://globalnews.ca/news/5995301/quebec-worker-dies-in-workplace-accident-on-cranberry-farm/, https://www.ohscanada.com/quebec-government-worker-dies-workplace-accident-cranberry-farm/

728. Climate change: Implication for food-borne diseases (Salmonella and food poisoning among humans in R. Macedonia). Viewed November 17, 2021. https://www.intechopen.com/chapters/38359

729. Climate change and the incidence of food poisoning in England and Wales. Viewed November 17, 2021. https://pubmed.ncbi.nlm.nih.gov/8530209/

730. Food safety climate change and the role of WHO. Viewed November 17, 2021. https://www.who.int/foodsafety/publications/all/Climate_Change_Document.pdf

731. Impact of climate on workers. Viewed November 17, 2021. https://www.cdc.gov/niosh/topics/climate/how.html

732. Climate change and the health of occupational groups. Viewed November 17, 2021. https://www.cmu.edu/steinbrenner/EPA%20Factsheets/occupational-health-climate-change.pdf

733. Hazard zone; the impact of climate change on occupational health. Viewed November 17, 2021. https://onlinepublichealth.gwu.edu/resources/impact-of-climate-change-on-occupational-health/

734. Psychological trauma and cybercrime. Viewed November 17, 2021. https://www.the-safetymag.com/ca/news/opinion/psychological-trauma-and-cybercrime/252447

735. why cyber security is the new health and safety. Viewed November 17, 2021. https://www.tripwire.com/state-of-security/featured/cyber-security-new-health-safety/

736. How cyber threats can eat up profits in the food and beverage industry. November 17, 2021. https://www.northbridgeinsurance.ca/blog/cyber-threats-eat-profits-in-food-industry/

737. How ready is your food business is ready for a cybercrime attack? November 17, 2021. https://www.newfoodmagazine.com/article/139380/cybercrime/

738. Imperial sugar dust explosion and fire. Imperial sugar dust explosion and fire. 2008. *CSB Accident Description and Analysis*. USA. Viewed November,18, 2021. https://www.csb.gov/imperial-sugar-company-dust-explosion-and-fire/

739. Inferno: Dust explosion at imperial sugar. CSB investigation and final recommendations. Viewed November 18, 2021. https://www.youtube.com/watch?v=Jg7mLSG-Yws https://www.youtube.com/watch?v=05NzONPeQnQ

References

740. Imperial sugar dust explosion at Georgia Sugar plant. Viewed November 18, 2021. https://www.youtube.com/watch?v=Nc4CImjZRcM

741. 2008 Georgia sugar refinery explosion. Viewed November 18, 2021. https://en.wikipedia.org/wiki/2008_Georgia_sugar_refinery_explosion

742. Maple Leaf case study: An example of crisis management. Viewed November 18, 2021. https://www.swlawyers.ca/wp-content/uploads/2019/06/Maple-Leaf-Case-Study_-_Colin-Stevenson.pdf

743. The future of food safety in Canada. Viewed November 18, 2021. https://www.foodsafety.ca/blog/top-5-food-safety-stories-2019

744. Food industry accidents. Viewed November 18, 2021. http://www.ehsdb.com/case-studies.php

745. A guide for conducting a food safety root cause analysis. Viewed November 18, 2021. https://www.pewtrusts.org/en/research-and-analysis/reports/2020/03/a-guide-for-conducting-a-food-safety-root-cause-analysis

746. Ontario's labor ministry issues 6 orders against Toronto food company after young worker dies. Viewed November 18, 2021. https://www.cbc.ca/news/canada/toronto/fatal-industrial-accident-food-company-ontario-labour-ministry-1.3750100

747. Maple Leaf has voluntarily pulled about 220 products made at the plant in one of the biggest food - Recalls ever in Canada, with direct costs to the company of about C$20 million ($19 million). Viewed November 18, 2021. https://www.reuters.com/article/us-meat-idUSN2526525120080826

748. The recent Supreme court decision on the deadliest foodborne disease outbreak in Canadian history – from the latest 'Food in Canada'. Viewed November 18, 2021. https://www.foodincanada.com/features/the-recent-supreme-court-decision-on-the-deadliest-foodborne-disease-outbreak-in-canadian-history-from-the-latest-food-in-canada/

749. Liquid nitrogen in the food and beverage industry. Viewed November 18, 2021. https://www.labour.gov.on.ca/english/hs/pubs/ib_liquidnitrogen.php

750. Accidents in the food-manufacturing small and medium sized Malaysian industries. Viewed November 19, 2021. https://www.researchgate.net/publication/41846223_Accidents_in_the_Food-Manufacturing_Small_and_Medium_Sized_Malaysian_Industries

751. Fast food restaurant customer injured in slip and fall. Viewed November 22, 2021. https://www.expertinstitute.com/resources/case-studies/fast-food-restaurant-customer-injured-slip-fall/

752. Food service accidents: Causes & prevention. Viewed November 22, 2021. https://study.com/academy/lesson/food-service-accidents-causes-prevention.html

753. The hazardous nature of food processing plants. Viewed November 22, 2021. https://www.foodmag.com.au/the-hazardous-nature-of-food-processing-plants/

754. Telling the story of temporary workers. Viewed November 22, 2021. http://www.grady.uga.edu/pdf/mcgill/research/docs/SimonCaseStudy.pdf

755. Fiera Foods 'partner' plant convicted over deaths of two temporary workers. Viewed November 22, 2021. https://www.thestar.com/news/gta/2021/06/07/fiera-foods-partner-plant-convicted-over-deaths-of-two-temporary-workers.html

756. Occupational accidents analysis in food and drink industry, a contribution to safety and risk management. Viewed November22, 2021. https://www.researchgate.net/publication/336554590_Occupational_Accidents_analysis_in_Food_and_Drink_industry_a_contribution_to_Safety_and_Risk_Management

758. Profiles in safety and health, the soft drink industry. Viewed November 22, 2021. https://www.bls.gov/opub/mlr/1992/04/art2full.pdf

759. Injuries, illnesses, and fatalities in food manufacturing. 2008. Viewed November 22, 2021. file:///C:/Users/enoroo/AppData/Local/Microsoft/Windows/INetCache/Content.Outlook/4F7X5154/injuries-illnesses-and-fatalities-in-food-manufacturing-2008.pdf

760. Sheri Sangji's story - UCLA chemical fire. Viewed November 22, 2021. https://www.youtube.com/watch?v=F6NEdcZY2WY

761. Learning From UCLA. Details of the experiment that led to a researcher's death prompt evaluations of academic safety practices. Viewed November 22, 2021. https://cen.acs.org/articles/87/i31/Learning-UCLA.html

762. Sheri Sangji case. Viewed November 22, 2021. https://en.wikipedia.org/wiki/Sheri_Sangji_case

763. Did lax laboratory safety practices kill this UCLA chemist? Viewed November 22, 2021. https://www.motherjones.com/politics/2012/07/ucla-sheri-sangji-patrick-harran/

764. Sarah Richardson. 2020. Is safe food good food? Looking beyond safety to regulate good food systems. A Phd dissertation, McGill University. Viewed November 23, 2021. https://escholarship.mcgill.ca/concern/theses/cf95jg71h

765. The impact of what we eat: Food as an ethical choice. Peter Singer. Julie Ann Wrigley global institute of sustainability. Arizona State University. Viewed November 23, 2021. https://www.youtube.com/watch?v=MC6SKkGhXJs

766. What is food ethic? Food ethic council. Viewed November 23, 2021. https://www.foodethicscouncil.org/learn/food-ethics/what-is-food-ethics/

767. Factors affecting food choice – CCEA. Viewed November 24, 2021. https://www.bbc.co.uk/bitesize/guides/z7fw7p3/revision/3

768. The Caribbean islands poisoned by a carcinogenic pesticide. Viewed November 24, 2021. https://www.bbc.com/news/stories-54992051

769. Pesticide residues in foods. Viewed November 24, 2021. https://www.who.int/news-room/fact-sheets/detail/pesticide-residues-in-food

770. Banana cultivation is pesticide-intensive. Viewed November 24, 2021. https://www.ewg.org/news-insights/news/banana-cultivation-pesticide-intensive

771. Food sovereignty in Canada. Movement growing to control our own food and agriculture. Viewed November 24, 2021. https://www.policyalternatives.ca/publications/monitor/food-sovereignty-canada

772. Food sovereignty. Viewed November 24, 2021. https://en.wikipedia.org/wiki/Food_sovereignty

773. List of food contamination incidents. Viewed November 25, 2021. https://en.wikipedia.org/wiki/List_of_food_contamination_incidents

774. Food safety incidents in China. Viewed November 25, 2021. https://en.wikipedia.org/wiki/Food_safety_incidents_in_China

775. Food adulteration and its problems (intentional, accidental and natural food adulteration). Viewed November 25, 2021. https://euroasiapub.org/wp-content/uploads/2016/10/13FMApril-3410.pdf

776. The 5 biggest food fraud cases ever pulled off. Viewed November 25, 2021. https://www.ideagen.com/thought-leadership/blog/the-5-biggest-food-fraud-cases-ever-pulled-off

777. Gray Allison and Hinch Ronald. 2019. *A Handbook of Food Crime*. Pages 44 & 69.

778. Recent advances in detection of food adulteration. Viewed November 25, 2021. https://www.sciencedirect.com/topics/food-science/food-adulteration

779. Pesticide Action Network. North America. OP food poisoning: Case study in India. Viewed November 25, 2021. https://www.panna.org/legacy/panups/panna-op-food-poisoning-case-study-india

780. Food poisoned with pesticide in Bihar, India: New disaster, same story. Viewed November 25, 2021. https://oem.bmj.com/content/71/3/228.2

781. Bihar school meal poisoning incident. Viewed November 25, 2021. https://en.wikipedia.org/wiki/Bihar_school_meal_poisoning_incident

References

782. What is a case study? Viewed November 26, 2021. https://www.questionsanswered.net/article/what-is-case-study?utm_content=params%3Ao%3D740012%26ad%3DdirN%26qo%3DserpIndex
783. Case study cholate products, Lessons learned from global food safety and fraud data and the guidance it can provide to the food industry. Viewed November 26, 2021. https://agroknow.com/wp-content/uploads/2020/07/Foodakai_Case_Study_Chocolate_Industry_Report.pdf
784. Effect of the food production chain from farm practices to vegetable processing on outbreak incidence. Viewed November 26, 2021. https://www.ncbi.nlm.nih.gov/pmc/articles/PMC4265071/
785. Accidents and accident prevention in agriculture a review of selected studies. Viewed 26, 2021. https://www.sciencedirect.com/science/article/pii/016981419290098K
786. Workplace and organisational factors in accident analysis within the Food Industry. Viewed November 26, 2021. https://www.researchgate.net/publication/223371353_Workplace_and_organisational_factors_in_accident_analysis_within_the_Food_Industry
787. Instructor guide: Using Chemical Safety Board (CSB) reports and videos. A resource for developing case studies to use in health and safety training. Viewed November 26, 2021. https://losh.ucla.edu/wp-content/uploads/sites/37/2021/02/CSB-Guide.pdf
788. Occupational hazards in seafood industry; a case study in Ernakulam District of Kerala. Viewed November 29, 2021. http://ijarse.com/images/fullpdf/1482765445_N137ijarse.pdf
789. Danger below deck: Study investigates at-sea processing accidents. Viewed November 29, 2021. https://www.nationalfisherman.com/alaska/danger-below-deck-study-investigates-at-sea-processing-accidents
790. Accident with fish processing equipment. Viewed November 29, 2021. https://www.maritimeinjurycenter.com/accidents-and-injuries/fishing-processing-equipment/
791. Maritime workers injuries caused by fish processing equipment. Viewed November 29, 2021. https://www.attorneystevelee.com/our-library/maritime-worker-injuries-caused-by-fish-processing-equipment/
792. Safety issues investigation into fishing safety in Canada. Viewed November 29, 2021. https://www.tsb.gc.ca/eng/rapports-reports/marine/etudes-studies/m09z0001/m09z0001.html
793. Gulf of Mexico seafood harvesters, part 2: Occupational health-related risk factors. Viewed November 29, 2021. https://www.mdpi.com/2313-576X/4/3/27/htm
794. Safety issues related to the maintenance operations in a vegetable oil refinery: A case study. Viewed November 29, 2021. https://www.researchgate.net/publication/262569012_Safety_issues_related_to_the_maintenance_operations_in_a_vegetable_oil_refinery_A_case_study
795. Please also refer to the manufacture of vegetable oils: A risky activity. Viewed November 29, 2021. https://www.aria.developpement-durable.gouv.fr/wp-content/uploads/2019/02/La-fabrication-des-huiles-v%C3%A9g%C3%A9tales-VP-_EN.pdf
796. Multiple tank explosions at refinery plant: A case study. Viewed November 29, 2021. https://www.chemistryviews.org/details/ezine/5012571/Multiple_Tank_Explosions_at_Refinery_Plant_A_Case_Study.html
797. Multiple tank explosions in an edible-oil refinery plant: A case study. Viewed November 29, 2021. https://onlinelibrary.wiley.com/doi/abs/10.1002/ceat.201300075
798. Food fraud and its impact on the flavors and fragrances industry. Viewed November 30, 2021. https://www.sigmaaldrich.com/CA/en/technical-documents/technical-article/food-and-beverage-testing-and-manufacturing/regulatory-compliance-for-food-and-beverage/food-fr

References

799. Flavorings-related lung disease: Exposures to flavoring chemicals. Viewed November 30, 2021. https://www.cdc.gov/niosh/topics/flavorings/exposure.html
800. A brief history of making and faking flavor. Viewed November 30, 2021. https://www.eater.com/sponsored/9443515/a-brief-history-of-making-and-faking-flavor
801. Flavor added: The science of flavor and the industrialization of taste in America. Viewed November 30, 2021. https://repository.upenn.edu/cgi/viewcontent.cgi?article=4501&context=edissertations
801. The inexorable rise of synthetic flavor: A pictorial history. Viewed November 30, 2021. https://www.popsci.com/history-flavors-us-pictorial/
802. Fast food nation: The dark side of the all-American meal. Page 124. Viewed November 30, 2021. https://books.google.ca/books?id=dU13X_AM_N8C&pg=PA124&lpg=PA124&dq=accident+with+flavor+industry&source=bl&ots=DpPmOQ-rDr&sig=ACfU3U1lQxusj2G0Pr2477F-
803. Work-related injuries and fatalities on dairy farm operations—A global perspective. Viewed November 30, 2021. https://www.researchgate.net/publication/248703755_Work-Related_Injuries_and_Fatalities_on_Dairy_Farm_Operations-A_Global_Perspective
804. Investigating slips, trips and falls in the New Zealand dairy farming sector. Viewed November 30, 2021. https://www.researchgate.net/publication/7614856_Investigating_Slips_Trips_and_Falls_in_the_New_Zealand_Dairy_Farming_Sector
805. Case study: Ammonia release incidents (2007–2017). Viewed November 30, 2021. https://www.technicalsafetybc.ca/case-study-ammonia-release-incidents-2007-2017
806. Ammonia refrigeration system safety for the food and beverage industry. Viewed November 30, 2021. https://berg-group.com/blog/ammonia-refrigeration-safety-food-beverage/
807. Occupational health and safety risk factors in beekeeping. Viewed November 30, 2021. https://www.researchgate.net/publication/316916226_Occupational_health_and_safety_risk_factors_in_beekeeping
808. Honey: Processing & hazards involved. Viewed November 30, 2021. https://www.pmg.engineering/honey-processing-hazards-involved/
809. Research offers insight into worker safety in egg production. Viewed November 30, 2021. https://www.ishn.com/articles/99257-research-offers-insight-into-worker-safety-in-egg-production
810. Poultry farm worker. Viewed November 30, 2021. https://www.ilo.org/wcmsp5/groups/public/---ed_protect/---protrav/-- safework/documents/publication/wcms_193147.pdf
811. Health risks for workers in egg production systems and methods of control. Viewed November 30, 2021. https://www.researchgate.net/publication/287919755_Health_risks_for_workers_in_egg_production_systems_and_methods_of_control
812. Poultry Processing. Viewed November 30, 2021. https://www.osha.gov/poultry-processing
813. International Labor Organization (ILO). Encyclopedia of occupational health and safety. Industries based on biological resources. Food processing sector. Viewed December 2, 2021. https://www.iloencyclopaedia.org/part-x-96841/food-industry/food-processing-sectors
814. Guide to Writing an OHS Policy Statement. Viewed December 6, 2021. https://www.ccohs.ca/oshanswers/hsprograms/osh_policy.html
815. Responsibilities of an organisation. Viewed December 6, 2021. https://www.ukessays.com/essays/organisations/responsibilities-of-an-organisation.php
816. Safety and health practices in the food industry and ergonomic interventions. Viewed December 6, 2021. https://www.longdom.org/open-access/safety-and-health-practices-in-the-food-industry-and-ergonomicinterventions-2165-7556-1000e146.pdf
817. EU legislation and functional foods: A case study. Viewed December 6, 2021. https://www.sciencedirect.com/topics/food-science/food-legislation

References

818. Elements of accident prevention plans. Viewed December 6, 2021. https://www.tasbrmf .org/learning-news/insiderm/home/safety-security/elements-of-accident-prevention -plans.aspx

819. Injuries in food manufacturing. Viewed December 7, 2021. https://www.tasanet .com/Knowledge-Center/Articles/ArtMID/477/ArticleID/329608/Injuries-in-Food -Manufacturing

820. Healthy eating: Food ethics and sustainability. Viewed December 7, 2021. https://www .qsrautomations.com/blog/restaurant-management/food-ethics/

821. Sustainability in the workplace. Viewed December 7, 2021. https://www.osha.gov/ sustainability

822. Occupational safety and health in the food and drink industries. Viewed December 7, 2021. https://www.ilo.org/wcmsp5/groups/public/---ed_protect/---protrav/---safework/ documents/publication/wcms_219577.pdf

823. How a cyberattack led to the shortage of cream cheese. Viewed December 21, 2021. https://lifestyle.livemint.com/food/discover/how-a-cyberattack-led-to-the-shortage-of -cream-cheese-111639363700971.html

824. Biopiracy: when indigenous knowledge is patented for profit. Viewed December 21, 2021. https://theconversation.com/biopiracy-when-indigenous-knowledge-is-patented -for-profit-55589

825. Food prices are about to make the largest jump in history. December 31, 2021. https:// canadiangrocer.com/food-prices-are-about-make-largest-jump-history

826. How Turkey's currency crisis is hurting your favorite Nutella. December 31, 2021. https://www.indiatimes.com/worth/news/how-turkeys-currency-crisis-is-hurting -nutella-557260.html#highlight_76674

827. 'Beware of oysters' and other remedies for donor threats to academic freedom. Viewed Jan 6,2022 newsletters@univcan.ca

828. What are the trends in health and safety for 2022? Viewed, Jan 6, 2022. https://www .thesafetymag.com/ca

829. Why safety professionals need to lead by example. Viewed Jan 10, 2022. https://www .thesafetymag.com/ca/topics/leadership-and-culture/why-safety-professionals-need-to -lead-by-example/320954

830. Inerting, purging and blanketing gases. Viewed Jan 11, 2022. https://www.coregas.com .au/applications/inerting-purging-and-blanketing

831. Grain, flour milling and animal feed. Viewed Jan 24, 2022. https://www.hse.gov.uk/ food/grain.htm

832. Fruits and vegetables. Viewed Jan 24, 2022. https://www.hse.gov.uk/food/fruitveg.htm

833. A handbook of sea food processing. Viewed Jan 25, 2022. https://novascotia.ca/lae/ healthandsafety/documents/SafeFishProcessHandbook.pdf

834. Food processing – Induced chemicals. Viewed Jan 25, 2022. https://www.canada.ca/en /health-canada/services/food-nutrition/food-safety/chemical-contaminants/food-pro- cessing-induced-chemicals.html

835. Hazards in the food processing industry. Viewed Jan 27, 2022. https://www.maintworld .com/HSE/Hazards-in-the-Food-Processing-Industry

836. A recipe for disaster: The top occupational health hazards in the food industry. Viewed Jan 27, 2022. https://www.dbocchealth.com/blog/occupational-health-hazards-in-the -food-industry/

837. Food processing machinery. Viewed Jan 27, 2022. https://www.hse.gov.uk/food/ machinery.htm

838. Evaluation of respiratory concerns at a snack food production facility. Viewed Jan 27, 2022. https://www.cdc.gov/niosh/hhe/reports/pdfs/2011-0037-3172.pdf

Index

Page numbers followed by f, t, and n indicate figures, tables, and notes respectively

Accident, 23, 122
  electrical, 140
  industrial, 225
  prevention and causation models, 24–25
  prevention management, 91–92
Agri-Food, 204
Agrobacterium tumefaciens, 71
Allergen, 69, 70
American Chemical Society, 133
Association of Workers Compensation Board of
    Canada (AWCBC), 166
Asthma, 73
Astrovirus, 63

Baby blue, 79
Bacillus cereus, 63, 65
Bacteria, 64
Band saws, 116–117
Beekeeping, 181–182
Beverage products industry, 185
Biological hazards, 62–67, 112, 169, 203, 230
  food allergy, sensitivity, and food intolerance,
    69–71
  foodborne illnesses by pathogenic
    microorganisms, 63–67
  food hygiene safety and sanitation
    certification, 74–78
  food safety of biotechnology-derived
    products, 71–72
  food safety priorities, 67–68
  hazards associated with live or dead
    animals, 67
  hearing problems, 73–74
  respiratory illnesses, 73
  skin disease/dermatitis, 72–73
Biosecurity, 207–208
Bioterrorism, 204
Bottled water alcoholic and non-alcoholic drink
    and beverage industry, 183–184
  brewing industry, 187
  coffee manufacturing, 190–191
  evolution of the industry, 184–186
  fire and explosion, 189
  fruit juice production and frozen concentrate,
    189–190
  hazard prevention in beverage and drink
    industries, 192–193
  hazards and their prevention, 198–199

injury management and prevention in food
    and beverage sectors, 193–197
main causes of injury, 186–189
potable spirits industry, 189
soft drink concentrates, 190
soft drinks industry, 187
spirit manufacturing hazards and
    prevention, 192
sugar-beets industry, 197–198
tea manufacturing, 191
wine industry, 192, 191t
working conditions, 198
Brucella spp., 63
Brucellosis spp., 63
Bureau of Labor Statistics (BLS), 167, 184
Burn hazard, 148

Campylobacter spp., 63
  Campylobacter jejuni, 65
Canada, 7, 123
  legislative levels in, 14–15
Canada Labor Code, 34
Canadian Food Inspection Agency (CFIA), 10,
    13, 14, 21, 71, 101, 102t, 208, 211
Canadian Food Safety Approach, 13–14
Canadian Food Safety Code of Practice, 101
Canadian General Standards Board (CGSB), 12, 17
Canadian Purchasing Profession, 44
Canning, 178–179, 179f
Cardiopulmonary resuscitation (CPR), 2, 31
Carpal Tunnel Syndrome, 137
CCP, *see* Critical control points
Center for Disease Control and Prevention
    (CDC), 182
CFIA, *see* Canadian Food Inspection Agency
Chemical energy, 124
Chemical hazard, 36, 53, 64, 78–80, 112, 121,
    203, 230
  fire safety, *see* Fire safety
  future transition to GHS/WHMIS after GHS,
    82–83
  natural toxicants/prevention, 80, 80t
  prevention, 80–81
  WHMIS, 81–82
Chemical Safety Board (CSB), 225, 236–238
Chilled and frozen products, 183
CIMS, *see* Continuous Improvement
    Management System

283

Index

Clean Air Act, 19
Clostridium spp.
    Clostridium botulinum, 64–65
    Clostridium perfringens, 65
    Clostridium porringers, 63
Cobalt-60, 128
Cocoa, 173
Comebacker sakazakii, 65
Compound annual growth rate (CAGR), 184
Confined space hazard, 149–150, 169, 190
    identifying and assessing, 156
Consumer Product Safety Commission (CPSC), 99
Consumers, 7, 12–14, 20–21, 65–69, 79, 100, 102,
        185, 226, 229–230, 233
    awareness, 15
    confidence, 17
    demands, 8
    expectation, 63
    food, 10
    fraud, 209
    goods, 162
    information, 20
    preferences, 144
    products, 127
    protection of, 9, 13–14, 19, 209
    role, 205–206
    threats, 71
    worldwide, 160, 219
Continuous Improvement Management System
        (CIMS), 231
Contractor, 28, 42–43
Conveyors, 116
COVID-19 pandemic, 214, 233–234
Critical control points (CCP), 17, 115
Cryptosporidium parvum, 63, 66
Cybercrime, 214–215
Cyclospora spp., 63
    Cyclospora cayetanensis, 65

Departmental safety committee, 2
Department of Energy and Defense, 99
Dermatitis
    allergic contact dermatitis, 72
    contact urticarial, 72
    high-risk occupations, 72–73
    irritant contact dermatitis, 72
    prevention, 73
Diacetyl, 180–181
Diamine Oxidase (DAO), 69–70
Dichlorodiphenyltrichloroethane (DDT), 78
Dust hazard, 37, 88

E. coli, 63–65
Electrical energy, 124
Electrical hazard, 140–141
    prevention, 141–142

Electrical safety, 158
    arc flash, 139
    power cord safety, 143–148
    preventing power tool hazard, 142–143
    prevention, 141–142
    protection against, 140–141
Emergency food services, 215–217
Emerging trends, 230–231
Employer, 28, 150
Engineering, 135
Enterotoxigenic *E. coli*, 63
Environment, health, and safety (EHS), 1
Environmental hazard, 83, 162–163, 233
Environmental Protection Agency (EPA), 19,
        78–79, 184
EPA, *see* Environmental Protection Agency
Ethics
    food ethic, 225–227
    food sovereignty, 225
    purposes of code of ethics, 224
    and waste management, 224
European Standardization Committee (CEN),
        114–115
Eurostat, 60

Failure Mode and Effects Analysis (FMEA), 18
FDA, *see* Food and Drug Administration
FDI, *see* Food and drink industry
Federal Emergency Management Agency
        (FEMA), 216
Federal Water Pollution Control, 19
Fire Loss Control Program, 87
Fire safety, 83–84, 158
    chemical inventory management, segregation,
        and storage, 92–93
    chemical storage safety, 90–92
    combustible dust and explosion hazards,
        88–90
    emergency fire procedures, 87–88
    fire classes, 84–85
    fire extinguisher classes, 85
    fire prevention management, 86–87
    fire prevention special hazard, 88
    inspection and maintenance of fire
        extinguishers, 86
    laboratory safety, 92
    special chemical hazards, 95–99
    type and size of fire extinguishers, 85
Flammable/explosive energy, 125
FLT, *see* Forklift truck
Food additive, 183
Food allergy, 69–71
Food and Agriculture Organization of the United
        Nations (FAO), 19–20, 215
Food and beverage industry, occupational injury
        and management in, 165–166

Index

285

bread, cake, and biscuit manufacturing, 172–173, 172t
canning, 178–179, 177–178t, 179f, 180t
catering and food service industry, 182
chilled and frozen products, 183
chocolate and sugar confectionary manufacturing, 173, 174t
common occupational injuries, 166
dairy, milk, and cheese manufacturing, 175, 176t
egg production, 180
flavor manufacturing, 180–181
food additive, 183
fruit and vegetable processing industries, 173–174, 174t
grain and flour milling industries, 170–171, 170–171t
health and safety temperature requirements, 183
honey production, 181–182
main causes of injury and accidents in, 166–168
meat, poultry, fish, and petfood industry, 168–170
oils and fat processing industries, 177, 178t
spice manufacturing, 182
tobacco manufacturing, 182
working in chill units and freezers, 183
Food and drink industry (FDI), 60
Food and Drug Administration (FDA), 9, 18–19, 99, 102
Food and Drug Cosmetic Act, 70
Food biosecurity, 213–214
Foodborne diseases, 20, 63, 65–66, 205
Foodborne illness, 10, 63–64, 66, 68, 101, 168, 205, 220, 236
Foodborne pathogens, 63–64
Food defense, 207–208, 206–207t
Food fraud and adulteration
in 21st century, 212
definition, 209–210, 210t
fight against, 209–210
food crime, 211
food fraud or error, 211
food fraud vulnerability, 211–212
Food hygiene safety and sanitation certification, 74–75
burn, 77–78
musculoskeletal injury, 75–76
transient emissions, 77
working with animals, 76–77
Food inspection, 18
Food intolerance, 69–71
Food legislation, 212
Food Products Division of the Industrial Accident Prevention Association (IAPA), 6

Food safety, 63, 65–66
of biotechnology-derived products, 71–72
priorities, 67–68
and quality, 13
and quality management systems, 16–20
and sanitation, 205–206
systematic control of, 17
Food Safety Act, 21–22
Food Safety Enhancement Program (FSEP), 14
Food Safety Management System (FSMS), 19
Food Safety Modernization Act (FSMA), 9, 206
Food security, 203–204
Food system industries, 167–168
Forklift truck (FLT), 111–112

Gas Flushing, 144
General Agreement on Tariffs and Trade (GATT), 12
Genetically modified (GM) foods, 70–72
GFCI, *see* Ground Fault Circuit Interrupter
Giardia lamblia, 63
Glass/bottling hazards, prevention management, 127
Globally Harmonized System (GHS), 36, 81–83
GMP, *see* Good Manufacturing Practices
Good Manufacturing Practices (GMP), 6, 11, 17–19, 205
Ground Fault Circuit Interrupter (GFCI), 142
Gulf of Mexico, 169

HACCP, *see* Hazard Analysis Critical Control Points
Hand–arm vibration syndrome (HAVS), 137
Hazard
accidental, 110
analysis, 18
assessment, 35–36, 50, 156
atmospheric, 56, 150
explosion, 84, 88, 96, 158, 170, 177
identifying and assessing, 119
impact, 37
inventory, 87
machine, 115
machine guarding, 122–123
occupational, 59, 130, 137, 161, 165–166, 193
optical, 38
potential, 121, 157, 179
prevention strategy/hygiene, 60
psychological, 112, 233
psychosocial, 161–162
safety, 44, 53, 114–115, 168, 186, 190
source of, 35, 60, 194, 232
specific equipment/machine/tools, 116–117
spirit manufacturing, 192
stress, 162–163

286

Index

transport, 120, 169–170, 172, 187
undue, 48
Hazard alert
explosive atmospheres, 152
flammable material, 152
ignition sources, 153
oxygen, 152
oxygen, too little or too much, 151
toxic atmospheres, 151–152
Hazard Analysis Critical Control Points
(HACCP), 6, 11–12, 15–20, 67, 75,
102–103, 205, 207, 232
HCP, *see* Hearing Conservation Program
Health and Safety Commission (CNESST), 170
Health and Safety Committee, 33–34
Health and Safety Executive (HSE), 73, 75, 97,
114, 116, 169–170, 173–175, 236
Health and safety policy
accident report form, 32
health and safety forms, 31–32
Health Canada, 14–15, 69, 184
Health hazard, 44, 69, 83, 124, 168, 209
Hearing, 73–74
Hearing Conservation Program (HCP), 135
administrative, 135
engineering, 131t, 135
Heat energy, 124
Hepatitis A, 63
High-hazard atmosphere, 151
Hippocrates (460 BC), 63
Histamine, 69–70
Honey, 181–182
HSE, *see* Health and Safety Executive

ILCI Loss Causation Model, 25, 25t
IMP, *see* Injury Management Policy
Inaugural Sustainable Labs Award, 223
Incident, 23
Industrial disasters, 25
Industrial hygiene
conducting inspection, 34
definition, 32
health and safety committee/joint health and
safety committee, 33–34
legislation, 40–42
personal protective equipment, 34–40; *see
also* Personal protective equipment
preventive maintenance, 42–44
procedure, 33
purchasing policy, 44–47
purpose, 32–33, 33t
working alone, 49–57
work refusal, 47–49
work stoppage, 49
Inert gas, 143
Injury

definition, 23
eye, 38
fatal, 6–7, 45
head, 39
main causes of, 114, 186–187
musculoskeletal, 75–76
nonfatal, 6
rates, 7
Injury Management Policy (IMP), 232
Integrated Inspection System (IIS), 14
Intentional (non-accidental) hazard sources, 60
International Board for Certification of Safety
Managers (IBFCSM), 1
International Code of Conduct on Pesticide
Management, 78
International Labor Organization (ILO), 193, 197
International Organization for Standardization, 19
International Risk-Based Food Inspection
System, 19–20
Irradiated food, 212
Irritant Contact Dermatitis (ICD), 72
ISO 22000, 19

JHSC, *see* Joint Health and Safety Committee
Joint Health and Safety Committee (JHSC),
33–34, 54

Knife hazard, 118–119

Lab safety program, 3
Lindane, 78
Liquid nitrogen (LN2), 148–149
Listeria monocytogenes, 63, 65
Lockout (energy control), 124–125
food irradiation process and hazard, 128–129
fragile devices, 127
glass/bottling hazards prevention
management in processing areas, 127
glassware, sharp, and bottling hazards, 126
handling sharp objects, 126–127
hazard prevention, 126
hazard's priorities, 126
hoists, 126
lockout/tagout exercise, 125, 125f
multiple workers/contractors and lockout/
tagout, 125
radiation/and food irradiation hazards,
127–128
Losses
and due diligence, 7–8
financial, 76, 234
job, 226
Low-hazard atmosphere, 151

Machine guard, 124
Machine hazards, 123

Index

287

Maple Leaf Foods Inc., 236
Material Safety Data Sheet (MSDS), 28, 31, 36, 40, 74, 81–83, 87, 96
Mixer hazards
    identification and assessing, 117
    prevention management, 117–120
    transport hazards, 120–124
Moderate-hazard atmosphere, 151
Modern safety management evolution, 8
Modified Atmosphere Packaging (MAP), 144, 146–147
Monsanto Company, 71
MSDS, *see* Material Safety Data Sheet
Mycotoxins, 64

Nanotechnology, 212
National Aeronautical and Space Administration (NASA), 17–18
National Fire Protection Association (NFPA), 84
National Institute for Occupational Safety and Health (NIOSH), 99, 181
National Institute of Safety and Health (NISH), 98, 181
National Safety Council (NSC), 30, 105
Nitrosamine, 79
Noise
    confined space entry program, 156
    construction, 156
    contents, 157
    hazardous atmosphere, 157
    identifying and assessing confined space hazards, 156
    location, 157
    risk of drowning, 155–156
    work activity, 157
Noise hazard/hearing problem, 129–135
    food and beverage industry, 133, 134
    hearing conservation program (HCP), 135, 131t
    hearing problem and noise level, 133–134
    identifying and assessing, 134, 130t
    noise preventive management and control, 134, 131t
    stunning electrical equipment, 135–139
Non-food system industries, 167–168
Novel food, 71

Occupational health and safety (OHS), 1–3, 5–6, 8–9, 11–12, 23, 26, 60, 81–82, 163, 165–168, 177, 200, 213, 219, 222, 224–226, 229–230, 233–235, 237
    general practices, 194–195
    model, 231–232
    specific practices, 195–197
Occupational Health and Safety Act (OHSA), 29, 33–34, 123, 126

Occupational Safety and Health Administrator (OSHA), 19, 99
Occupational Skin Disease (OSD), 72
OHS, *see* Occupational health and safety
Organic foods, 212
Organization, 2, 18
Organizational hazard, 112
OSHA, *see* Occupational Safety and Health Administrator

Parasites, 64
Personal protective equipment (PPE), 2, 5, 34–35, 158–159
    body protection wear, 36
    chemical hazards, 36
    devices for chemical hazards, 36
    devices for dust hazards, 37
    devices for impact hazards, 37
    dust hazard, 37
    employee training, 40
    foot protection, 38
    hazard assessment, 35–36
    head protection, 39
    impact hazard, 37
    optical hazard, 38
    prevention, 39
    requirements, 35
    respiratory protection, 39–40
Personal risk, 5
Pew Charitable Trusts, 220
Physical hazards, 83, 99–100
    classification of, 101, 102
    commercial mixers, 117–120; *see also* Mixer hazards
    current regulations around the world, 102
    effective, 101
    electrical shock, 154–155
    fall hazards assessments, 108
    falling objects, 154
    falls from height and principles of working at height, 107–108
    in food and drink, 100–101
    in food and drink manufacturing, 100
    food processing machinery, tools, and equipment safety, 114–117
    forklift truck, 111–112
    hazard prevention, 117
    identifying and assessing fall hazards, 107
    ladder hazard prevention management, 109–110
    lockout, 124–129; *see also* Lockout
    loose and unstable materials, 153–154, 153t
    mechanical equipment, 154
    poor visibility, 155
    prevention of, 102–103, 103t
    preventive management, 103–104

preventive measures, falls from height, 108–109
scaffold safety, 110
slip, trip, and fall hazards, 104–107, 154; *see also* Slip, trip, and fall hazards
struck by objects, 112–114
substances entering through piping, 155
temperature extremes, 155
workplace transport, 110–111
Pillsbury Company, 18
Pipelines, 137–139
Potential energy, 125
Power cord safety
  compressed gases, 144–148
  hazard preventive management, 144
  inerting and purging, 143
Pressure vessel systems, 137–139
Priority allergens, 69
Priority areas, 231
Produce Safety Rule (PSR), 9
Production departmental responsibility, 2–3
PSR, *see* Produce Safety Rule
Purging, 143, 157

Quality assurance (QA), 18
Quality control (QC), 18
Quality management program (QMP), 14
Quebec, 166

Radiolytic Products, 128
Recall and traceability, 20
Rehabilitation, 200–201
Residual energy, 154
Respirator, 39–40
Respiratory illnesses, 73
Rhinitis, 73
Root Cause Analysis (RCA), 236
Rotating energy, 124
Rotavirus, 63

Safe Food for Canada Act, 8
Safety, 23
  chemical storage, 90–92
  comprehensive responsibility, 1–2
  definition, 1
  department, 2
  electrical, 115, 139–148, 158
  fire, 158
  goggles, 37
  laboratory, 3, 92
  machine guarding/hazard and, 122–123
  promotion and recognition, 56–57
  purchasing policy and, 121–122
  scaffold, 110
Safety Data Sheet (SDS), 31, 82–83, 87, 96
Safety management system, elements of, 8
Safe Water Drinking Act, 184

Salmonella spp., 63–65
  Salmonella typhi, 63
Scaffold safety, 110
Scalding, 77
SDS, *see* Safety Data Sheet
Self-reported work-related illness (SWI) survey, 59
Sensitivity, 69–71
Shigella spp., 63, 65
Slip, trip, and fall hazards, 104–107
  contamination, 107
  environmental factors, 107
  flooring, 104–105, 105f
  footwear, 105–107
  worker's performance, 107
Soft drink industry, 185
Specific hazards and elements, 160–163
  environmental and stress hazard, 162–163
  ergonomic hazards, 163
  hazard associated with nanotechnology, 160
  preventive management and control, 161
  psychosocial hazards/stress, 161–162
  sanitation and sanitary occupational hazards, 161
  temporary and seasonal workers, 160–161
  welding, 160
Spices, 182
Standard Council of Canada (SCC), 12
Staphylococcus aureus, 63, 65
Streptococci, 63
Stunning electrical equipment, 135
  cleaning and storage, 136
  hazard prevention, 135–136
  pressure vessels and pipelines, 137–139
  vibration, 136–137
  work equipment safely, 136
Sustainability an agri-food concern
  hazardous waste management, 222–223
  in laboratory, 223–224
  sustainability pillars, 221–222; *see also* Sustainability pillars
Sustainability pillars
  economic sustainability, 221–222
  environmental sustainability, 221
  social sustainability, 222

Temporary Permit, 74
Terrorism, 207–208, 213–214
Thermal hazards, 148–149
  confined space hazard, 149–150
  cryogenic hazard prevention, 149
  hazard alert, 151–153; *see also* Hazard alert
  hazardous atmospheres, 150–151
  risks, 150
Total Quality Management System (TQMS), 18
Toxoplasma gondii, 63, 65
Traceability

Index

definition, 21
recall and, 20–21
Trichinella spp., 63
   Trichinella spiralis, 64–65
TQM, *see* Total Quality Management System

Unintentional (accidental) hazard sources, 60
United States Department of Agriculture
   (USDA), 19
University of Regina, 7
USA, 6, 10, 12–13, 19, 24, 26, 30, 78, 83–84, 96,
   99, 104, 109, 123, 167, 169, 184, 194,
   204–205, 209, 220, 225, 230, 237–238
USA House of Representatives, 21
US Chemical Safety and Hazard Investigation
   Board, 88
U.S. Consumer Product Safety Commission, 109
US Department of Labor Occupational Safety
   and Health Administration, 90

Vibration, 136–137
Vibrio spp., 63, 65
   Vibrio cholera, 63

Viruses, 64

WFP, *see* World Food Program
WHMIS, *see* Workplace Hazardous Material
   Information System
WHO, *see* World Health Organization
Whole-body vibration, 137
Workers, 150
Work permit, 159
Workplace Hazardous Material Information
   System (WHMIS), 15, 28–29, 32,
   81–83
   controlled products, 82
Workplace hazards, 39, 53, 111, 182, 234
Workplace inspection, 50–52
Workplace transport, 110–111
World Bank Group, 177
World Food Program (WFP), 216
World Health Organization (WHO), 12, 15, 20,
   65, 206, 215
World Trade Organization, 12

Yersinia enterocolitica, 63, 65